THE RHETORIC OF RISK

Technical Documentation in Hazardous Environments

THE RHETORIC OF RISK

Technical Documentation
in Hazardous Environments

Beverly Sauer
Carnegie Mellon University

Routledge
Taylor & Francis Group
New York   London

First published by Lawrence Erlbaum Associates, Inc., Publishers
10 Industrial Avenue
Mahwah, New Jersey 07430

Transferred to digital printing 2010 by Routledge

Routledge

711 Third Avenue,
New York, NY 10017

2 Park Square, Milton Park
Abingdon, Oxon OX14 4RN, UK

Cover design by Kathryn Houghtaling Lacey

Library of Congress Cataloging-in-Publication Data

Sauer, Beverly J.
   The rhetoric of risk : technical documentation in hazardous environments / Beverly Sauer.
     p.   cm.
   Includes bibliographical references and index.

   1. Risk communication.   I. Title.

T10.68 .S38   2002
363.11'2—dc21   2001040673
                                                                 CIP
 ISBN 13: 978-0-8058-3685-1 (hbk)
 ISBN 13: 978-0-8058-3686-8 (pbk)

To Elizabeth and Caroline, with love

Contents

List of Figures   xiii

Editor's Introduction   xvii

Introduction: The Rhetoric of Risk   1

    Rhetorical Invention in the Context of Risk   3
    Finding Out the Available Means of Persuasion   7
    Integrating Stakeholder Knowledge in Expert Judgments About Risk   11
    Bridging the Boundaries Between Risk and Rhetoric   14
    The Cycle of Technical Documentation in Large Regulatory Industries   17
    The Framework of the Book   18
    Acknowledgments   21

**I   The Problem of Technical Documentation
in Hazardous Environments**   25

1   Regulating Hazardous Environments:
The Problem of Documentation   27

    Pyro-Technics: Regulating an Explosive Environment   34
        The Problem of Standardizing Experience   37
        The Problem of Wording   43
        The Problem of Regulatory Revision and Review   47

The *Difficulties of Compliance*   50
    The *Tension Between Strict Enforcement and Day-to-Day Compliance*   51
    The *Problem of Complying "In Time"*   55
    The *Costs and Benefits of a Well-Regulated Environment*   57
The *Problem of Documentation*   58
    The *Problem of Timely and Adequate Documentation*   59
    The *Problem of Unwarrantable Failure*   61
The *Nature of Technical Documentation in Hazardous Environments*   64

2   Moments of Transformation: The Cycle of Technical
    Documentation in Large Regulatory Industries   65

    The *Need for a Rhetorical Framework*   67
    The *Cycle of Technical Documentation in Large Regulatory Industries*   72
    Six Critical Moments of Transformation   75
    A *Collaborative Notion of Expertise*   78
    The *Rhetorical Force of Documents Within the Cycle*   85
    The *Material Consequences of Manipulating Risk on Paper*   87
        Shifting the Focus of Institutional Attention   88
        The *Consequences of Managing Risk on Paper*   90
    The *Rhetorical Transformation of Experience*   91
        The *Rhetorical Transformation of Experience in Accident Investigations*   92
        The *Retransformation of Experience in Training*   93
    The *Uncertainty of Knowledge in Large Regulatory Industries*   97

3   Acknowledging Uncertainty: Rethinking Rhetoric
    in a Hazardous Environment   99

    Uncertainty at the Highest Level of Exigence: Imminent Danger   103
    The *Dynamic Uncertainty of Hazardous Environments*   107
    The *Variability and Unreliability of Human Performance*   110
    The *Uncertainty of Premium Data*   112
    Uncertainty in Social Structure and Organization   116
    The *Rhetorical Incompleteness of a Single Viewpoint*   122

II Moments of Transformation   127

4   Reconstructing Experience: The Rhetorical Interface
    Between Agencies and Experience   129

    The *Structure of Situated Knowledge in Hazardous Worksites*   133
        The *Literal and Figurative Structure of Work in Hazardous Worksites*   133
        The *Vocabulary of Situated Viewpoints*   135
        The *Rhetorical Uncertainty of Documentation in Hazardous Worksites*   136

Documenting Local Experience: Kenny Blake's Narrative   138
Constructing the Agency's Perspective: Outby the Disaster   140
   Mapping Embodied Positions   142
   Reconstructing Time   143
   Categorizing Rhetorical Positions   144
   Reconstructing the Collective Experience of Risk   145
   Constructing Collective Agreement   147
   Drawing on Previous Experience   148
   Evaluating the Outcome   150
The Potential for Multiple Viewpoints   152

5   Learning From Experience: Enlarging the Agency's
Perspective in Training and Instruction   154

*The Problem of Anomalous Behavior: Rethinking Instruction
   as Hierarchical Procedure*   156
*Constructing the Proper Perspective: MSHA Reconstructs
   Miners' Experience*   159
*The British Miner's Perspective: Free to Speculate, Miners Can
   Articulate Gaps in the Agency's Construction of Experience*   163
*FATALGRAMS in U.S. Training: Encouraged to Speculate, Miners
   Retransform the Agency's Limited Perspective*   166
*The Welsh Miner's Perspective: Faced With Complexity, Individuals
   Can Manage a Surprising Number of Diverse Perspectives*   175
*Revisiting a Feminist Perspective: Can Rhetoric Accommodate Multiple
   Perspectives?*   178

6   Warrants for Judgment: The Textual Representation
of Embodied Sensory Experience   181

The Problem of Roof Support in U.S. and British Mines   185
The Nature of Warrants Grounded in Experience   189
   Embodied Sensory Experience (Pit Sense)   189
   Engineering Experience   191
   Scientific Knowledge   191
The Effect of Warrants on Risk Decisions and Risk Outcomes   193
The Rhetorical Incompleteness of Written Instructions and Procedures   199
   The Rhetorical Incompleteness of Written Instructions   200
   The Interdependence of Scientific Knowledge and Local Experience   203
   The Rhetorical Presence of Tacit Knowledge   204
The Textual Dynamics of Disaster   207
The Limits of Science in the Context of Risk   210
Implications for Rhetorical Theory   213

III   Documenting Experience   217

7   Embodied Experience: Representing Risk
     in Speech and Gesture   219

     Hand and Mind: What Gestures Reveal About Thought   224
     The Role of Gesture in Speakers' Representations of Risk   227
          Mimetic and Analytic Viewpoints   228
          Multiple-Viewpoint Representations   230
     Mimetic Viewpoints in Miners' Representations of Risk   232
     Analytic Viewpoints in Miners' Representations of Risk   238
     Simultaneous Viewpoints in Miners' Representations of Risk   243
     Missing Viewpoints in Miners' Representations of Risk   244
     Sequential Viewpoints in Miners' Representations of Risk   249
     Interpreting Miners' Representations of Risk: The Problem of Audience   252
     Appendix A: Characterizing Viewpoints in Miners'
          Representations of Risk   254

8   Manual Communication: The Negotiation
     of Meaning Embodied in Gesture   256

     The Effect of Viewpoint in Miners' Representations of Risk   259
          Viewpoint Redefines the Relationship Between Speaker and Audience   260
          Viewpoint Shapes Semantic Content   262
          Viewpoint Helps Speakers Elaborate the Meanings Conveyed in Speech   266
          Viewpoints Help Miners Integrate Theory and Practice   268
          Viewpoint Transforms a Speaker's Understanding of Experience   271
     The Role of Gesture in the Rhetorical Construction of Meaning   272
          Miners Use Gesture to Interpret Gesture   273
          Speakers Use Gesture to Amplify Component Features and Concepts   274
          Gestures Influence the Production of New Meanings in Gesture   276
     Miners Use Gestures Rhetorically in Collaborative Interactions   279
          Speakers Use Gesture to Explore Meanings Embodied in Gesture   279
          Speakers Use Gesture as Queries to Invite Response   279
          Speakers Co-Construct Knowledge in Gesture   280
     Implications for Writers   281

IV   Transforming Experience   285

9   Capturing Experience: The Moment of Transformation   287

     Controversy and Uncertainty Following the Southmountain Disaster   292
     Legal Uncertainties Cloud Agency Investigations   294
          Preventing Self-Incrimination   294
          Unraveling Institutional Authority and Liability   295

Locating Responsible Agents   296

Corroborating Factual Findings   299

Uncertainties in the Process of Rhetorical Transformation   300

  *Differences in Vocabulary Reflect More Fundamental Differences
    in the Relationship to Risk*   301

  *Differences in Language Reflect Underlying Differences in the Ways
    That Miners and Investigators Warrant Judgments About Risk*   303

*The Transcripts Indicate That Miners Gesture, but Interviewers Do Not
  Employ Conventions of Coding That Would Allow Us to See What
  Individuals Express With Their Hands*   307

*Two-Dimensional Maps Cannot Represent the Spatial and Temporal
  Complexity That Is Possible When Speakers Coordinate Speech
  and Gesture*   311

The Complete and Forthright Testimony of Witnesses   317

10   Conclusion: The Last Canary?   320

The Function of Written Documentation in Hazardous Environments   324

Towards a Better Understanding of Workplace Discourse   328

Industrial Strength Documentation in an Electronic Culture   330

  *Expert Systems Are Constructed on a Foundation
    of (Frequently Unarticulated) Tacit Knowledge*   330

  *The Uncertainties in New Technologies Are Less Visible, but the Problem
    Is a Question of Scale, Not Kind*   332

  *The Move Off Shore Has Relocated Many of the Problems
    of Transformation to a More Difficult Rhetorical Context*   333

  *Electronic Commerce Has Not Entirely Eliminated the Problems
    of Industrial Labor, Even Within Its Own Institutions*   333

Implications for a Theory of Rhetoric   334

References   336

Author Index   357

Subject Index   361

List of Figures

Figure 1.1. This sample program information bulletin alerts the coal mining industry to an increase in fatal accidents related to haulage. The cluster of accidents alerts the agency to an underlying problem in safety. Source: McAteer (1996).

Figure 1.2. Memo from William Bruce, Chief, Ventilation Division, regarding MSHA's draft revision of the mine standards for methane-producing mines (June 10, 1983). Source: Bruce (1987, p. 392).

Figure 1.3. Bruce includes a copy of the gassy mine standards. In this second document, he circles typographical errors and other changes that have reduced worker safety. Source: Bruce (1987, pp. 420-421).

Figure 1.4. Bernard's memo demonstrates the presence of many different voices in the text. Source: Bernard (1987, p. 480).

Figure 2.1. Risk management framework. Source: Presidential/Congressional Commission on Risk Assessment and Risk Management (1997, p. 7).

Figure 2.2. The cycle of technical documentation in large regulatory industries.

Figure 2.3. Six critical moments of rhetorical transformation.

Figure 2.4. Mental models approach: Influence diagram for risk produced by radon. Source: Bostrom, Fischhoff, and Morgan (1992). (In Morgan, Fischhoff, Bostrom, & Atman, 2001, p. 49). Reprinted with the permission of Cambridge University Press.

Figure 2.5. Many writers limit the scope of their analysis to a limited set of documents.

Figure 2.6. Perception influences the outcomes of risk decisions.

Figure 3.1. The Physical Barrier Analysis Matrix. This document seems constructed to support the conclusion that physical barriers are "not possible or practical" in most of the cases reported on the form. Source: U. S. Department of Labor, Mine Safety and Health Administration (1988a, p. 63).

Figure 3.2. Sample mine training bulletin. This document shows trainers how to structure accident report narratives. Source: U. S. Department of the Interior, Bureau of Mines (1993, p. 55).

Figure 3.3. The Rhetoric of Risk focuses on the rhetorical transformation and retransformation of experience at the boundaries between agencies and experience.

Figure 4.1. This map shows the areas inby and outby the fire at the Wilberg mine.

Figure 5.1. FATALGRAMs depict an oddly comic distortion of the laws of time and space. Source: U. S. Department of Labor, Mine Safety and Health Administration, Metal and Nonmetal Mine Safety and Health (1991b).

Figure 5.2. MSHA's FATALGRAM foregrounds two miners (on their knees) and presents two others as shadow-like objects in the background as a massive block of rock falls on their head. Source: U. S. Department of Labor, Mine Safety and Health Administration, Metal and Nonmetal Mine Safety and Health (1991c).

Figure 5.3. In this FATALGRAM, both observer (photographer) and object (miner) died together in a mine accident. Source: U. S. Department of Labor, Mine Safety and Health Administration, Metal and Nonmetal Mine Safety and Health (1990).

Figure 6.1. This mine safety pamphlet shows one mining inspector as he assumes three different roles in one inspection. Source: U. S. Department of Labor, Mine Safety and Health Administration, National Mine Academy (1990).

Figure 6.2. In the Bilsthorpe disaster, the roof fall occurred above the level of the bolts, beyond the limits of either technical control or scientific prediction. Source: Langdon (1994).

Figure 6.3. This American diagram presents an idealized picture of the strata or rock layers above the coal seam in the mine. Source: U. S. Department of Labor, Mine Safety and Health Administration, National Mine Health and Safety Academy (1989).

Figure 6.4. Unlike the American diagram, this British diagram depicts both vertical and horizontal pressures in the mine roof and ribs (pillars of coal left standing to support the mine roof). Source: British Coal, Nottinghamshire Area, Lound Hall Training Centre (n.d.).

Figure 7.1. E3 holds tightly to an imaginary wheel in this mimetic gesture.

Figure 7.2. E3 holds the wheel intensely. In this mimetic viewpoint gesture, he reenacts his physical struggle to control the machine with his entire body.

Figure 7.3. E5 uses her entire body to describe the extent and magnitude of a roof fall.

Figure 7.4. E5 imitates her boss's gestures. In the narrative, she imitates his voice. This speaker uses her entire body (note shoulder position) and head to represent another speaker talking.

Figure 7.5. E5 imitates the gestures of her manager. Although she talks about installing roof bolts, she does not employ the characteristic torquing motions (like screwing in a light bulb) that experienced roof bolters employ when they describe how to install a roof bolt. Her gestures demonstrate that manager and miner do not share similar viewpoints in relation to risk and safety.

Figure 7.6. E5 depicts classic "roof fall" gesture.

Figure 7.7.   E5 looks into the space created between her hands. In this case, E5's hands depict the two-foot pads of the t-bar which pinned her to the rib and nearly crushed her.

Figure 7.8.   E5 traces shape of an arch with flat palm of hand, fingers splayed. In a mine, these steel arches are overhead—like the metal supports in a tunnel. As she describes the arches, E5 assumes a viewpoint that is above the arches, looking down upon them, as she traces their shape with her hands. To assume a mimetic viewpoint of the same space, miners hold both hands overhead in the shape of an arch.

Figure 7.9.   In this sequence, E5 represents two viewpoints simultaneously in speech and gesture. In her gesture, E5 holds an imaginary object in her hands. She assumes the character of her sister, who was holding onto the plate during a roof cave-in. In her speech, she describes the event from the distanced viewpoint of an observer.

Figure 8.1.   Libby's original strata/compression gesture.

Figure 8.2.   Strata variant 1: This gesture retains the parallelism that characterizes strata, but the gesture does not highlight the compression component of this gesture.

Figure 8.3.   Strata variant 2: In this variant, Libby's colleague depicts the movement of the strata as parallel plates that shift, but her gesture does not show the compression that Libby depicts in her gesture.

Figure 8.4.   Strata variant 3: This gesture loses some of the visual character of the strata component as Libby's colleague amplifies the movement characteristic of compression.

Figure 8.5.   Libby's colleague explores the meaning of compression in a new 13-second gesture that demonstrates how speakers take up meanings embodied in gesture. In the background, a second colleague co-constructs a smaller version of the same gesture.

Figure 8.6.   Libby's original spacing gesture.

Figure 8.7.   Spacing variant 1: This gesture takes up the notion of size but produces a mismatch in semantic content.

Figure 8.8.   Spacing variant 2: This gesture is noticeably wider and more expansive than Libby's original gesture. The gesture was produced by flipping the wide strata gesture 90 degrees.

Figure 8.9.   Libby's colleague (front, foreground) uses gesture as a query to invite response.

Figure 8.10.   Miners co-construct similar gestures within a fraction of a second. These imitations are so closely coordinated that they seem deliberately synchronized. These gestures show how miners are attuned to the meanings expressed in both speech and gesture. When speakers co-construct gestures, the process of uptake is nearly instantaneous.

Figure 9.1.   Six critical moments of transformation.

Editor's Introduction

The opening paragraphs of chapter 1 of Beverly Sauer's *The Rhetoric of Risk* powerfully frame the gravity of the issues she addresses. For centuries mining has been known to be a dangerous occupation that regularly kills and maims workers. Despite almost a century of regulation in the United States, miners are still regularly killed and maimed.

Miners work in threatening, constantly changing, uncertain physical environments. Government regulation is the work of symbols, reports, regulations, and training manuals. How can the work of paper pushers affect this most embodied and bodily-threatening of labors? How can the work of symbol makers ever capture the realities of the miners, miners' judgments, and the mines? What important knowledge gets lost as the moment-by-moment physical realities of ordinary operations and moments of disaster are translated into the quiet of offices and hum of computers? How can what is learned and regulated in an office be made real, present, and meaningful in the dark of the mines? How can this process of translation from physical danger into reflective comment, rule and guideline making, and training and then back into miners' *in situ* judgement and decision making be improved to lead to a more intelligently safe practice? These are the weighty questions Sauer takes on and brings remarkable clarity to.

As editor of the Rhetoric, Knowledge, and Society series, I have had the opportunity to introduce many truly wonderful books, but none I think so graphically shows the importance and delicacy of written language in the world. The records, reports, science, and training manuals have provided major tools to develop mining technology. Mine safety texts have provided ways to reflect on practice, find shortcomings in practices, calculate possible dangers, standardize preferable practices, inform all above and below ground of useful knowledge. But mines and mining are more real and complex and unknowable than texts can yet capture and regularize. Without miners' embodied knowledge texts are incomplete and dangerously misleading. And without being integrated into miners' embodied sense of the physical world they move in, texts are of no use at the moments when they are needed, because no one can check a book as the roof of the shaft is collapsing.

Significantly, Sauer finds the fulcrum of her investigation in gesture—where embodied sense emerges into the world of representation and symbols. It is through gesture that the miners say what they sense and start to articulate those things they experience, and it is through gesture that the abstractions of the world can be brought into the material space of bodily being.

While the poignancy and importance of lives at risk give compelling meaning to Sauer's work, the book also carries significant general meanings for rhetoric and communication. The book provides a deep and complex example of the way texts enter into complex systems of material and social practice. It explores the boundary between embodied experience and abstract representation. And it starts to unpack how gesture sits between experience and representation and can inform both. In doing all these things, Sauer's book pulls rhetoric from the place of words, the *agora,* to the place where physical danger meets the complex social organizations of workers, unions, industry, government, law, engineering, and science. Although Sauer frames her argument within classical rhetoric, she implicitly challenges rhetoric to consider problems unimagined in the classical canon, even as she challenges technical communication to consider the complexly situated rhetorical nature of risk communication in the workplace.

Charles Bazerman
University of California, Santa Barbara

Introduction: The Rhetoric
of Risk

The Shuttle Challenger disaster, the crash of an Amtrak train near Balti-
more, the collapse of the Hyatt hotel in Kansas City, the incident at Three
Mile Island, and other large-scale technological disasters have provided powerful examples of the ways that communication practices influence the
events and decisions that precipitate a disaster. These examples have raised
ethical questions about the responsibility of writers within agencies,
epistemological questions about the nature of representation in science, and
rhetorical questions about the nature of expertise and experience as grounds
for judgments about risk. The answers to these questions frequently reflect
more general concerns about the nature of modernity, the role of language as
a determinant of cultural practices within a postmodern society, and the nature and distribution of risk in a technological society.

The present project grew out of these investigations, but my own work has
taken a much different focus—the result of a single memo from William
Bruce to Madison McCulloch.[1] In this memo, Bruce (1987) argues that errors
introduced into mine safety standards were not only an embarrassment to the
agency, but they had actually diminished worker safety. That memo changed
my career: It explains how a woman living in Maine became engaged with
the politics and production of documentation in one of the world's most dangerous industries. I began the research to determine how gender and power
might be inscribed in the official communication practices that affected min-

---

[1]Bruce, 1987. The memo is reproduced in chap. 1.

ers' lives.[2] I have ended this project with a profound respect for the men and women who have struggled for more than a century to meet this nation's energy needs. Talking with men and women at all levels who deal with risk judgments on a daily basis, I have confirmed my belief that these individuals have much to contribute to an understanding of the events and decisions that precipitate a disaster—if we can develop interpretation and transcription practices to capture their understanding in writing.

This is a book about rhetorical practices, not mine safety, though mine safety is central to the questions I am asking. Rather, mine safety provides a rich technical and historical context where problems of rhetorical agency, narrative, and the negotiation of meaning have visible and tragic outcomes. When I am asked, "What's at stake?", I frequently answer, "Miners' lives." But that answer does not address what is at stake in rethinking technical documentation from the perspective of those who work in hazardous environments.

In "Sense and Sensibility: How Feminist Approaches to Technical Documentation Can Save Lives in the Nation's Mines" (Sauer, 1993), I argued that communication practices that documented conditions "for the record" might fail to effect change in risky environments. In *The Rhetoric of Risk*, I examine the formal and informal communication practices in US, British, and South African mines in order to understand why some documents might be said to "fail" in the context of risk. What are the visible (visual and verbal) forms of communication in hazardous environments? To what extent do these forms provide readers with the information they need to assess and manage risk? What is the rhetorical function of these documents within large regulatory industries? What rhetorical strategies do individuals use when they describe their observations and experience in these environments? What happens when writers attempt to document these observations and experiences in writing? What makes documentation so difficult?

By exploring the strategies that individuals employ to make sense of the material, technical, and institutional indeterminacies of their work, I hope that my research will help rhetoricians rethink their (frequently) unquestioned assumptions about workplace discourse and the role of writers in hazardous worksites.[3]

Readers familiar with rhetorical theory will recognize the influence of two important mentors in my work. Richard Young's work in invention has

---

[2]Cf. Gaventa, 1980; Lukes, 1989. I am grateful to Howard Segal, who introduced me to these authors.

[3]One salient example is the unquestioned textbook assumption that well-written instructions can produce predictable and certain outcomes.

helped me frame these questions as a problem of rhetorical invention. J. V. Cunningham's influence helped me build the foundations of my claims brick by brick. If my method is eclectic in drawing on many different sources to test and evaluate any single piece of evidence, that eclecticism reflects my assumption that rhetoricians must move outside of disciplinary and institutional boundaries if they are to discover the full range of rhetorical strategies available to rhetors in any given situation. Aristotle called this the problem of invention.

RHETORICAL INVENTION
IN THE CONTEXT OF RISK

In defining the kinds of questions that rhetorical theory might address, Aristotle defined rhetoric as the "art of finding out the available means of persuasion."[4] Aristotle acknowledged that rhetoric dealt with questions of uncertainty in deliberation, judgment, and evaluation. As a theory of public discourse, Aristotle's Rhetoric did not define how individuals might find the correct answer (as physicians might prescribe a remedy) but rather how individuals might employ a theoretical framework to discover arguments that might be effective in public deliberation and judgment.[5] Aristotle thus distinguishes between those proofs which are givens and those proofs (or arguments) that the rhetor constructs to persuade audiences in particular situations. This distinction is fundamental to the notion that evidence alone will not persuade audiences; instead, rhetors must have at their command a range of rhetorical strategies that address the knowledge, understanding, values, belief systems, fears, hope, and shame of the audiences they seek to persuade. Aristotle recognized that foreign audiences might not share the same set of *enthymemes* or commonplaces that he describes in his Rhetoric.[6] In unfamiliar domains, rhetors must first find out an audience's attitudes and beliefs in order to move the audience to agreement.

Studies of science have demonstrated the degree to which evidence in science is also shaped by a shared set of beliefs and practices. Following Bloor (1991), social scientists have argued for a more reflexive approach to ques-

---

[4]Aristotle, 1991, p. 37.

[5]Aristotle, 1991.

[6]Kennedy (1991) defines an *enthymeme* as "a rhetorical syllogism, i.e., a statement with a supporting reason introduced by for, because, or since or an if . . . then statement. In contrast to a logical syllogism, the premises and conclusion are ordinarily probable, not necessarily logically valid. A premise may be omitted if it will be easily assumed by the audience" (Kennedy, 1991, p. 315).

tions of controversy in science; and they have argued that judgments about risk can profit from an integrated approach that draws upon the knowledge and understanding of lay audiences in the deliberation of questions of policy and value. But the notion that all knowledge is ultimately socially constructed does not address the rhetorical problems that policymakers face when they attempt to predict (and prevent) disaster: What counts as persuasive knowledge in the context of risk? Are there rhetorical principles that can guide decision makers as they attempt to document rapidly changing conditions in hazardous environments? How do written and oral communication practices affect the negotiation and construction of knowledge in risky environments?

Previous rhetorical studies of the rhetoric of science, technology, and risk have, in large part, followed Aristotelian models and have employed Aristotelian vocabularies to describe the role of *ethos*, *pathos*, *logos*, *stasis*, and *kairos* (to name but a few terms) in risk communication and deliberation.[7] Rhetoricians have analyzed the rhetoric of environmental policy, medicine, laboratory science, nuclear power, computer science, and engineering; and they have played an important role in understanding the problems of communication in risky environments.[8] They have shown how rhetorical theory can contribute to an understanding of communication practices in science,[9] and they have used feminist theory to articulate how the discourses of science and technology reflect silent and salient power structures that deliberately or inadvertently silence the voices of women and others in matters of science and technology.[10]

By demonstrating the social constructedness of knowledge claims, rhetorical theorists have helped us understand the nature of the invisible negotiations and power relationships that influence knowledge in science.[11] Rhetorical theory has provided a powerful tool for understanding the rhetorical

---

[7]See especially Dombroski, 1992; Herndl, Fennell, & Miller, 1991; Miller, 1989; Miller, 1994; Miller, 1998.

[8]Bazerman & Paradis, 1991; Cohn, 1987; Dombroski, 1992; Frank & Treichler (Eds.), 1989; Herndl, Fennell & Miller, 1991; Killingsworth & Palmer, 1992; Winsor, 1990.

[9]Fahnestock, 1996; Gross, 1990; Myers 1990; Prelli, 1989; Selzer, J. Ed., 1993; Simons, 1980; Stewart & Feder, 1986.

[10]Brasseur, 1993; Cohn, 1987; Gershung, McConnell-Ginet, & Wolfe, 1989; Hacker, 1990; Keller, 1985; Lay, Gurak, Gravon, & Myntti (Eds.), 2000; Sauer, 1993; Sauer, 1994. Merchant's (1983) feminist accounts of mining were important in helping me think about the relationship between nature and science in early modern mining.

[11]Activity theorists have extended the notion of text and action to investigate how texts and actors interact in the material world of work. Social scientists have adopted the language of rhetoric to examine the social history of science as an institution and the force of culture in shaping scientific communities. See Engeström, 1996; Hutchins, 1996; Scribner, 1997.

underpinnings of the human sciences that support quantitative and qualitative studies of risk—most famously McCloskey's analysis of the *Rhetoric of Economics* and more generally in what is now known as the Rhetoric of Inquiry.[12] This research has been particularly important in showing how rhetorical practices affect "the content and conduct of communication—as much as the communication—of research" and how—on a practical level—text and activity interact in the workplace.[13] More recently, Bazerman (1999) provides a model for the kind of detailed contextual work that we need to understand how technologies achieve both symbolic and material success—finding their "fit" within an already established system of values, discourse patterns, and human action.[14]

My own work shares with Aristotelian theorists a belief that rhetoric is an inventional art—an art of "finding out the available means of persuasion."[15] It couples this Aristotelian perspective with a postmodern and feminist awareness that existing communication practices within industries are shaped and constrained by political and economic assumptions that may inadvertently silence or render invisible the kinds of information that decision makers need to assess and manage risk in hazardous environments. The problem of rhetorical invention—revisited as problem of discovering or "finding

---

[12]McCloskey, 1985 & 1992; Nelson, Megill, & McCloskey (Eds.), 1987.
[13]Nelson, Megill, & McCloskey 1987, p. ix.
[14]Bazerman (1999) writes:

> For any technology to succeed (that is, to establish an enduring place within the world of human activities), it must not only succeed materially (that is, produce specified and reliably repeatable transformations of matter and energy); it must also succeed symbolically (that is, adopt significant and stable meanings within germane discourse systems in which the technology is identified, given value, and made the object of human attention and action). (p. 335)

Bazerman's work extends the meaning of persuasion to include the "entire range of actions occurring across all discourse networks" (p. 341). Persuasion in this sense is not confrontational, but may include acts of cooperation, negotiation and assent. As Bazerman argues, as technologies increase in complexity, they are accompanied by increasingly complex "discursive networks necessary for the production, maintenance, and use of the technology: laboratory notebooks, installation and repair manuals, patents, monthly bills for service, publicity pamphlets, technical journals, city ordinances, and stock reports" (p. 336). In the *Languages of Edison's Light*, Bazerman (1999) demonstrates how individual elements in the discursive system must persuade individuals to coordinate their work with one another, invest in stock offerings, produce a new part according to specifications, or write positive stories that sell the product (p. 241). In each case, the criteria for success depends upon the tasks to be accomplished by the utterance. His own work demonstrates the range of acts, texts, and activities that produce a successful technology.
[15]Cf. Aristotle, 1991.

out" the range of rhetorical practices in hazardous worksites—is thus simultaneously a theoretical problem and a methodological problem for rhetorical
theorists: How do we discover the means of persuasion that are not recorded
in writing, inscribed in textual practices, and authorized as conventional
within the disciplines and institutions we choose to study? To paraphrase
Lynch (1985), the issue is not *which* representation to analyze but rather how
to determine what counts as a representation or as the visible form of a representation, in the setting of that representation's production.[16]

For rhetoricians, the problem of invention is not to make visible the already conventionalized and schematized categories of neo-Aristotelian theory *within* documents whose governing assumptions are derived from those
categories, but rather (a) to investigate the full range of genres and communication practices that arose in response to particular problems of work, risk,
authority, uncertainty, and disaster and (b) to make visible those margin-
alized forms of representation that might not be visible with conventional
methods of analysis. This is the central methodological problem of feminist
theory and the crux of postmodern cultural studies—to make visible the silences or invisible knowledges that are not present in written texts.[17]

For my own purposes, I have turned to psychological and linguistic studies
of gesture in order to penetrate the silences in technical documentation in
hazardous environments.

My own research suggests that gestures play an important, though largely
unexamined role, in workplace communication. Workers use gesture to communicate their understanding of their environment, but the information
conveyed in these gestures may not be captured in writing if researchers do

---

[16]Lynch (1985) writes:

The issue of "unique adequacy" in the use of a material record of work is not a question of
empirical grounding of a theory. Any of the varieties of records which are available at
the scenes of work, or which are produced through the constructive involvement of the
analyst within the scene (tape recordings, ethnographic field notes, films, photographs,
interviews, questionnaires, etc.) is adequate as "grounds" for particular claims and arguments on which "data" reveal. What is raised as a problem here is not a matter of the validity or reliability of such records of work. Instead the problem is a matter of what
counts as a record of work, or as the visibility of the work, in the setting of that work's ac-
complishment. It further involves issues of what ways of addressing such records reveal
the work's observable and reportable detail in a way that would not initially idealize
these details through the devices of extrinsic descriptions. (pp. 9–10)

[17]Perelman and Olbrechts-Tyteca (1969) reenergized the study of Aristotelian rhetoric and
argumentation with their New Rhetoric, but their work does not invent a new rhetoric. Much
of the rhetoric of science has borrowed from Aristotelian frameworks in order to demonstrate
the presence of rhetoric in scientific texts and activities.

not have an adequate system of documentation for recording and analyzing the presence of information embodied in gesture. My analysis of accident investigations demonstrates that investigators follow legal conventions when witnesses gesture (see chapter 9). At these moments, the text acknowledges that the witness "indicated" something, but the text does not document what the witness indicated. My interviews with miners suggest that these moments might provide information not visible in the written text, but we cannot determine whether this information can contribute to risk management and assessment until we can actually see and interpret what miners are representing with their hands.

FINDING OUT THE AVAILABLE MEANS OF PERSUASION

Analysis of gesture helped me recapture some of the embodied knowledge that is lost when writers attempt to capture local knowledge in writing, but gesture is only one of the rhetorical strategies potentially available to individuals when they talk about their work.

To discover the full range of rhetorical strategies that individuals employ when they observe, analyze, and assess risk, I interviewed miners in the United States, Great Britain, and (most recently) South Africa during a five-year period from 1992 to 1997. Many of these interviews took place within mines and training sites. Mine management has been extremely generous in providing me with both the time and the access to miners. In one large British mine, miners were given time off (on a Friday night) and were paid a regular shift's work to speak with me. I observed and participated in apprenticeship training programs in England and one-day refresher training in the United States. (At one mine, I passed the roofbolting supervisor's training program.) Given the constraints of production, such time is particularly generous and illustrates the concern with safety among both managers and miners.

One concern in these interviews, of course, is the degree to which miners who were paid to talk with an American interviewer felt free to discuss problems of safety and risk. Miners' comments about the interviews suggest that they were nervous at first to talk with a woman presumed to represent management interests. But many miners seemed to relax as the interview progressed. They frequently phrased critical comments with the warning "I'll probably get drawn and quartered by 'X' "—their manager or supervisor, but criticism seems to be accepted within safety circles, particularly when the aim is improving safety.

To validate the results of these interviews, I also interviewed miners and miners' wives in their homes, in pubs, in union meetings, in cars, and—in one case—in a taxi driven by a former miner. In these interviews, miners were more cautious, particularly in England, where miners targeted as troublemakers might lose their jobs because of the tight economic climate and bad feelings that followed the mine closures and privatization of British mines. (In one case, a miner spoke with me in a pub in a distant city because his wife feared that he would lose his job.) One particularly vocal union safety official was unemployed following the pit closures. He openly criticized management practices, but he had not worked in several years.

Women against Pit Closures provided me with another perspective on safety. Miners' wives play an important role in British mine culture. Unlike miners, who may fear losing their jobs, women can more freely protest pit closures and economic changes in the mines. They also speak freely about the anxieties they face every day when their men are underground, the risks of mining, and the signs of risk they see when their men return home: The dusty clothes and headaches that signal the conditions for an explosion, for example, or men too tired to think about safety.

In the United States, women miners also provided me with a different perspective on safety, and I am grateful to miners in the Coal Employment Project for including me in their local meetings. Many of these women entered the mines in the early 1970s. Last hired, they were the first fired as U.S. mines reduced their workforce, but they remain committed to mine safety and to training women to enter the workforce as equal partners with men.

In addition to miners, I conducted semi-structured interviews with engineers, safety trainers, and writers at the U.S. Mine Safety and Health Administration; the U.S. Bureau of Mines; the (former) British Coal Training Centre at Eastwood Hall, England; Her Majesty's Health and Safety Executive in Bootle, England; the U.S. Bureau of Mines; and the Council for Scientific and Industrial Research (Miningtek) in Johannesburg, South Africa, in order to understand the rationale for documentation practices and procedures. I interviewed managers, undermanagers, safety engineers, and specialists in electrical hazards, explosions, ventilation, and roof support at both large and small mines in the United States, Great Britain, and South Africa. I am particularly grateful to Bob Painter and other personnel at the U.S. Mine Safety and Health Administration (MSHA) for helping me understand the relationship between accident reports, policy, and procedures, and to Chris Mark and George Schnakenberg at the U.S. Bureau of Mines (now NIOSH) for help in this project.

Finally, I examined agency reports, training manuals, safety training research, public inquiries, accident investigation reports, video training materi-

als, and engineering reports describing all facets of safety training and instruction in the United States, Great Britain, and South Africa in order to
compare how culture affected documentation practices. Many of these materials were dispersed when British Coal privatized mines in 1994, but they
have been collected in small libraries at colliery training centers. They enabled me to examine the nature of technical documentation under British
Coal. But they also demonstrated how the content of local training programs
can become quickly outdated and limited in scope without centralized and
ongoing research and administration.

   In the five years since I began the project, I myself had seen so many
changes that it seemed difficult to make any generalized statement about technical documentation in mining disasters. Originally, I had intended to compare U.S. and British safety training programs at a time when British Coal
training programs were thought to be superior to U.S. mine safety training programs. In 1994, however, British Coal privatized its mines and disbanded its
central training center. Miners in the north were particularly hard hit. The
John T. Boyd Report had recommended mine closures on the basis of the
mine's history of labor relations and its potential for introducing new technologies that could reduce costs.[18] This meant, first, that mines with an active union presence were threatened with closure and, secondly, that miners could be
pressured to increase production to keep mines operating. In practice, this
meant that union miners were less likely to agitate for changes in the safety culture of the mine when the life of the pit was on the line. In response to the
Boyd report, British mines also introduced the American system of roofbolting
to save costs. When Britain's first major fatality occurred in a mine that had recently introduced roof bolts (the Bilsthorpe disaster), British miners blamed
management for cutting costs at the expense of miners' lives.

   Mining was changing in other, more subtle ways, as new technologies and
training replaced traditional practices. In one mine, I videotaped the last canary in a British mine. This canary came to symbolize the emphasis on embodied experience that characterizes my research. Like many British miners,
this canary had been forcibly retired (in British terms, "made redundant") in
the newly privatized mines. For 200 years, miners had depended on canaries
and flame-safety lamps to sense changes in the atmosphere that signaled the
presence of Black Damp and explosive methane. Canaries could be replaced
by methanometers that could produce—in theory—more scientifically accurate readings of mine gases. As the mine manager left with the canary in a
cardboard box, he made a fatal pronouncement in a heavy Yorkshire accent:
"Thus ends two hundred years of industrial history."

---

[18]John T. Boyd Company, Department of Trade and Industry, 1993.

In the United States, major changes had also occurred since the beginning of the project. During the Reagan Administration, mines had introduced new mining technologies, including longwall mining. These technologies promised increased production and lower costs. But these new technologies also introduced new problems of safety—problems that could not be addressed by new regulation because the Reagan Administration had imposed a freeze on all new regulation. When the freeze was lifted, MSHA struggled to implement new regulations. Under the Clinton Administration, Davitt McAteer, a strong labor advocate, assumed the post of Chief of Safety. When I first conducted interviews at MSHA in 1992, McAteer had been a clear outsider.

Historically, mine safety standards had been separate from the more general standards for occupational safety promulgated under the U.S. Occupational Safety and Health Administration (OSHA). In 1998, the Bureau of Mines was reorganized, and mine safety research was absorbed by the National Institute for Occupational Safety and Health (NIOSH). Bureau personnel were relocated to new projects—like diesel emissions—that seemed distant from the Bureau's historical concerns about dust, ventilation, ground control, and explosions.

In 1992, I had focused my research on U.S. mine safety training programs. In 1994, I expanded my research to compare U.S. and British mine safety training programs. Although U.S. mines were three times more dangerous than British mines (prior to 1994), conditions in British mines changed rapidly in 1994 with the denationalization of British Coal. At the same time, safety standards in British mines were threatened by cheap coal from South Africa and Eastern Europe. Safety is expensive in an industry where production is measured in tons per second. To remain competitive, the newly privatized mines were forced to push production and reduce outlays for safety training. Sitting in a room with 20 novice miners in a training center in Doncaster, I wondered which of the eager men in front of me would lose fingers or limbs in a mining accident. I could not think of technical documentation as an abstract ideal divorced from the realities of production.

Like the global economies they fueled, mining communities throughout the world are tight-knit and closely connected networks. American miners introduced me to British and South African miners. When I spoke with miners in the north of England, they had already checked me out to see whether I was "safe." Throughout my research, I have taken seriously the provisions of the Human Subjects' Protection that governs my interviews with miners. As a result, individuals' names have been changed and identifying features have been removed to insure miners' privacy and anonymity when requested, but I have attempted to maintain distinctions between British, U.S., and South

African miners where such distinctions reflect differences in culture, language, geology, regulatory regimes, or technical and management practices.

Readers more comfortable with statistical methods may perhaps be uneasy with what they see as the attempt to build theory from isolated instances of the oral, written, and gestural production of meaning in these chapters. If, on the other hand, we view each instance as one individual's momentary solution to a problem of communication, we can then see how each isolated instance provides a new possibility for extending our understanding of the rhetorical strategies available to individuals in hazardous worksites. With a limited number of examples, we cannot determine the frequency of this instance within an extended discourse community. Instead, we must see the discovery as both epistemic and inventional. Each time we attempt to interpret and document a new instance, we extend our knowledge of workplace discourse.

INTEGRATING STAKEHOLDER KNOWLEDGE IN EXPERT JUDGMENTS ABOUT RISK

Risk specialists and rhetoricians share a common interest in discovering how their intended audiences think about risk. As Kunreuther, Slovic, and MacGregor (1996) demonstrate, the most successful risk decisions result from "an atmosphere of trust between the proponent and the host community."[19] To develop this trust, they encourage agencies to develop a "broad-based participation process" that involves stakeholders at all phases of problem definition and risk management and assessment within large regulatory agencies.[20]

Risk specialists have created a variety of frameworks to depict the process of risk management and assessment within large regulatory agencies (chapter 2).[21] But they disagree about the role that stakeholders play in expert judgments about risk.[22] Many risk specialists applaud the attempt to involve

---

[19]Kunreuther, Slovic, & MacGregor, 1996.

[20]Kunreuther et al., 1996.

[21]Presidential/Congressional Commission on Risk Assessment and Management, 1997.

[22]The 1997 Presidential/Congressional Commission on Risk Management and Assessment demonstrates the tensions that occur when agencies value stakeholder involvement but fail to define a set of rhetorical strategies that might help them represent stakeholder knowledge in regulatory decision making. In its final report (1997), the Commission proposed a generalizable framework to help agencies integrate stakeholder values into the process of problem definition and risk management within large government agencies. This framework depicts the process of risk management and assessment as an iterative process of problem definition and evaluation that allows stakeholders to serve as partners at any stage in the process. (Chapter 2 describes the agency's framework in detail.)

stakeholders, but they doubt that stakeholders can participate in the regulatory process without some sort of technical assistance. They argue that stakeholders "should not participate directly in the assessment itself" to avoid risk assessments becoming "too politicized."[23]

Proponents of increased stakeholder involvement have countered that experts are no less susceptible to error than lay audiences, and they frequently make errors in judgments. Citing Kahneman and Tversky, Shrader-Frechette (1990) argues that experts are likely to share many of the biases that affect lay judgments about risk, including sampling errors, availability errors (assessing the frequency of a class, or the probability of an event, by the ease with which instances or occurrences can be brought to mind), and anchoring errors (making estimates on the basis of adjusting values of an initial variable).[24] Shrader-Frechette argues for regulatory reform in order to "protect us from the most dangerous consequences of expert error and to insure us that the lay people most likely to be affected by a risk have a larger voice in making public policy regarding it."[25]

Both sides in this debate believe that research in communication can provide a tool to "gain the confidence of stakeholders, incorporate their views

---

[23]Presidential/Congressional Commission on Risk Management and Assessment, 1997, p. 185.

[24]Kahneman & Tversky, 1981, p. 46 (In Shrader-Frechette, 1985, p. 16).

[25]Shrader-Frechette, 1990, p. 18. Shrader-Frechette suggests that we must reform regulatory decision making in "at least" three different ways:

First, instead of having experts perform a single study, we need to develop alternative technology assessments or environmental impact analyses, weighting them on the basis of different value systems and different epistemic or methodological value judgments. Second, we need to debate the merits of these alternative analyses, each with its own interpretational and evaluative weights. In this way citizens can decide not only what policy they want, but also what value systems they wish to guide their decisions. Third, in areas where assessors obviously have more technical knowledge, e.g., of probabilities, we need to weight expert opinion on the basis of past predictive successes. In other words, we need to calibrate the scientists and engineers who provide information relevant to policy choices. Let's examine some of the reasons for each of these moves.

Since no necessary connection exists between Pareto optimality (the central concept of benefit-cost analysis) or Bayesian rules and socially desirable policy, it would be helpful if there were some way to avoid the tendency to assume that economic methods or Bayesian rules, alone, reveal socially desirable policy. Alternatively weighted assessments would enable persons to see that sound policy is not only a matter of economic calculations but also a question of epistemological and ethical analysis, as well as citizens' negotiations. (pp. 26–27)

This analysis begs the question of whether citizens, experts, and regulatory agencies will be able to create new collaborative mechanisms that achieve the processes she recommends.

and knowledge, and influence favorably the acceptability of risk assessment and risk management decisions."[26] But they disagree about the kinds of solutions that can produce change. They encourage risk managers to listen to constituents to improve risk communication, but their arguments often reinforce the notion that experts and lay audiences are separate communities that operate in complementary but different spheres of influence.

Morgan et al. (2002) have proposed a "mental models" approach to risk communication to overcome some of the problems of integrating stakeholder knowledge into regulatory risk assessment.[27] The term "expert" in the phrase "expert model" does not connote an individual agent. Instead, the term serves as an adjective that denotes the character of the knowledge represented in the model. In rhetorical terms, an expert model is the end process of highly inventional thinking. To develop an expert model, researchers interview experts (what we might call "authorities on the subject") and summarize their beliefs in an influence diagram. The expert model summarizes and integrates many different types of expertise in a single formulation. In simple terms, an influence diagram attempts to capture a snapshot of all the events, decisions, and conditions that produce or shape the outcome of risk decisions. Risk specialists can use these models to compare an individual stakeholder's knowledge with the idealized expert model. They can then construct risk communication messages that provide people with the "information they need in a form that fits their intuitive ways of thinking."[28]

Although much of the information in the model may be extraneous to actual risk decision making, risk specialists recognize that lay audiences need background information to understand numerical estimates of risk.[29] When risk specialists incorporate influence diagrams in the model, the expert model can show causal relationships and correlations that influence the final outcome. The ideal expert model thus shows how each node in the model is more than just an isolated piece of information. Individuals can see how the information represented in the node can influence the outcomes of risk decisions. The process involves much iteration and testing: "Getting questions to work for both lay people and experts can require several iterations, involving

---

[26]Presidential/Congressional Commission on Risk Management and Assessment, 1997, p. 39.

[27]Morgan, Fischhoff, Bostrom, & Atman, 2002.

[28]Morgan, Fischhoff, Bostrom, Lave, & Atman, 1992, p. 2050.

[29]Morgan et al. (1992) argue that lay audiences need to understand the physical processes that create risk. They want to develop competence so that they can formulate their own opinions about future options. If they know nothing, risk communications will be incomprehensible.

both subject matter and communication specialists."[30] Morgan et al. (2002) warn that the model frequently uncovers many more misconceptions and gaps than can be "filled" with a reasonably short risk communication message. They thus advise researchers to set priorities and identify those factors that will most significantly affect an individual's risk judgment. When issues involve scientific uncertainties, risk specialists must decide whether to include a balanced approach that represents all sides in the controversy; whether they should provide information and allow audiences to draw their own conclusions; or whether they should represent information that will achieve the desired outcome.

BRIDGING THE BOUNDARIES BETWEEN RISK AND RHETORIC

As Selber (1999) points out, researchers in technical communication have resisted "rationalistic" approaches that are "blatantly system centered."[31] Thus, despite what seems like a natural alliance in investigating the nature of deliberation, communication, and probabilistic reasoning in the context of high levels of ethical and political uncertainty, rhetorical theorists have not taken advantage of the insights and research findings of risk specialists to understand the nature of decision making in uncertain, dynamic, and hazardous environments.

At the most obvious level, risk specialists can help rhetoricians elicit what audiences know and how they structure their understanding of complex issues. They can also help rhetoricians think more systematically about the material processes that produce risk and hazard in the workplace. Morgan et al. (2002) write:

> The expert model is an attempt to pool in a systematic manner, everything known, or believed, by the community of experts that is relevant for the risk decisions the audience faces. . . . Despite the name, the model need not exist in any one expert mind. Even were a single expert to have such comprehensive knowledge, this expert modeling procedure is a reactive one; it forces the experts to think more systematically about their beliefs than they might have otherwise. As a result, they may have different beliefs at the end of the interview process then they had at its beginning.[32]

---

[30]Morgan et al. 2002, p. 91.
[31]Selber, 1999, p. 468.
[32]Morgan et al. 2002, p. 23.

But risk specialists can also benefit from an understanding of the rhetorical practices that influence how we document what audiences know.[33]

First, a careful descriptive analysis of workplace discourse can contribute to the development of a generalized expert model in many important and complementary ways: Because the interview process can affect a subject's beliefs and values, researchers must be particularly careful when they attempt to elicit their lay audience's beliefs and misconceptions.[34] If interviews change a subject's knowledge, then interview results will produce a skewed conception of what audiences need to know to understand risk. When interviewers employ a conversational style, for example, they can introduce new words or ideas,[35] but they can also overlook critical concepts. If the purpose of the interview is to discover people's mental models so that communicators can construct risk messages to affect a wider audience, the interview process must not create misconceptions or reshape the subject's mental model. When researchers employ a variety of methods, they must insure that the different forms of inquiry (questionnaire or interview, for example) do not affect their results. Morgan et al. (2002) note that "designing good questions is not easy."[36]

Second, research in rhetoric can help risk specialists identify particular rhetorical strategies that might influence the outcomes of their research. It can draw attention to features of texts and narratives that influence interpretation and meaning; and it can help researchers understand how an agency's documentation practices may inadvertently silence information critical to a speaker's understanding of risk.

Finally, researchers in rhetoric can help risk theorists develop a contextual framework for theoretical models of risk decision making. As Fischhoff (1996) argues in an article entitled "The Real World: What Good Is It?", laboratory experiments in risk decision making often bear little resemblance to real-world decision making. In laboratory experiments, researchers deliberately seek to limit the number of distinct options, clear consequences, and intervening uncertainties. When real teenagers talk about their decisions, he concludes, the intellectual demands of their decision-making tasks "overwhelmed the theoretical capacity of economics."[37]

---

[33]Most writers resist the notion that writing can be separated from the processes of invention and knowledge production. They see language as inextricably tied to semantic content and meaning. For writers, language structures thought. They believe that writers must understand how specific features of rhetorical practice influence the structure of beliefs and the organization of ideas within any knowledge domain.

[34]Morgan et al. 2002, p. 24.

[35]Morgan et al. 2002, p. 24.

[36]Morgan et al. 2002, p. 86.

[37]Fischhoff, 1996, p. 21.

As *The Rhetoric of Risk* demonstrates, the ideal expert model exists in a fragile tension with the pressures of production and the economics of decision making in hazardous worksites. Idealized rhetorical models also exist in tension with the capacity of individuals to document the knowledge needed to assess and manage risk. Real environments are dynamic, uncertain, and complex; real writers work under stress to meet deadlines. These individuals may not identify themselves as writers by occupation or training, but they are nevertheless responsible for documenting events and conditions that might produce disaster.

Ideally, individuals list all feasible courses of action, evaluate outcomes, determine the various probabilities, and decide on an optimal course of action based upon the probabilities of a outcome.[38] They weigh options, consider the net benefits, and evaluate the likely consequences before they make their decisions. In real time, few individuals have the time or resources to predict all of the elements that might produce disaster.[39] When a crisis occurs, individuals need adequate documentation to help them make decisions, and they need appropriate strategies to persuade others to act to reduce and manage risk.

Rhetoricians cannot create a single document to present all of the information that audiences need, but they can help individuals and agencies understand how discourse practices create gaps in the official record that paralyze agencies and leave workers unprotected.

The rhetorical problem is to discover the full range of communication practices, both documented and undocumented, (a) so that individuals can make strategic interventions to evoke more complete (more "adequate") accounts of the complex interaction of events, decisions, and conditions that precipitate disaster and (b) so that they can understand what is lost or rendered invisible in written documentation. In arguing for a more complex understanding of documentation practices, I am not arguing that complexity alone will facilitate decision making. Decision makers need tools to sort through the complexity, and they need rhetorical knowledge to understand how audiences construct and negotiate meaning as they communicate with others in their work.[40]

Although rhetoricians do not ordinarily distinguish between knowledge and its representation, rhetorical knowledge (knowledge about documentation practices) cannot tell us which specific standards or which specific forms

---

[38]Fischhoff, 1988.

[39]Fischhoff, 1988.

[40]Cf. Flower, 1994. I am grateful to Linda Flower for her suggestions in this chapter in regard to the negotiated construction of meaning in the discourse of mining.

of testimony will provide the best level of protection for workers. Like Aristotle, we must leave the "full examination" of such practices to political science.[41] At the same time, we believe that analysis of the rhetoric of risk can help agencies discover what kinds of information are important, how they discover this information in the course of their work, and how they document this information in writing.

This process can also help risk specialists develop new strategies to learn more about what stakeholders know so that they can construct appropriate risk communication messages that build upon and augment the knowledge and experience of lay audiences.

THE CYCLE OF TECHNICAL DOCUMENTATION IN LARGE REGULATORY INDUSTRIES

My research enabled me to develop a framework for understanding the Cycle of Technical Documentation in large regulatory industries.

This framework enables us to identify Six Critical Moments of Rhetorical Transformation in large regulatory industries. At these moments, writers must extract information that is presented in one rhetorical modality (oral testimony, for example) and literally change the form so that the information can be re-represented for a different audience: (1) when oral testimony and embodied experience are captured in writing; (2) when the information in accident reports is re-represented in statistical records; (3) when statistical accounts are re-represented as arguments for particular policies; (4) when policies and standards are transformed into procedures; (5) when written procedures are transformed into training; and (6) when training is re-represented to workers at local sites. (Chapter 2 describes the cycle in detail.)

Because an investigation of all six moments of transformation is impossible in a single volume, The Rhetoric of Risk focuses on two specific moments of transformation and re-transformation at the boundaries between agencies and experience: accident investigations (when oral testimony and embodied experience are captured in writing) and training (when procedures are re-represented to workers). As I discuss in chapter 2, this focus reflects a deeper theoretical concern with the specific rhetorical transformations that take place at the rhetorical interface between agencies and the material sites they seek to regulate. I thus distinguish the rhetorical transformation of experience and its retransformation in instruction and training from the more visi-

---

[41]Aristotle, 1991, p. 53.

ble rhetorical transformations that take place as writers negotiate and construct new knowledge *within* organizations at other moments in the cycle.

The chapters that follow use this framework to examine how agencies attempt to reconcile diverse viewpoints to make sense of accidents (chapters 4
and 5); how embodied sensory experience is rendered invisible in writing
(chapter 6); and how speakers' gestures help them understand the temporal
and spatial complexity of a hazardous environment (chapters 7 and 8). In
chapter 9, I analyze 31 oral interviews with miners following the Southmountain disaster in Norton, Virginia (January, 1993), in order to show how
embodied experience and gesture are documented in writing at one critical
moment of transformation.

As I argue in the chapters that follow, we cannot simply settle disputes by
affirming that each individual's subject position constructs an alternative—if
rhetorically incomplete—version of reality. What if individuals do not have
sufficient rhetorical knowledge to document that experience in writing?
What if they cannot interpret an individual's tone or gesture? How can they
draw upon that experience to warrant judgments about the health and safety
of workers? How can others evaluate that knowledge if it is not documented
in conventional (written) forms?

The Rhetoric of Risk does not presume to answer all of these questions in a
single volume. Instead, the book maps a new territory for investigation and
suggests new methods for investigating those questions.

THE FRAMEWORK OF THE BOOK

The book itself has four parts. *Part I: The Problem of Technical Documentation
in Hazardous Environments* lays out a general framework for understanding the
problems of technical documentation within large regulatory agencies. Chapter 1 describes the problems of regulating an uncertain and hazardous material environment and the need for technical documentation to warrant judgments about risk. As chapter 1 demonstrates, regulatory agencies must create
standards that can be applied across multiple sites. But standards are not
bright lines. To standardize experience, regulatory agencies must wrestle with
problems of precision in wording to avoid ambiguity and misinterpretation
that can produce disaster. But too much emphasis on problems of wording
and interpretation can paralyze agencies and leave workers unprotected.

Chapter 2 describes the cycle of technical documentation within large
regulatory agencies. This framework allows us to identify Six Critical Moments of Rhetorical Transformation when knowledge in one modality (oral

vs. written, spoken vs. gestural) is transformed into a new modality for new audiences. This framework allows us to see how individual documents have a rhetorical force that influences future policy and procedures. It shows how writers can manipulate risk on paper, shifting the focus of institutional attention and influencing future policy.

Chapter 3 argues that the material and institutional uncertainty of hazardous environments challenges conventional notions of writing in the workplace and forces us to rethink documentation practices in the context of risk. This chapter describes five sources of uncertainty that affect technical documentation in hazardous environments: (1) the dynamic uncertainty of hazardous environments; (3) the variability and unreliability of human performance; (3) the uncertainty of the agency's notions of "premium data" (categories of analysis); (4) uncertainty in social structure and organization; and (6) the rhetorical incompleteness of any single viewpoint. As this chapter argues, hazardous environments are complex, dynamic, and uncertain. Individuals differ in their ability and experience. But documentation practices can also affect the certainty of knowledge within institutions—even at the highest level of exigence in the case of imminent danger. In these environments, conventional notions of coherence, consistency, and hierarchy may work against "good" representations of risk.

These introductory chapters provide a framework for examining specific instances of written documentation in large regulatory industries.

Part II: Moments of Transformation shows how information critical to risk assessment may be reconstructed and rendered invisible *in writing* at specific moments of transformation within the cycle. Chapter 4 describes how two different writers attempt to capture experience at one specific moment of transformation. This chapter contrasts a miner's narrative, inside of the hazardous environment he describes, with the agency's reconstruction of the accident, above and outside of the mine. When John Nagy attempts to reconstruct events in the Wilberg disaster, his account presents only a limited view of the complex viewpoints that individuals must assume to work safely in dynamic and hazardous environments. This chapter argues that writers must pay increased attention to the ways that individuals manage multiple viewpoints because agencies rely on accident narratives to determine the cause or causes of a disaster.

Chapter 5 shows how individuals can retransform the agency's narrow construction of events (what the agency calls the "proper perspective"). This chapter examines the problem of anomalous behavior (when experienced miners act in an apparently irrational or life-threatening manner). As this chapter argues, anomalous behaviors can be viewed as problems of attention or focus. In each case, workers placed themselves in danger because they failed to turn their attention to other aspects of their work. This chapter raises questions about the narrow focus of instruction and training, but it also suggests that agencies can encourage more flexible thinking and speaking. As this chapter argues, even the most limited documents can function epistemically if individuals are encouraged to speculate freely in the context of instruction.

Chapter 6 shows how speakers draw on three different types of warrants to ground judgments about risk: embodied knowledge ("pit sense"), engineering experience, and scientific knowledge. This chapter describes the rhetorical incompleteness of conventional notions of instructions and procedures, the interdependence of scientific knowledge and local experience, and the rhetorical presence of tacit knowledge. This chapter argues that conventional notions of technical documentation in the workplace have not taken into account the nature of knowledge grounded in experience.

Part III: Documenting Experience explicitly draw upon psychological studies of gesture in order to document how individuals represent their work in speech and gesture. Chapter 7 provides instances that demonstrate the range of viewpoints individuals can represent in speech and gesture. This chapter argues that speech and gesture together provide a richer representation of risk than either speech or gesture alone. But knowledge embodied in gesture can be lost when writers lack the strategies to document this knowledge in writing.

Chapter 8 argues that speakers can deploy two different viewpoints simultaneously in speech and gesture to produce complex representations of risk that integrate theory and practice. When individuals shift viewpoints deliberately, they can transform their understanding of experience. Individuals can use gesture as an active inventional strategy to explore, query, and co-construct knowledge in collaborative interactions. This chapter argues that gesture is both epistemic and inventional. By deliberately shifting viewpoints, speakers can develop a new understanding of their work.

Part IV: Transforming Experience takes a second look at the negotiations that occur at the boundaries between agencies and experience. Chapter 9 examines the presence of embodied knowledge and gesture in 31 depositions following the Southmountain disaster in Norton, Virginia. This chapter shows how writers sometimes indicate the presence of gesture and embodied experience in written transcripts, but these indications are neither consistent nor systematic. Without a method for capturing this experience, knowledge critical to the understanding of risk may be lost in the written record.

Chapter 10 explores the implications of this research for future research in rhetoric. This analysis suggests that rhetoricians and historians can also benefit from the methods of analysis developed in this project.

ACKNOWLEDGMENTS

As the previous discussion indicates, this book owes a great deal to many people whose ideas and input contributed to the shape of the final argument. I am grateful, first, to Rachelle Hollander and the National Science Foundation for providing the economic support for my research and for helping me in the earliest stages of my career. I am also grateful to Charles Bazerman, my editor and friend, who shares with me a common heritage in the work of J. V. Cunningham, our dissertation adviser. This book would not have taken shape without his support and guidance at all stages in the process.

Martha Alibali first suggested the idea of using gesture as a lens for understanding miners' representations of risk. This project would not have been possible without her continuing conversations and support. I am also grateful to David McNeill, Susan Goldin-Meadow, and other members of the Department of Psychology and Linguistics at the University of Chicago for help with the analysis of gesture. I am especially grateful to David McNeill for his help with my interpretation of Libby's gestures in chapters 6 and 7.

Alan Irwin, Steve Woolgar, and other members of CRICT at Brunel University helped me think about the tensions between local and expert knowledge. Their initial support was particularly helpful as I attempted to make sense of British miners' narratives during the earliest phases of my research.

My colleagues Linda Flower, Barbara Johnstone, and David Kaufer in the Department of English at Carnegie Mellon have read my manuscripts and have kept my focus on rhetoric, even when I was most caught up in the details of roof bolts and narratives of disaster. William Keech, Chair of Social and Decision Science at Carnegie Mellon University, gave me extremely detailed comments that helped me bring the project to completion. Baruch Fischhoff has helped me understand the perspective of the risk community, and for his continuing support, I am extremely grateful. Bob Painter, of the Mine Safety and Health Administration, read an early draft of the manuscript with a miner's keen eye for detail. Jean Vettel, an undergraduate at Carnegie Mellon University, read the book with extraordinary care and helped me make the final revisions to the manuscript. I am extremely grateful to her for helping me in this project. Altogether, this work would not have been possible without the support of my colleagues at Carnegie Mellon University; their comments and editorial suggestions have been invaluable.

I am grateful to the Department of English for providing a home for research like mine, and to Carnegie Mellon in general for providing an environment that allows me to work across disciplines in collaboration with so many interesting colleagues. This work would also not have been possible, of course, without the support and assistance of graduate students like Terri Palmer, Geoff Sauer, Susan Lawrence, Laura Franz, Naomi Shabot, and, most recently, Angela Meyer, who corrected the color contrast and resolution of the digitalized pictures of miners that appear in chapters 7 and 8. Laura Martin, Amy Cyphert, and Anne Garibaldi worked diligently to analyze and understand gestures, supported by Carnegie Mellon University's Student Undergraduate Research Grants. Their enthusiasm and curiosity helped make the project a success.

Stephen Brockmann, R. Craig Windham, David Nagel, and Rebecca Burnett have provided the continuing emotional support that has enabled me to bring the project to completion. I am also grateful to Caroline Acker, in the Department of History at Carnegie Mellon, and to Jane Bernstein, Chris Neuwirth, and Peggy Knapp for their friendship in the Department of English. Richard Young's influence in my work may not be visible, but it is a continuing presence.

In the field of rhetoric and professional communication, I have been blessed with much support. I am grateful to Carolyn Miller, Mary Lay, David Russell, David Wallace, Jimmie Killingsworth, Laura Gurak, Dorothy Winsor, Charie Thralls, Paul Anderson, Pat Sullivan, Steve Witte, Cheryl Geisler, Chris Haas, Jack Selzer, and Davida Charney, to name only a few. Nell Ann Pickett and Beth Tebeaux first brought me into the profession as a member of the National Council of Teachers of English committee on scientific and technical writing; their enthusiasm for technical communication has made my career possible.

Because my work is cross-disciplinary, I have also been blessed with wonderful colleagues in the Department of History at Carnegie Mellon: Ed Constant, David Hounschell, Mary Lindemann, Judith Modell, Joel Tarr, and Joe Trotter. Conversations and interactions with historians and their graduate students have helped shape my work. I am grateful to the members of Society for the History of Technology (SHOT) and the Society for the Social Study of Science (SSSS) for helping me locate my work within the history and sociology of science.

This book would not be possible, of course, without the dozens of miners who have contributed their time and effort to help me in so many ways. Many of these men and women cannot be acknowledged because their comments must be kept anonymous. I am particularly grateful to Dave Douglass in Doncaster, England; to Andre deKock and Jan Oberholzer at the Council for Scientific and Industrial Research in Johannesburg, South Africa; to Chris Mark and George Schnakenberg at the U.S. Bureau of Mines; to Bob Painter and George Fezak at the U.S. Mine Safety and Health Administration, and to Cosby Totten and the women of the Coal Employment project, to name a few. This is not a book about mining, but my experience with mining has changed my life. Miners have kept my work grounded, and have encouraged me to make visible the incredible intelligence and linguistic elegance of their representations of their work. The pictures in this book represent only a small fraction of the knowledge of risk that might be available to researchers if we could capture their knowledge in writing.

Finally, to my daughters, Elizabeth Levy and Caroline Levy, I can only say thank you for bringing so much joy into my life. Your presence in my life has helped me in more ways than you can know.

I

THE PROBLEM OF TECHNICAL DOCUMENTATION IN HAZARDOUS ENVIRONMENTS

1

Regulating Hazardous Environments:
The Problem of Documentation

What is required is a paradigm that on the one hand acknowledges the inevitable interaction between known and unknown, and on the other hand respects the equally inevitable gap between theory and phenomena.[1]
—Fox Keller, (1985)

I do not want to sell coal that is stained by the blood of Kentucky coal miners.[2]
—Carroll, 1976. (In Gibson, 1977)

As miners dig, they continually reshape the underground landscape. Every cut changes the immediate shape of the walls, floor, and roof (ceiling) of the mine. As miners remove rock from the face, pressures from the overburden—the mass of mountain above the immediate mine roof—produce subtle changes in the rock strata. If the rock itself is unstable or if miners fail to support this rock, these changes can produce sudden and violent falls of rock that may crush miners without warning. Tiny cracks in the rock may also liberate methane that can burn as a layer of blue flame above miners' heads. If mines are not properly ventilated, if miners do not replace their picks frequently, if cutting picks hit the hard layer of rock rather than the soft coal layer, if electrical equipment sparks, or if miners increase the force of the cut to increase production, frictional heat produced by the cutting pick can ignite methane

---

[1]Keller, 1985, p. 139.
[2]*Oversight Hearings on the Coal Mine Health and Safety Act of 1969 (Excluding Title IV)*, 1977, p. 4.

in a violent chain reaction that can turn tiny particles of coal dust into a deadly explosion.

The earliest printed mining texts acknowledge the risks inherent in mining. In the *De Re Metallica* (1556/1950), Georgius Agricola admits that the dangers of mining are of such "exceeding gravity" and so "fraught with terror and peril" that he would "consider that the metals should not be dug up at all, if such things were to happen very often or if miners could not safely guard against danger by any means":[3]

> The critics say further that mining is a perilous occupation to pursue because the miners are sometimes killed by the pestilential air which they breathe; sometimes their lungs rot away; sometimes the men perish by being crushed in masses of rock; sometimes, falling from the ladders into the shafts, they break their arms, legs, or necks; and it is added there is no compensation which should be thought great enough to equalize the extreme dangers to safety and life.[4]

Four hundred years later, miners continue to die.

Despite improvements in mining technology and engineering, 20th century miners still confront risks from gas, falling rock, and black lung disease.[5] In 1983, Richard Trumka, President of the United Mineworkers of America (UMWA), testified before Congress about what he called the continuing "slaughter and maiming" of coal miners in the U.S. Between 1900 and 1968, he argued, over 71,000 coal miners had died and over 300,000 miners had been totally disabled with black lung disease.[6] "Untold thousands of coal miners" had been injured, he argued, because the Federal government had no means of enforcing mine safety standards and mining regulations.

---

[3]Agricola, 1556/1950, p. 6. Agricola believed that miners could avoid both economic and physical risk if they observed the indices or signs that predicted risk in the environment. He thus set about to describe the various practices that expert miners could employ to locate profitable veins and stringers, work cooperatively with others in the Gesellschaft (miners' union), and control risk in the mines. Agricola's work provided practical guidance and clear illustrations for more than 400 years. Herbert Hoover translated the work with his wife Lou Henry Hoover in 1912 so that he could use the work to develop Chinese mines, which lacked the hydraulic equipment that provided power in U.S. mines.

[4]Agricola, 1556/1950, p. 6.

[5]The history of mining provides gruesome evidence that accidents were both frequent and deadly despite continuing improvements in mining technology and engineering. In some cases, in fact, improvements in technology—the hydraulic drill, for example—brought increased danger as miners struggled with heavy equipment underground. Yet few miners argue for a return to the golden age of picks and shovels.

[6]*Hearings before the Committee on Education and Labor, Subcommittee on Health and Safety*, 98th Cong., 1st Sess., 1983, p. 2. See also Lewis, 1947.

The history of Federal mining legislation demonstrates the continual struggle to create enforceable standards that would protect the health and safety of workers. Although Congress had introduced a bill to create the Federal Bureau of Mining as early as 1865, the earliest mining legislation lacked "any right or authority with the inspection or supervision of mines. . . ."[7] In 1910, Congress established the Bureau of Mines within the Department of the Interior to investigate mining methods, mining operations, and new technologies to improve mine safety. This legislation was intended to "attack" hazards in the mining industry, but Federal inspectors lacked authority to inspect or supervise mines.[8] In 1941, Congress gave Federal inspectors the right to enter and inspect all bituminous, anthracite, and lignite mines.[9] But the Mine Safety Code (established by the Federal Bureau of Mining) served only as a guideline for Federal inspectors and "compliance by operators was purely voluntary."[10] As a result, inspectors had no legal means of enforcing compliance with standards.

In 1951, following the death of 119 miners in West Frankfurt, Illinois, Congress gave the Bureau of Mines a "reinspection closing order" (Public Law 552) that enabled inspectors to prevent some repeat violations of the Bureau recommendations. But the legislation fell short of President Harry Truman's recommendations submitted to Congress, and miners continued to die.[11] Public Law 89-376 (1966) extended Federal authority to small mines that had been excluded in previous legislation. Because this law was intended to attack major disasters, it covered only 10% of all occurrences that might cause a fatality or accident. Ninety percent of accident occurrences were still left to be covered by State law and the Bureau's voluntary safety code.[12]

Then, on November 20, 1968, a spark ignited gas at the Number 9 Farmington mine in Marion County, West Virginia. In that one disaster alone, 78 miners died. Trumka recounts the horror in vivid terms:

> There was a dangerous accumulation of methane in that mine that day. I would like for you to remember that that was the cause of the disaster, an accumulation of methane, coupled with the lack of rock dust, coupled with the

---

[7]U. S. Senate, (1977, May 16).

[8]U. S. Senate, (1977, May 16).

[9]U. S. House of Representatives, (1969, October 13). Public Law 49, known as Title I of the Federal Coal Mine Health and Safety Act was first introduced as Senate Bill S. 2420 on May 16, 1939 and passed by the Senate in 1940. See U. S. Senate (1977, May 16).

[10]U. S. Senate, (1977, May 16).

[11]As Truman signed the law, he lamented that "the legislation falls short of the recommendations I submitted to Congress to meet the urgent problems in this field." U. S. Senate (1977, May 16).

[12]U. S. House of Representatives, (1969, October 13).

accumulation of coal dust. . . . A spark ignited that gas and turned the Farmington mine into an inferno; 78 miners lost their lives that day. They were cooked. They were cooked externally. They were fried like you could fry an egg. . . .[13]

The 1968 Farmington disaster marked a turning point in U.S. policy. In the wake of public outcry over Farmington, Congress adopted the Federal Coal Mine Health and Safety Act of 1969 (Public Law 91-173). The legislation was drafted in two weeks over Christmas vacation.

The Mine Act (as it is now referred to)[14] represented a major victory for labor in its struggle to hold management accountable for safety underground. The Act provided the first mandatory Federal standards, the first mandatory inspections, and the first civil penalties for violations of the Act's standards. The Act authorized a separate enforcement agency, the Mine Enforcement and Safety Administration (MESA). The Act promised miners "the safety that they deserved in their workplace, a healthful workplace that they deserved, free of dust, free of noise, free of dangerous accumulations of coal dust and methane"—with compensation for black lung disease and civil and criminal penalties to force compliance.[15] Under the Act, the Bureau of Mines retained its responsibility for mine safety research, funded through a separate budget.

With the passage of the 1969 Mine Act, miners were no longer dependent upon the good will or voluntary compliance of owners and management.[16] The Act specified that management must provide training for both new and experienced miners, and it specified a mechanism for enforcement and penalties for violations. Most important, the Act held management ultimately lia-

---

[13]*Hearings before the Committee on Education and Labor, Subcommittee on Health and Safety*, 98th Cong., 1st Sess., 1983, p. 4.

[14]The Act was originally called the Coal Act because metal and non-metal mines were covered under separate legislation. In 1977, Congress authorized new legislation that united metal and non-metal mines under a single Mine Act: Public Law 91-173, as amended by Public Law 95-164.

[15]*Hearings*, 98th Cong., 1st Sess., 1983, p. 4.

[16]A section describing the background of legislation preceding the Federal Mine Safety and Health Act of 1977 reports that compliance with the Bureau's code of practice was as low as 33% prior to passage of the 1969 Mine Act. The writers note:

The 1966 amendments only reached a small portion of the causes of fatalities and accidents occurring in mines. The larger number of such occurrences lay outside and beyond the reach of the Federal statute, and was left by Congress to be embraced by State laws and the Bureau of Mines Advisory Coal Mine Safety Code. (The Federal Mine Safety and Health Act of 1977).

ble for willful violations of safety. The Act was further revised in 1977 in response to another series of disasters that underscored the inadequacy of safety and enforcement.[17]

The 1977 Act consolidated metal and coal mines under a single set of standards and transferred the functions of the former MESA (under the Bureau of Mines) to the newly created Mine Safety and Health Administration (MSHA).[18] The 1977 Act thus placed the responsibility for worker safety within the Department of Labor and removed the potential conflict of interest that might occur if MESA inspectors interrupted production to enforce health and safety standards.

But miners continued to die.

In 1987, Congress convened oversight hearings to determine why agencies had failed to publish regulations to protect workers' safety and health. These hearings revealed problems with the regulatory process at all levels, from compliance and enforcement to problems with training, ambiguities in language and interpretation, and a process of regulatory review and revision that paralyzed the agency and left workers unprotected.

Labor and management agreed that the regulatory process needed to be improved, but they disagreed about the locus of liability and the cause of the problems. In testimony before Congressional oversight hearings, the Bituminous Coal Operators' Association and American Mining Congress (1977) argued that regulations were hard to interpret because they lacked "a clear and sensible meaning"; yet if operators failed to abide by the strict letter of "this compendium of legal detail," they were cited for "violation" of the stan-

---

[17]The most famous of these is the Scotia disaster, where 23 miners and three Federal inspectors died in March, 1976. The Act notes: "At Scotia, in March, 1976, twenty three miners and three Federal inspectors died in two explosions of accumulated methane gas when the mine safety enforcement effort was unable to detect and address chronic conditions of inadequate ventilation in that mine."

[18]The Federal Mine Safety and Health Act of 1977 reads:

SEC. 302. (a) There is established in the Department of Labor a Mine Safety and Health Administration to be headed by an Assistant Secretary of Labor for Mine Safety and Health appointed by the President, by and with the advice and consent of the Senate. The Secretary, acting through the Assistant Secretary for Mine Safety and Health, shall have authority to appoint, subject to the civil service laws, such officers and employees as he may deem necessary for the administration of this Act, and to prescribe powers, duties, and responsibilities of all officers and employees engaged in the administration of this Act. The Secretary is authorized and directed, except as specifically provided otherwise to carry out his functions under the Federal Mine Safety and Health Act of 1977 through the Mine Safety and Health Administration.

dards.[19] Operators charged that it was unreasonable to hold management liable for mine safety violations when MSHA and management disagreed about the meaning and interpretation of regulations. The number of regulations and their complexity made unreasonable demands upon foremen, inspectors, supervisors, and management, who were forced to interpret and apply general standards in highly local and unpredictable situations in the field.

The United Mineworkers countered that tougher standards and consistent enforcement would prevent disaster. They criticized MSHA for allowing mines to violate safety standards and change plans. They argued that such flexibility encouraged abuse and would return miners to the hazards they faced under voluntary compliance.

Senator Edward Kennedy (1987) charged that the agency spent too much time "reviewing the re-review."[20] Kennedy called the process a "regulatory nightmare" that "provide[d] little chance for regulation ever to be finalized."[21] United Mineworkers President Joseph Trumka (1987) was more direct. He argued that the agency's proposals for new mine ventilation standards had missed the "simple changes in work procedures" and "simple measures" that could have prevented disaster.[22] In calling upon agencies to write clearer, more enforceable standards, unions and Congressional overseers sought to move the site of risk decision making away from the pressures of production and the economies of mine management in local mines.[23]

---

[19]*Oversight Hearings on the Coal Mine Health and Safety Act of 1969 (Excluding Title IV)*, 98th Cong., 1977, p. 333.

[20]Kennedy asks, ". . . why does MSHA have to go back and take another look at regulations that were proposed by individuals from important safety areas who, as I understand, are the best that you had in the Department? Why do you have to spend more time in re-reviewing the re-review? That is something that I just have some difficulty in understanding." *Oversight of the Mine Safety and Health Administration*, 100th Cong., 1st Sess., 1987, p. 520.

[21]*Oversight of the Mine Safety and Health Administration*, 100th Cong., 1st Sess., 1987, p. 512.

[22]*Oversight of the Mine Safety and Health Administration*, 100th Cong., 1st Sess., 1987, p. 718. Like the accidents at Pyro and Farmington, the disasters at Grundy, Scotia, and Wilberg revealed disparities in the local enforcement and suggested that MSHA and mine operators had watered down regulations, softened enforcement efforts, and endangered miners' lives. Following the hearings, MSHA instituted changes in training and enforcement throughout the agency and promulgated specific regulations to address problems discovered during the accident.

[23]In a detailed critique of MSHA's policies, Richard Trumka charged that more than 2,156 miners had died during the period in which MSHA was reviewing and re-reviewing standards. "Over half of those mine disasters have occurred since 1980," he wrote, "a rate 60% more frequent than during the 1970s" (*Oversight of the Mine Safety and Health Administration*, 100th Cong., 1st Sess., 1987, p. 706). Trumka argued that the investigative reports on those accidents

In criticizing the agency for failing to write clear standards that might protect miners' lives, the hearing raised questions about the problems that agencies faced as they attempted to regulate hazardous environments: Why did management deliberately and willfully fail to comply with standards even as conditions deteriorated? How did the failure to document changing conditions affect the outcome of the disaster? Why was it so difficult to write clear and enforceable standards?

In this chapter, I describe three rhetorical problems that make regulation so difficult: the problem of standardizing experience, the difficulties of day-to-day compliance with standards, and the problem of documentation. To set the stage for this analysis, I begin with a particularly egregious case of noncompliance—an explosion at the Pyro Mining Company's Number 9 Slope, William Station Mine in Sullivan, Union County, Kentucky (generally referred to as Pyro), on September 13, 1989. Given the levels of methane released during mining operations, the explosion at Pyro seems inevitable in retrospect. Although there are as many accounts of blame as there are agents involved in the decision, the Pyro accident supports the union's argument that stricter enforcement might have prevented disaster. But Pyro also demonstrates how written regulations can become a dangerous instrument if operators misinterpret or misapply poorly written and outdated standards inappropriately in new contexts.

In acknowledging the rhetorical complexity of regulating hazardous worksites, this analysis does not excuse management for its failure to detect the signs of imminent danger; nor does it attempt to excuse such failures on the grounds that all judgments are relative. Instead, I want to consider Pyro as an example of the intentional risk taking that makes standards so difficult to write, enforce, and apply in hazardous worksites.[24]

---

provided more than enough information to warrant a change in standards—particularly where a pattern of violations identified mines like Pyro with a high likelihood of disaster.

[24]I borrow the term *regulate* from MSHA's *Report of Investigation* (Childers, 1990, p. 48) for the Pyro accident to underscore the degree to which human decision-makers (regulators) work in tandem with mechanical systems (also known as regulators) to *regulate* conditions in the environment. To regulate airflow, ventilation, and water levels, decision-makers calibrate measuring devices, monitor changing conditions, and refigure physical structures in the environment. Federal regulatory agencies work through inspectors to enforce regulations. General standards (regulations) set limits to unsafe practices and conditions, but human regulators must continually work to bring risky worksites into compliance. Operators sometimes deliberately and knowingly continue to operate (They "mine past the regulator") even when conditions exceed minimum safety standards. Because mines are complex geographic structures, each decision produces changes throughout the system that must be re-calibrated and re-regulated to manage risk.

PYRO-TECHNICS: REGULATING AN EXPLOSIVE ENVIRONMENT[25]

The term *regulation* has many different meanings in hazardous worksites. At the agency level, regulators construct written regulations that define safety standards and set limits to risky practices and procedures. At the local level, human decision-makers (regulators) work in tandem with mechanical systems (also known as regulators) to monitor and control (regulate) conditions in the environment. To determine whether monitors provide accurate readings, human regulators must continually calibrate and re-calibrate mechanical regulators. Some mechanical regulators automatically shut down production when sensors detect dangerous conditions. But the decision to stop production is by no means automatic.

Even when regulations specify clear limits like the standards for methane, human regulators must exercise judgment when unfavorable conditions begin to exceed minimum safety standards. Management and miners themselves sometimes illegally bypass regulators to maintain levels of production in an industry where production is measured in tons per second. Workers and management may also bypass regulators if they believe that regulators are too sensitive to local conditions—as one might carelessly and deliberately remove the battery from a fire alarm that is too sensitive to cooking gases in a kitchen. In the Southmountain disaster in Wise County, Norton, Virginia, on December 7, 1992, someone placed a greasy rag on a methane monitor because water spray from the mining machine was creating false alarms.[26] In the Wilberg fire in Emery County, Utah, on December 19, 1994, individuals bypassed the safety switch on a compressor. In both cases, the decision to bypass the regulator produced disaster.[27]

On September 13, 1989, 10 miners died in an explosion at the Pyro Mining Company in Sullivan, Kentucky. At the time of the explosion, Pyro was under a five-day, mandatory inspection program, based upon two previous explosions within the last five years. The mine had a history of repeated violations and had been cited in the previous inspection for dangerous accumulations of methane.[28] One of the citations also stated that no air movement could be de-

---

[25]Childers et al., 1990, p. 48. "Mining past the regulator" is the title of one section of the report.

[26]Thompson et al., 1993.

[27]Huntley, Painter, Oakes, Cavanaugh, and Denning, 1992.

[28]The bleeder system for Longwall Panel "M" functioned properly until January 4, 1989, when MSHA inspectors from the Madisonville District Office issued two Section 104(a) citations for accumulations of methane. Methane concentrations of 2.5% and 3.0% were found in the connecting entries between the 2nd Main North Bleeder Entries and the 7th West Entries. (Analysis of bottle samples taken revealed 3.40% and 4.92% respectively.)

tected in the connecting passageways.[29] On the day of the explosion, Federal investigators were on the surface, preparing to inspect the mine.[30]

What seems so problematic in retrospect is the degree to which Pyro operators worked in conditions that were so volatile and dangerous that it seems almost ludicrous to talk in terms of reasonable action. The mining company had violated all of the major regulations relating to methane control: (a) The volume and velocity of air were not sufficient to exhaust methane [30 CFR § 75(301)]; (b) no preshift exam was conducted on the day of the explosion [30 CFR § 75 (303)]; (c) the approved ventilation system and the dust control plan were not being followed (30 CFR § 75(316); and (d) changes in the ventilation plan substantially affected the health and safety of miners [30 CFR § 75(322)].

In January, 1989, MSHA inspectors had cited Pyro for imminent danger (stopping work until dangerous conditions were abated) when 4.8% methane was discovered in one area of the mine. (The legal limit is 1.0%.) The mine was permitted to resume production when methane levels were reduced to 2.5% "as long as methane checks were made in the affected area at the designated time intervals."[31] But the mine continued to encounter mechanical problems that disrupted the flow of fresh air and created dangerous accumulations of methane.[32] Despite the presence of Federal inspectors at the mine,

---

[29]Childers et al., 1990, p. 40.

[30]A complete MSHA Safety and Health Inspection of the entire Pyro No. 9 Slope, William Station Mine had been conducted from June 6, 1989, through July 5, 1989. During the inspection, 23 citations and five orders were issued. During the day shift of the explosion, federal inspectors were on the surface preparing to inspect the mine (Childers et al., 1990, p. 3).

[31]On January 5, 1989, MSHA inspectors from the Morganfield Field Office issued a Section 107(a) imminent danger order along with a Section 104(a) citation for a violation of the approved ventilation plan when 4.8% methane (analysis of bottle samples revealed 4.11%) was found in this same general area of the bleeders.

Specifically, the methane was detected at the junction of the No. 1 Entry of the 8th West Entries and the No. 6 Entry of the 2nd Main North Bleeder Entries. When methane levels were reduced to 2.5%, the Section 107(a) imminent danger order was modified to allow production to resume as long as methane checks were made in the affected area at the designated time intervals (Childers et al., 1990, p. 40).

[32]According to the report, "poor roof conditions affected air flow and squeeze conditions," and water in the bleeder entries blocked air flow through the bleeder system and forced the majority of return air from the longwall face to return down the no. 1 entry of the 5th West (Childers et al., 1990, p. 42). Boreholes to control water in the bleeders also had major problems: "This borehole missed the underground workings as did a second borehole which was drilled during August 1988. A third borehole was started on September 1, 1988, but this hole was abandoned when the drill bit used to enlarge the diameter of the borehole became lodged in the hole. The fourth borehole was started on October 26, 1988, and finished on February 11, 1989, but was not energized since no water was present at the pump inlet (Childers et al., 1990, p. 46).

methane monitors had been removed from the mining machine, and mining operations had advanced beyond the limits of the regulator—a piece of equipment that worked to hold back deadly gases that might accumulate in the non-working sections of the mine.[33]

Methane concentrations were so high on the days prior to the accident (6% and 9%, respectively) that it is difficult to understand why no one documented these accumulations in the approved record book or took action to correct the conditions [30 CFR §75(324)]. These concentrations fall within the explosive range of methane—the level at which gases will ignite even without a source of ignition. The concentration of methane also reduces the level of oxygen available to miners in respiration.[34]

In hindsight, the situation at Pyro supported union arguments that tougher enforcement and stronger regulation might have prevented disaster. But Pyro also demonstrated that new technologies would constantly render the standards of the Mine Act obsolete. In the anti-regulatory atmosphere of the Reagan Administration (which had imposed a moratorium on new regulations), written standards could become a dangerous instrument if inspectors strictly adhered to the letter of outdated regulations, misinterpreted ambiguous standards, or failed to apply regulations appropriately in new contexts.

Like other disasters, Pyro revealed these lessons in hindsight. Following a disaster, agencies will argue for new regulations, corporations will increase safety training, and individuals will follow instructions and pay closer attention to the signs of risk in their environment. After an accident, both workers and management take extra precautions. Such extra care reduces accidents and fatalities, and thus the likelihood of accidents in the future. In time, however, workers and management may develop a false sense of secu-

---

On June 8, 1989, the water was 1000 feet long and 2 feet deep. The submersible pump installed in March, 1989, was energized on June 2, 1989, but this pump failed to pump any water (p. 47). The mine experienced problems in the pumping rate until September 13, 1989, the day of the accident (Childers et al., 1990, p. 47).

[33]During the latter part of mining on Panel "O," Pyro maintained a regulator to control the amount of air flowing across the working face and to maintain pressure to hold back gases that might accumulate in non-working sections of the mine (Childers et al., 1990, p. 48). Sometime between August 31, 1989, and September 2, 1989, the mining machine mined past the regulator. The longwall equipment also had a methane monitor, but at the time of the explosion, the methane monitors had been removed along with the equipment they were mounted on (Childers et al., 1990, p. 10).

[34]Several miners I have interviewed have joked grimly that the concentration of methane probably affected the level of oxygen to the brain and reduced the thinking capacity of decision makers who decided to keep working under such dangerous conditions.

rity and once again test the limits of safe practice—increasing the likelihood of disaster.

Agencies create standards because experience is a poor teacher. But standards are not bright lines. Well-written standards help individuals identify moments of possible danger, but workers are still dependent upon information from colleagues and management to help them determine which practices are the best to control and manage danger. In rhetorical terms, standards help workers identify moments of exigence—moments of crisis that demand action. But standards do not always define how individuals should respond to those crises.

The Problem of Standardizing Experience

Like other scientific and technical communities, government agencies "shape written knowledge" through textual practices like accident reports, investigations, and standards.[35] But standards blur the distinctions between "expert" science and lay understanding.

Standards are heterogeneous and complex texts that attempt to formulate in writing the accumulated body of scientific and local knowledge about preventing disaster in hazardous worksites. Standards reflect reasonable and practicable solutions to complex technical problems in the workplace. But they are also the outcome of many deliberations about the value of human life and the distribution of risk among affected populations.

---

[35]Historical studies of the evolution of rhetorical practice within specific scientific communities reveal "how much work, thought, intelligent responsiveness to complex pressures, and fortunate concatenations of events" go into the process that ultimately "constitutes" scientific discourse (Bazerman, 1988, p. 14). Moments of conflict in science reveal the power of rhetoric in shaping scientific knowledge. At these moments of conflict, rhetoric in science becomes most visible in the strategies that scientists draw on to argue their positions. (See Dear, 1991; Gross, 1990; Prelli, 1989.)

Researchers in the sociology of science have not explicitly identified their work as studies of historical narratives in science, but their accounts of individual laboratories and communities reveal the ways that the material world becomes invisible in the traces and inscriptions that constitute scientific fact (Latour, Woolgar, and Salk, 1986).

Markley (1993) critiques both the method and the scope of these studies because, he claims, researchers have focused on the "contested area where science seeks to write out of existence the knowledge claims of competing accounts of representation and reality" (p. 19). As a result, sociologists of science have not explored the ways in which scientific discourses are "interpenetrated by the very semiotic systems . . . that they appropriate and seek to de-legitimize" (p. 18). His work suggests that researchers must investigate the ways that competing representations of reality affect the discourses of science.

To construct standards, agencies solicit input from many sources: scientific theory, local practice, engineering experience, and common-sense knowledge.

Written standards represent the attempt to standardize the historical experience of individuals accumulated over time. Some standards reflect scientific theories about gas behavior, combustion, or rock mechanics. Standards like 30 CFR § 75(301-5), for example, specify the maximum concentrations of explosive gases that can accumulate in a section. Other standards specify general procedures like air sampling without citing specific techniques or codes of practice [30 CFR § 75(301-6)]. Some standards specify testing methods developed by scientific testing institutes like American Society for Testing and Materials (ASTM) [30 CFR § 75(302-3)]. And some specify the sequence of procedures that must occur in pre-shift examinations and inspections [30 CFR § 75(303)].

Many standards reflect highly local and frequently common-sense practices and procedures. These standards standardize the experiences of individuals accumulated in narrative histories and oral accounts of risk. Some standards reflect reasonably practicable solutions to the problem of safety and thus may codify outdated industry practices. Other standards simply reflect common-sense guidelines formulated as law. Thus, the miner's warning, "never go under unsupported roof," becomes 30 CFR §75(202) of the Mine Act:

(b) No person shall work or travel under unsupported roof unless in accordance with this subpart.[36]

Congress recognized that the standards promulgated under the original 1969 Mine Act would be continually outdated unless the legislation could

---

[36]Federal Mine Safety and Health Act of 1977, 30 CFR § 75(202), p. 500. I refer to the July 1, 1990, version of standards in this passage.

Because the Mine Act is a living document, it is possible to speak of specific sections without identifying the year in which these provisions were created. Thus, the 1969 Act and its amendment (1977) will always be the current set of regulations in force at any one time. If regulations change, the Act is updated with the new standard or standards. The old standards no longer exist in the public record because outdated standards could endanger the health and safety of workers. The research for this project is based upon the standards available in 1990, except when I cite a specific iteration of the Act. The reader should be warned that 1990 standards may no longer be in force in the current iteration of the Act. It is also technically incorrect to speak of a 1990 Act. Citations in the text are intended to help orient the reader to the specific wording that might have caused confusion at specific moments in time. The citations are not intended as legal citations to authority.

change in response to new information, new technologies, and changing mining practices. To keep standards up-to-date, both the original Mine Act (the Federal Coal Mine Health and Safety Act of 1969) and the 1977 Federal Mine Safety and Health Act (Public Law 91-173, as amended by Public Law 95-164) provide a mechanism for regulatory revision and review to insure that no provisions promulgated under the Act might reduce the health and safety of miners. Section 101 of the Act (mandatory safety and health standards) specifically authorizes the Secretary of the Interior to "develop, promulgate, and revise as may be appropriate, improved mandatory health or safety standards for the protection of life and prevention of injuries in coal or other mines" [Public Law 91-173. §101(a)]. Under the provisions of the Act, regulatory agencies authorized by the Secretary can propose new rules and revise existing standards without having to redefine the purpose and intent of the original Act.

The standards promulgated under the Mine Act thus create what one agency researcher called "a living document" that can change in response to the specific needs of management and labor.[37] Thus, while we can speak of the specific provisions of the original 1969 Act as enacted by Congress, many of the Act's original standards have been revised and updated to reflect new research, new policy, or a better understanding of mine safety. Agency personnel refer to the Mine Act (or simply "the Act") without reference to a specific year or date. Outdated provisions are quickly excised from the living record; they are hard to find except in archives. The Mine Act reserves specific sections for specific kinds of regulation, but the details of regulatory practice change over time. Thus, while Section 57 always refers to diesel standards, specific provisions within the Act will change as researchers develop a better understanding of the hazards of diesel exhaust underground. Section 73 is reserved for ventilation. Section 103 authorizes inspections, investigations, and record keeping. Section 104 refers to citations and orders.

Ultimately, any standard promulgated under the Act is evaluated according to the level of protection it provides for the health and safety of workers. According to the Act, "No mandatory health or safety standard promulgated under this title shall reduce the protection afforded miners by an existing mandatory health or safety standard" (Public Law 91-173, Title I-General; §B). In Section 2 (Findings and Purpose), Congress "declares" that "the first priority and concern of all in the coal or other mining industry must be the health and safety of its most precious resource—the miner" [(Public Law 91-173 §2(a)]. Documentation plays an important role in the process. The Act

---

[37]Schnakenberg (Personal interview, 2001).

specifies that the Secretary must hold hearings and evaluate evidence so that the agency can "adequately assure" that the new standard will not decrease the level of safety currently guaranteed by the provisions of the Act.

Unfortunately, the meaning of the Act and its specific standards are not always apparent in day-to-day practice. The Act acknowledges that no single standard can be generalized to apply to diverse environments and mining operations. Management or workers can petition to modify particular standards if they can demonstrate that the new practice will not diminish the health and safety of workers [§101 (c)]. The Act also acknowledges that conditions in the mine must ultimately dictate safe practice. Thus, the Act provides clear engineering standards for regulating methane and air flow in the mine, but the Act also allows mines to employ highly subjective standards if air currents cannot be controlled or measurements cannot be taken—as in some anthracite mines—as long as mines take into account the "health and safety of workers."[38] Unfortunately, what constitutes the health and safety of workers is not always apparent, and regulators disagree about the meaning and interpretation of standards in different mining contexts. Federal agencies thus reinterpret standards in policy memos, compliance guidelines, and program information bulletins like the sample in Fig. 1.1. When accidents occur, agencies must alert miners and management to the conditions and practices that produced the disaster.

To write clear and enforceable standards, writers within agencies must balance the tensions between production and open-ended scientific inquiry, rationality and "regulatory reasonableness" (Blau, 1969)—satisfying the needs of stakeholders with diverse political and economic agendas. To be perceived as rational, Blau argues, agencies must provide "a rule-making record that satisfies the reviewing court that all relevant considerations have been explored."[39] But they must also provide "single correct solutions" to difficult problems in local contexts. To protect their decisions against possible reversal in the political sphere, agencies frequently create "massive technical reports and analysis."[40] But these reports create more, not less, controversy as

---

[38]The 1990 version of the Federal Mine Safety and Health Act of 1977, Section 303, for example, specifies that "all active workings" must be ventilated by a current of air containing "not less than 19.5 volume per centum of oxygen, not more than 0.5 volume per centum of carbon dioxide, and no harmful quantities of other noxious or poisonous gases" (Public Law 91-173, as amended by P. L. 95-164). The law further specifies that the minimum quantity of air reaching the last open crosscut shall be 9000 cubic feet a minute and that the minimum quantity of air in any coal mine reaching each working face shall be 3,000 feet per minute. The law allows regulatory agencies to exceed these specifications to protect the health and safety of miners (CFR 30 § 75-301).

[39]Blau, 1969, p. 39. See also Bardach & Kagan, 1982; Bryner, 1987.

[40]Blau, 1969, p. 39.

U.S. Department of
Labor

Mine Safety and Health
Administration
4015 Wilson Boulevard
Arlington, Virginia
22203-1984

ISSUE DATE: September 16, 1996

PROGRAM INFORMATION BULLETIN NO. P96-19

FROM: MARVIN W. NICHOLS, JR. *Marvin W. Nichols Jr.*
                    Administrator
                    for Coal Mine Safety and Health

                                                    Kenneth J. Howard

             KENNETH   T.   HOWARD
             Director of Technical Support

SUBJECT: The Danger of Oxygen Deficiency In Underground
                    Coal   Mines

Scope
This bulletin applies to underground coal mine operators and Mine Safety and Health Administration
(MSHA) personnel.

Purpose
The purpose of this bulletin is to increase awareness of the detrimental effects associated with low oxygen
concentrations, emphasize the importance of testing for oxygen concentrations at certain locations, and
identify some of these locations in underground coal mines.

Information
Methane liberation in the absence of effective ventilation can present serious hazards to miners. In addition
to being an explosive gas, an accumulation of methane in high concentrations can result in a mine
atmosphere that is deficient in oxygen. Atmospheres with oxygen concentrations below 19.5 percent can
have adverse physiological effects, and atmospheres with less than 16 percent oxygen can become life
threatening.

Methane may be released from a variety of sources in a mine, including strata in the roof or floor, as well as
the coal seam itself. Since methane continues to be released after mining is completed, effective ventilation
or sealing of mined-out areas is critical. Consequently, to verify that ventilation is effective and seals are
maintained, weekly examinations by mine operators, as well as other examinations and surveys by MSHA
personnel are required. To avoid exposing themselves to hazardous situations, each person making these
examinations should have a thorough understanding of how the ventilation system is intended to function
and how and where low levels of oxygen may exist; and they should be able to recognize situations
conducive to accumulation of methane or other contaminant gases. Some of the more likely areas where
oxygen-deficient atmosphere may be encountered are bleeder entries, worked-out areas, approaches to

FIG. 1.1.   This sample program information bulletin alerts the coal mining
industry to an increase in fatal accidents related to haulage. The cluster of accidents alerts the agency to an underlying problem in safety. Source: McAteer
(1996).

Methane is lighter than air and tends to migrate upwards. Methane may accumulate when the ventilation system is not moving air with sufficient velocity to mix with the methane, or the air is not reaching into cavities or other high locations, such as those created by roof falls. At times, methane accumulations may reach concentrations high enough to displace oxygen and create an irrespirable atmosphere that is immediately life threatening.

Recently, a fatal accident occurred when two foremen were making examinations in a bleeder system. As the two were traveling over fallen rock and into the roof fall cavity area, which had been supported after the fall, they encountered a methane-rich oxygen- deficient atmosphere. Both foremen immediately lost consciousness and collapsed. One foreman apparently fell to an area with a higher oxygen level near the mine floor where the ventilation air stream existed. He regained consciousness and traveled outby to seek help. The other died due to lack of oxygen.

Miners and MSHA personnel should understand the cause of this tragic accident and be aware that the harmful effects of entering an oxygen-deficient atmosphere can be so immediate that it is impossible to retreat to safety. Miners, and especially examiners, should exercise caution wherever high methane concentrations are suspected, particularly when entering or examining roof cavities. Miners and MSHA personnel should also fully understand the operation of multiple gas detection devices and the significance of their visual and audible alarms. Such areas should be approached slowly to permit adequate time for gas detection devices to react. High levels of methane or other contaminant gases should be a warning that the oxygen level may be lower than normal.

Background

Oxygen is depleted from the atmosphere by oxidation of coal, wood, or other organic materials. Oxygen deficiency can also occur as contaminant gases liberated from the coal strata displace the oxygen. Atmospheres with oxygen concentrations below 19.5 percent can have adverse physiological effects. The following are the likely effects of depressed oxygen levels in air:

Percent Oxygen in Air	Effect
17	Faster, deep breathing
15	Dizziness, buzzing in ears, rapid heartbeat
13	May lose consciousness with prolonged exposure
9	Fainting, unconsciousness
7	Life endangered
6	Convulsive movements, death

Historically, emphasis has been placed on testing for oxygen near mine floors and in low areas to detect gases that are heavier than air, such as carbon dioxide. However, gases that are lighter than air, such as methane, tend to migrate upward; and oxygen-deficient atmosphere may be found near the roof or other high spots. For example, methane seeping from feeders in cavities such as those created by roof falls will migrate upward, displace the air, and create an oxygen-deficient atmosphere if ventilation is ineffective. Similarly, methane that is layered near the roof or high areas can migrate updip for a considerable distance, even against air currents, and accumulate in these areas. These situations, however, can be prevented with ventilation that causes the air to move with a velocity sufficient to mix with the methane or other gases and render harmless atmosphere.

Typically, multiple gas detection devices give visual and audible warning signals when a predetermined gas concentration is encountered. Users of these devices should be familiar with their operation and should realize that there is a delay between the time the gas is encountered and the device responds. This delay period varies with the type, age, and condition of the device. It should be noted that when multiple gas concentrations are encountered, an alarm will sound when one of them reaches the predetermined

FIG. 1.1. (Continued)

concentration that activates the alarm. If a second gas reaches its predetermined level, the alarm may continue to sound and an additional visual warning signal will be activated. For example, a visual and audible signal may activate when methane concentration in excess of 2.0 percent is detected; then, if an oxygen concentration of less than 19.5 percent is detected simultaneously, another visual warning will activate and the audible alarm may remain unchanged.

Authority
Section 107(a) of the Federal Mine Safety and Health Act of 1977;
30 CFR 75.321(a)(1) and 323(e).

Issuing Office and Contact Person
Coal Mine Safety and Health, Safety Division
Salwa El-Bassioni, (703) 235-1915

Technical Support
John Urosek, (412) 892-6934

Distribution
Program Policy Manual Holders
Underground Coal Mine Operators
MSHA Special Interest Groups

FIG. 1.1. *(Continued)*

unions decry the lack of enforceable standards and management complains that general standards fail to take into account important differences in various types of mining.

In 1987, the U.S. Senate conducted Congressional Oversight hearings following a series of mine disasters involving longwall mining and ventilation. In these hearings, congressional overseers scrutinized the rough drafts of standards in order to recommend changes in MSHA's procedures for writing and revising safety regulations. As these documents demonstrate, agencies draw heavily on previous experience—documented in accident reports and investigations—to determine how much leeway they can allow operators to make on-site decisions about compliance and safety. The documents presented as evidence in these hearings thus enable us to see how the rhetorical problems of wording and interpretation reflect deeper disagreements about the locus of safety and liability in a hazardous worksite.

The Problem of Wording

Standards must be precise enough to prevent misinterpretation and ambiguity, but they must be broad enough to apply generally to all mines. The stakes are high because inaccurate and poorly worded standards can endanger miners' lives.

Documents submitted as testimony in Congressional hearings show how minor editorial changes can diminish worker safety. The first document presented as evidence (Fig. 1.2) is a memo from William Bruce, Chief, Ventilation Division, regarding MSHA's draft revision of the Mine Standards for methane-producing mines (June 10, 1983). To demonstrate his point, Bruce includes a copy of the gassy mine standards. In this second document, he circles typographical errors and other changes that have reduced worker safety (Fig. 1.3).

Many of Bruce's comments reveal a grammarian's annoyance at poorly worded language. Bruce faults the document's lack of parallelism and criticizes other errors that he believes were "introduced capriciously" or as the result of word-processing errors.[41] He complains that "this kind of error is difficult to understand since the proper wording is contained on a flexible disk for the word processing units."[42]

Bruce's comments underline the ways that simple changes in language can seriously jeopardize miners' lives. The revision changes "or" to "and" in two different standards.[43] The authors of the revision explain: "Wording was changed slightly to be as consistent as possible with wording of draft revised standard 58.21-640(a)." But these slight revisions have important consequences for safety. In the original, a reading of 1.0% methane at any *one* site would require immediate action to reduce methane. In the revision, methane must be present in concentrations greater than 1.0% at all four sites of a working mine ("the back, face, rib, *and* floor"). Bruce argues that this revision substantially reduces the standard of safety provided in the original.[44]

The problem of wording ultimately reflects deeper concerns about how to produce technical documentation that can improve the health and safety of workers. As Pyro demonstrates, imprecise and poorly worded standards can derail the agency's intentions and leave workers unprotected. But too much focus on precision can also prevent the agency from promulgating standards to regulate new technologies.

---

[41]Bruce (1987) comments on typographical and grammatical errors in marginal comments: "Wording was capriciously changed" (p. 406); "Changed. Not Parallel" (p. 400).

[42]Bruce, 1987, p. 421.

[43]Gassy Mine Standards, section 58.21-639 (U).

[44]The revision reads: "If 1.0 percent or more methane is present in air returning from a working place, or is present in the air not less than 12 inches from the back, face, rib *and* floor of a working place, changes or adjustments shall be made immediately so that the concentration of methane is reduced to less than 1.0 percent" (Italics added). (U.S. Congress, House, Oversight of the Mine Safety and Health Administration, p. 418). Reviewers note that this change allows the standard to be as consistent as possible with the wording of draft revised standard 58.21-640 (a).

U. S. Department of Labor   Mine Safety and Health Administration
P O Box 25367
Denver, Colorado 80225

SAFETY AND HEALTH TECHNOLOGY CENTER
Ventilation Division

June 10, 1983

MEMORANDUM FOR: MADISON MCCULLOCH
Director, Technical Support
Arlington, VA

THROUGH: A. Z. DIMITROFF
Chief, Safety and Health Technology Center

FROM: WILLIAM E. BRUCE
Chief, Ventilation Division

SUBJECT: Draft Revision of Gassy Mine Standards
for Metal and Nonmetal Mines

I have attached Category III of the proposed new gassy mine standards
for metal and nonmetal mines. Technical and typographical errors were
noted throughout the entire package which consists of a preamble and
five individual mine categories. Category III, with handwritten comments
thereon, is attached as an example of problems which, no doubt, pervade
the entire package.

Most likely, the errors could have been avoided had the total committee
membership been provided the opportunity to make one final review of the
package prior to its release to interested parties. The errors were
introduced into the package during final revisions and typing carried out
in Arlington. As it happened, there was no opportunity for the full
committee membership to participate in that final review, and although
not of our making, the errors may prove to be an embarrassment to the
committee and to Government workers in general. I personally did not
receive the final standards package until noon on June 8, 1983, and the
package was reportedly advertised in the Federal Register on June 10,
1983. With such a short period available for reviewing the 271-page
document, there was no chance for the committee membership to review
the entire package and to correct the deficiencies. This release of the
document to the Federal Register, without a final committee review, was
contrary to repeated assurances made to me that the committee would
receive one last review immediately prior to publication in the Federal
Register.

Of greater significance than the uncalled for typographical errors, however,
is the potentially diminished worker safety that was introduced as a result
of some of the changes which are noted in the attachment. I hope that my
comments will result in changes in the present practice which apparently does
not allow for a final review of draft standards by standards committee members.

Attachment

FIG. 1.2. Memo from William Bruce, Chief, Ventilation Division, regarding MSHA's draft revision of the mine standards for methane-producing
mines (June 10, 1983). Source: Bruce (1987, p. 392).

58.21-648

VENTILATION

57.21-48 Mandatory. Line brattice or other suitable devices shall be installed from the last open crosscut to a point near the face to assure positive air flow to the face of every active underground working place, unless the Secretary or his authorized representative permits an exception to this requirement.

58.21-648 (U) Draft Revision

Line brattice or fans with ducting shall be used to provide positive air flow across each working face.

Explanation: Revised wording of this standard was to clarify "other suitable devices" by specifying fans with ducting. Reference to placement of fans and ducts or brattice was eliminated as draft revised standard 58.21-634 requires a minimum amount of air flow to be maintained regardless of fan or brattice placement. Additionally, this revision reflects current application of ventilation practices in the industry.

> Formerly the committee version of 4-20-83 said " .... positive air flow across each face within a working place."
>
> The reworded version will allow any face in a set of advancing multiple entries to be totally unventilated, thus allowing a methane buildup and thus jeopardizing worker safety.
>
> The Green River MSHA personnel may have a problem with the wording relative to pillar extraction, but that can be handled and still retain the wording.

FIG. 1.3. Bruce includes a copy of the gassy mine standards. In this second document, he circles typographical errors and other changes that have reduced worker safety. Source: Bruce (1987, pp. 420–421).

46

The Problem of Regulatory Revision and Review

Documents presented to Congress reveal how the viewpoints of multiple stakeholders are literally written into the regulatory process. These documents provide a limited view of the hundreds of revisions and re-revisions that occur within agencies. In actual practice, a single standard can be revised as many as 40 times as it passes between lawyers, safety experts, and technical advisors.[45] While the process allows agencies to incorporate comments from stakeholders, the process creates controversy and fails to satisfy the needs of diverse stakeholders. Ultimately, Congressional overseers concluded, the process paralyzes the agency and jeopardizes workers' lives.

In 1987, MSHA proposed a series of draft revisions to ventilation standards in methane-producing mines. MSHA circulated copies of the draft revisions along with the original standard and comments justifying the proposed change. An excerpt from the document appears in Fig. 1.4. As the document circulated among interested parties, each new reader commented upon revisions and responded to comments of previous reviewers.

As the excerpt in Fig. 1.4 illustrates,[46] the layering of responses within the document becomes increasingly more complex as respondents react to comments from previous commentators.[47] As the number of commentators and interpretations increases with each new review, it becomes increasingly difficult to separate writer, respondent, and audience within the text.

To challenge a revision, reviewers must demonstrate the flaws in their opponent's interpretation of the evidence. Each commentator raises new questions about the uncertainty of new technologies and the efficacy of standards that were in effect at the time of previous disasters. In one memo, Bernard (1987) challenges the results of a technical review of the issues raised in previous memos. Bernard's handwritten comments above the typed text make explicit his alternative conception of Utah mining law.[48] Bernard draws upon a previous accident to challenge the agency's representation that these events were "rare."[49] He argues instead that the same standards were in effect in a previous accident "when MSHA and Co. personel [sic] were killed [in]

---

[45]Personal interviews, 1992.

[46]Bernard, 1987, p. 484.

[47]Beason (1987) objected to changes in standards on technical grounds, arguing that "the use of vent tubing has in the past proved disastrous" (Fig. 1.3). Bernard (1987) responded to Beason that "this language recognizes advancements in technology which are currently in use at one . . . mine in proposed category II. . . ." Beason forwarded the response to other members of the gassy standard committee, taking the liberty to write in some comments (Bernard, 1987).

[48]Bernard, 1987, p. 478.

[49]Bernard, 1987, p. 483.

ventilation control throughout the mine can be achieved with

ventilation tubing in the shaft or drift. This standard must be

viewed in conjunction with the remainder of the final rule which

contains requirements for air flow, monitoring for methane,

actions levels, etc.

In the Cane Creek explosion, gases liberated during blasting

were drawn into an underground shop where ignition sources were

present. The Belle Isle mine explosion occurred when an outburst

Referring to Calix explosion June 8, 1979

of gas, liberated by blasting, was ignited by sources in the

Not in The Deaths, Fire Belle Isle March 5, 1968

active workings. Neither of these explosions can be attributed

to the handling of intake and return air in a single mine

The Reason was Because intake + Return airways were Not provided

opening. Both instances, however, serve as evidence of the

Thus The Methane Inidated The Mine Back To The Ignition Source (Texas Gulf

destructive forces generated in the mine opening when an

explosion occurs and the need for secondary escapeways from the

mine.

No metal and nonmetal rulemaking effort has ever addressed

the issue of providing sufficient numbers of mine openings to

assure that pressures generated by an explosion would not damage

Explosions destroy Vent Tubing Thus CO, CO₂ kill Miners. The purpo:

a shaft. If this effort was made, each mine would have to be

OF Vents in Curtain walls is To Release The explosive pressures - vent

evaluated individually based upon the configuration of the mine.

Tubing would be emmidiatly destroyed - Thus distroying All Ventilation

The resulting requirements would then become immediately outdated

into the mine.

as the mining front was advanced and the mine size increased, so

it would be nearly impossible to develop such requirements.

The commenter is indeed correct in his contention that

secondary openings to the surface are necessary for escape

purposes. This protection is provided in Subpart J, Travelways

FIG. 1.4. Bernard's memo demonstrates the presence of many different voices in the text. Source: Bernard (1987, p. 480).

and Escapeways, Section 57.11050. It was never intended for inclusion in Subpart T, Methane Standards. In most metal and nonmetal mines, these additional openings to the surface are also incorporated into the mine ventilation scheme; therefore, intake and return air passes through separate openings. The Agency recognizes that these separate openings for intake and return air are usually the most efficient and effective way to ventilate an underground metal or nonmetal mine but does not intend that a ventilation requirement for substantial separation of fresh air and gas-laden air be used to provide this escape capability.

In summary, the majority of the commenters examples and opinions on this issue concern escapeways which <u>are not</u> <u>appropriately addressed</u> by methane ventilation standards.

↓ *TRUE They Are Not Appropriatly-addressed*

Issue #2. The commenter objects to the language of Section 57.22224, Auxiliary Equipment Stations, for gilsonite mines because the restriction for placement of potential ignition sources was changed from "within 100 feet of a mine opening" to "within 50 feet of a mine opening". He contends that this change does not recognize the explosibility and density of gilsonite. He substantiates his concerns by citing hazards which <u>could exist</u> in a <u>mill area</u> and an accident that occurred near the mine

Gilsonite in the Mine, Same As in the Mill -

opening when an <u>ore car was dumped</u> and a spill occurred.

The standard prohibits the installation of battery charging stations, compressor stations and electrical substations in the underground portion of the mine or within 50 feet of the mine opening. No reference is <u>made to mill installations.</u>

Unsafe in Mill- Also- Mine The Mill Also Transports
Ore by Air Lift Lines - and Fills - Ore Bins.

FIG. 1.4. *(Continued)*

West Virginia."[50] In another memo, Beason (1987)[51] argues that the new ventilation practice has already "proved disasterous [sic]."[52] Beason bases his assessment on previous accident investigations and recommendations from the American Mining Congress.

Although writers disagree about the need for new standards, they use many of the same types of arguments to justify their positions. In each document, respondents draw upon previous disasters to warrant their concerns about the new draft revisions. Comments sometimes reflect the respondent's personal experience of how prudent mine operators will act in the presence of risk. Thus, when Bernard asserts that "it is reasonable to assume that a prudent mine operator" would make smart choices about the location of a fan "absent the regulation,"[53] Beason counters: "If all operators were prudent—we wouldn't need any regulations, but history proves that wrong!"[54]

Many of the arguments are hypothetical. In one section of this memo, agency writers acknowledged that individuals might encounter hazardous conditions if they entered a mine following an explosion, but these writers assure their audience that previous standards had provided "protection . . . on these rare occasions."[55]

Not surprisingly, the quality of previous accident reports affects the meaning that reviewers are able to draw from these events. When accident reports provide inadequate information, agencies and individuals must spend time reviewing and re-reviewing the cause of accidents to make sense of the disaster. Without adequate evidence, these individuals cannot argue successfully for policies to prevent accidents in the future.

THE DIFFICULTIES OF COMPLIANCE

Once standards are promulgated, mines must interpret and apply standards in local contexts. Because many situations of compliance with standards seem so clear-cut, we may fail to appreciate the difficulties of compliance in a hazardous environment.

We comply with the law when we stop at stop signs; we comply with doctors' orders when we take our medicine as ordered; we comply with instructions when we maintain a constant room temperature, adjusting the furnace

---

[50]Bernard, 1987, p. 483. Bernard (1987) argued that "protection is provided on these rare occasions by the requirements of Sections 103 (j) and (k) of the Act which assures appropriate control in emergencies and by the mine standards of Part 49" (p. 490).

[51]A member of the Gassy Mine Standard Committee Moab field office.

[52]Beason, 1987, p. 461.

[53]Bernard, 1987, p. 490.

[54]Bernard, 1987, p. 490.

[55]Bernard, 1987, p. 490.

or air-conditioning as the weather changes. In the first case, we follow a clear rule; in the second case, we must determine whether conditions warrant our actions (e.g. when we are ordered to take pain medication "as needed"). In the third case, we may find ourselves technically out of compliance even as we increase our efforts to bring conditions into compliance. If our actions fail to solve the problem, we risk disaster. But many times, we succeed in bringing conditions into compliance without a problem. When the likelihood of a bad outcome seems remote, we develop a sense of complacency, testing the limits of compliance until we once again produce disaster.

Accidents like Pyro demonstrate how workers and management may deliberately continue to work in highly volatile conditions because they have miscalculated the time and resources they need to maintain production and bring the conditions into compliance. Ideally, individuals recognize rapidly changing conditions in the environment, follow safe practices, and control and manage hazards. But when conditions change rapidly, the law exempts those individuals "whose presence in such area is necessary, in the judgment of the operator or an authorized representative of the Secretary of Labor to eliminate the condition described in the order."[56] These workers work in rapidly changing conditions within a highly explosive and dangerous environment. Miners themselves accept this risk when they believe they can control hazards as they work.[57]

The Tension Between Strict Enforcement and Day-to-Day Compliance

The day-to-day problems of managing a hazardous environment reveal the tension between a strict enforcement of a standard and the day-to-day problems of compliance in a hazardous workplace.

Ventilation standards are some of the most specific in the Mine Act. The Mine Act specifies that mines must maintain a system of airflow that can dilute and exhaust methane and other toxic gases. The Mine Act also specifies the maximum level of methane (generally 1.0%). All mines must submit a

---

[56]The Federal Mine Safety and Health Act of 1977, Section 104 (d) (1), p. 751. The law further specifies that this withdrawal order exempts any public official whose official duties require him to enter such area; any representative of the miners in such a mine who is, in the judgment of the operator or an authorized representative or who is accompanied by such a person and whose presence in such area is necessary for the investigation of the conditions described in the order; and any consultant to any of the foregoing (The Federal Mine Safety and Health Act, Section 104d (2)-(4), p. 752). The citation is from the July, 1990, version of the Act. [Online] Available: http://www.msha.gov/REGS/ACT/ACT1.HTM#13act.

[57]Personal interview with the author.

ventilation plan for approval before they can start mining to demonstrate how they will comply with the law. They must check methane levels at regular intervals and shut down electrical equipment when levels exceed the maximum levels allowed by law. The standard provides a strict level of safety, but mines do not always withdraw miners and close shop every time the level of methane exceeds 1%. Someone must continue working to bring the mine into compliance.

So-called "gassy" mines like Pyro require constant regulation because they emit high levels of methane as a byproduct of normal mining operations. Natural changes in pressure and temperature affect the flow of air in and out of a mine. As air heats up underground, it rises to the surface, creating a natural airflow like a chimney. (Ancient mines used to light fires underground to heat up air and create a natural draw; these fires frequently ignited methane and caused explosions.) As miners remove rock and advance into the mine, they must work to keep air moving into the working face. To maintain a safe environment, miners must constantly adjust and re-adjust an underground system of permanent stoppings and temporary curtains (called "braddices") that redirect fresh air to each section in the mine.[58] These practices make compliance a difficult and time-consuming project. To save time, miners may take shortcuts that can affect ventilation throughout the entire mine.

To control the concentration of methane, mines have developed systems of ventilation that provide an influx of fresh air that flushes toxic gases. In addition to ventilation fans on the surface that force fresh air into the mine, mines create permanent tunnels (bleeders) that move fresh air into the mine and exhaust methane and other toxic gases through tunnels known as returns. If the system is inadvertently reversed, return air filled with methane and other toxic gases can accumulate in abandoned or unworked sections of the mine, creating the potential for a violent explosion.

Because mines are complex systems, normal mining practices can easily upset the fragile system of ventilation in a gassy mine even when individuals are making a good faith attempt to achieve compliance. When Pyro miners traveled from one section to another, for example, they passed through "mandoors" in stoppings or took down curtains to move equipment to the working face.[59] Because ventilation curtains interfered with production, miners did not always replace curtains except when high concentrations of methane were detected. One particular mandoor was so critical for regulating ven-

---

[58]According to the citation, this particular stopping line separated the No. 4 return from the longwall bleeders which contained the methane monitor (Childers et al., 1990, p. 50).

[59]Childers et al., 1990, p. 49.

tilation that simply leaving the mandoor open constituted a major violation of the mine's ventilation plan. When one miner closed the mandoor, airflow was restored and the violation was removed.[60]

Conditions in an unworked or abandoned section of the mine can also increase the potential for disaster.[61] At Pyro, water blocked ventilation systems and increased the accumulation of methane in one abandoned section. On June 8, 1989, the water accumulation in this section was 1000 feet long and two feet deep. A submersible pump installed in March, 1989, was finally energized on June 2, 1989, but this pump experienced problems in the pumping rate until September 13, 1989—the date of the accident.[62]

Economic pressures to increase and maintain production can influence decisions about safety at all levels. At the time of the explosion, Pyro was in the process of dismantling a previous worksite (Panel "O") and transporting it to a new location (Panel "P").[63] Management had originally planned to wait to begin production until the new panel was connected to the larger system of ventilation. But they decided to begin production on the new panel even though the ventilation systems were separated by about 150 feet. In January, 1989, MSHA cited Pyro for failing to follow mining projections. The final report blamed management for failing to recognize the effects of the continuing changes in the ventilation plan and for changing the ventilation system while miners were working.[64]

---

[60]One mandoor was so critical that MSHA issued a 104(a) citation (imminent danger) for violation of the ventilation plan. The final report reads: "With the mandoor open, return air from No. 4 Unit traveled through the door toward the longwall face. Ramsey closed the mandoor and the citation was abated. O'Leary talked to McDowell on the surface and told him to be very careful about leaving the mandoors open. McDowell acknowledged the fact that leaving the mandoors open would short circuit the air toward the longwall" (Childers et al., 1990, p. 44).

[61]Childers et al., 1990, p. 42.

[62]The report notes: "This borehole missed the underground workings as did a second borehole which was drilled during August 1988. A third borehole was started on September 1, 1988, but this hole was abandoned when the drill bit used to enlarge the diameter of the borehole became lodged in the hole. The fourth borehole was started on October 26, 1988, and finished on February 11, 1989, but was not energized since no water was present at the pump inlet" (Childers et al., 1990, p. 46).

[63]Since longwall panel "O" had finished production, work had progressed steadily to dismantle and recover the longwall mining equipment. As the equipment was recovered, it was transported to longwall panel "P" where it was being reassembled at the new face (Childers et al., 1990, p. 12).

[64]MSHA criticized management for failing to recognize the "sensitivity" of ventilation systems (Childers et al., pp. 128–9). When the mine drilled a borehole to control water, they also experienced mechanical problems (Childers et al., 1990, p. 47).

Even when we believe that Federal inspectors *ought* to recognize danger, accidents like Pyro demonstrate the degree to which even experienced inspectors can miss the signs of danger if agencies do not constantly revise regulations to keep up-to-date with new technologies.[65] In the Pyro disaster, both panels involved in the accident employed new longwall mining techniques. According to the agency's own *Internal Review* (Tattersall, 1991), MSHA inspectors were unfamiliar with the recently introduced method of longwall mining employed in the mine and thus were not aware these technologies might require more conservative management.[66] The *Internal Review* noted that the writers who drafted the ventilation standards in 1969 were not "fully mindful" of the increased ventilation requirements of longwall mining. As a result, they did not anticipate that longwall mines might liberate "substantially higher" levels of methane than conventional mining methods. (The report concluded, however, that there was "no evidence of any wrongdoing by MSHA officials" at the time.)[67]

In hindsight, accidents like Pyro reveal the degree to which individuals place their trust in management to bring conditions into compliance—in sufficient time—when conditions exceed minimum safety standards. As the following testimony from the Southmountain disaster demonstrates, miners trust that management will ultimately comply with standards even when they are working in highly volatile conditions. Davis testified:

> I just knowed that he come down there and he told us, he said, you know, Brian and all of us was standing there, and he come back down and told us he said, now, hit methane up at five right, you know, we's turning the break. And he

---

[65]This does not rule out other explanations for their inability to detect problems at the site. A discussion of the relationship between management, labor, and agencies is beyond the scope of the present analysis.

[66]Following the accident, MSHA conducted its first internal review to assess its own liability for decisions and practices that contributed to the accident. (See Tattersall, 1991.)

[67]The review concludes,

Because longwall mining methods were not prevalent when the mandatory ventilation standards in Subpart D were written in 1969, it is unlikely that the drafters of these standards were fully mindful of these methods when creating ventilation requirements for underground coal mines. For example, the minimum air quantity requirements in section 75.301 are not adequate minimum requirements for longwall mining sections, which means that ventilation plans must specify greater amounts. In 1969 when the Coal Act was drafted, one million cubic feet of methane liberation in a 24-hour period was considered high, primarily because mining was done by either conventional methods or with a continuous miner. Today, with longwalls, substantially higher methane is not unusual (Tattersall, 1991, p. 84).

said, get them curtains up. Well, it wasn't two minutes, well all of us worked on the curtains, you know, and it was allright (sic).[68]

Ultimately, the problem of compliance is a problem of time. When disaster is imminent under Section 104 of the Mine Act, mines *must* take action to protect the safety of workers—withdrawing all miners, if necessary, until conditions abate or can be brought back to standard. But calculating imminence is a difficult problem—particularly when individuals are dependent upon their own experience to assess and predict disaster.

The problem of compliance thus recasts the problem of regulation (What are the hazards? How great is the risk? What shall we do?) into a question of time: Will operators have enough time to abate conditions before they produce disaster?

The Problem of Complying "In Time"

The problem of predicting imminent danger demonstrates how the problem of time affects decisions about compliance.

The Mine Act defines several types of imminent danger. If inspectors or representatives of the Secretary of Labor find that an imminent danger "exists," they can order "forthwith" that the mine operator withdraw all persons from the section until the imminent danger no longer exists (Section 104a). MSHA can also order a withdrawal even if there is no immediate danger if mines fail to abate conditions cited in a previous order (Section 104b).[69]

The notion of imminence defines the highest exigence in terms of time. But the Mine Act does not specify what constitutes imminence. As a result, industry and labor disagree about how serious and immediate the danger must be before inspectors order miners to withdraw from the mine. Industry argues that the term "imminent" should be limited to those situations where the danger is either "immediate" or "so serious" that it will cause "immediate" physical harm.[70] Unions counter that miners must not be exposed to a greater

---

[68]Davis, 1993, p. 15. Spelling *Sic.*

[69]Inspectors can also order a withdrawal if they find another violation caused by the unwarrantable failure of the operator to comply within 90 days of the first inspection.

[70]Hurst, 1977, p. 410. In testimony before the U.S. House of Representatives, industry officials argued that the term "imminent danger" should not be expanded to include "non-existent conditions" that could only "potentially" affect the health and safety of miners. The report states:

Even though imminent danger was intended to include only those situations which posed an immediate danger or serious physical harm, the scope of applicability has not

danger than a "rational man with free choice would expose himself to."[71] Must the conditions be present and visible or merely "probable"? If conditions are merely "probable," how great should the magnitude of risk be? Is a one-in-four probability "imminent"? A one-in-two? How much time do we have before we decide to withdraw miners entirely?[72]

To determine whether danger is imminent, individuals must first identify hazards, determine the magnitude of the hazard effects (How great is the likely harm?), and judge the likelihood that conditions will "significantly and substantially"[73] affect the health and safety of miners within the time required to abate conditions. Next, they must specify practices that will bring the mine into compliance with standards. Finally, they must determine whether these practices will abate conditions in time to prevent disaster. In a brief filed by the Council of Southern Mountains, Inc.,[74] Galloway and McAteer proposed a simple test for imminent danger: How long would it take to abate conditions? How likely is the feared harm to occur before the condition or practice is abated if normal mining operations continue?[75] Would the inspector remain in the mine under such conditions?[76]

Unfortunately, individuals perform poorly as intuitive statisticians in assessing the magnitude of even the most common risks (though researchers disagree about the relative performance and the kinds of situations that individuals perform well in).[77] In addition, the meaning of terms like "possible" and "likely" is highly dependent upon context: "It's possible I'll be late" represents a different degree of magnitude than "It's possible a woman will be president."[78] As a re-

---

been so confined. It has been construed to include potential, non-existent conditions which could only possibly have created death or serious physical injury if normal mining operations had been permitted to continue. (*Oversight hearings on the Coal Mine Health and Safety Act of 1969 (Excluding Title IV)*, 95th Congress, 1st Sess., 1977, p. 410)

[71]Because the determination of imminent danger is highly subjective and intuitive, unscrupulous managers can manipulate statistical estimates to continue production. *Oversight hearings on the Coal Mine Health and Safety Act of 1969 (Excluding Title IV)*, 95th Congress, 1st Sess., 1977.

[72]Chap. 3 describes the problem of imminent danger in more detail.

[73]See below, fn. 84.

[74]Speaking as *amicus curae* on Pittsburgh and Midway Coal Mining Company vs. Mining Enforcement and Safety Administration and the UMWA, 1976.

[75]*Oversight hearings on the Coal Mine Health and Safety Act of 1969 (Excluding Title IV)*, 95th Congress, 1st Sess., 1977, p. 673.

[76]*Oversight hearings on the Coal Mine Health and Safety Act of 1969 (Excluding Title IV)*, 95th Congress, 1st Sess., 1977, p. 679.

[77]Fischhoff, 1988, p. 167.

[78]Morgan & Henrion, 1992.

sult, both expert and lay audiences differ widely in the numerical probabilities they attach to such terms as "good chance" and "possible."[79]

The Costs and Benefits of a Well-Regulated Environment

The question of time also affects the kinds of actions individuals propose to manage hazards in the workplace. "It's possible that the mine will explode" (as a general statement of possibility) does not demand the same level of action as the time-constrained warning: "It's possible that this mine will explode in 10 minutes." All mines fail eventually. For workers in hazardous environments, the question "when?" is essential to saving miners' lives.[80]

Mine operators have many options for creating a safe work environment. Mines could theoretically create tunnels as strong and enduring as highway tunnels, but the costs and the hazards of creating tunnels outweigh the advantages to safety. Alternatively, they can opt for the minimum level of roof support, monitoring changing conditions, and adding additional roof support as needed. They can keep the margin of safety close to the levels described in the standards, or they can work to create an environment that is so well-regulated that they create a large margin of safety.

Many mines opt for a large margin of safety because it is less costly over the long run and because it produces fewer occasions for disaster. If management chooses to maintain a small margin of safety, they must continually work to monitor conditions, and they create many more occasions for failure. To maintain a safe working environment, they would need an educated workforce that can spot changes in conditions. They would need flexible instruction and training that can respond to new conditions, and they would need individuals who could communicate changes quickly and efficiently. If mines opt for a large margin of safety, they do not need to invest as much in day-to-day monitoring underground, but they must plan carefully and anticipate potential changes in the geography that might require changes in practice.

Over time, however, even well-regulated systems fail. The most carefully constructed tunnels are subject to erosion, subsidence, and wear. Roof supports eventually fail. Metals corrode. Mines are dynamic systems that cannot continue to operate on the same course forever. If workers and management become complacent in the absence of any perceived hazard, they may begin to take short-cuts, cutting the margin of safety until they reach disastrous levels. The series of mine disasters that culminated in the Farmington disaster

---

[79]Morgan, 1998.
[80]VanDeMerwe, 1995.

shows how slowly management responded to diminishing levels of safety in the mines and—ironically—how quickly regulators were able to create the Mine Act when public outrage over the death of 178 miners forced regulators to act to improve safety and enforcement.

Changes in operation can also create the potential for disaster. When Pyro management decided to move the longwall, they disrupted a well-regulated system. They could no longer rely on previous interpretations of the mine regulations; instead, they were forced to reinterpret standards in light of the changes they had introduced. Regardless of whether they willfully ignored signs of imminent danger or simply misinterpreted the law, management failed to recognize that a change in operation required higher levels of caution and a higher margin of safety.[81]

To insure that decisions about safety are not merely good guesses, Federal law holds all parties accountable for the quality of their documentation. When management fails to document the reasons for their action, they can be held liable for "unwarrantable failure."

THE PROBLEM OF DOCUMENTATION

When speakers determine when and how to manage hazards, we expect them to have sufficient evidence to justify their decisions and we expect them to exercise due diligence in their reasoning.[82] As Toulmin suggests, it is ultimately the quality of the arguments and the rhetorical status of the evidence that determines what sort of probability terms speakers are entitled to use—not the presumed uncertainty of the claim. Despite the uncertainty of claims based on probability, Toulmin argues, claims based on probability are no less persuasive than those couched in certainty, for it would be highly irrational for individuals to believe that danger was likely and not act—assuming they were able—to reduce visible hazards that might create disaster. The expectation that speakers have sufficient evidence to support probability estimates enables us to hold individuals legally and morally accountable for the quality and completeness of their evidence. Toulmin writes:[83]

---

[81]In retrospect, it's hard to imagine that management believed they had time to bring the mine into compliance with ventilation standards. As Aristotle (1991, p. 74) points out, people do not normally argue for the disadvantageous, unjust, or shameful. But their judgments can be influenced by greed or opinion.

[82]Secretary of Labor, Mine Safety and Health Administration (MSHA) v. Emery Mining Corporation, p. 3.

[83]As Toulmin (1995) argues, "The most reasonable estimate a man can make of the probability of some hypothesis [is] that which is warranted by the evidence, and the terms 'bearing,'

We may not say, "I shall probably come," if we have strong reasons for thinking that we shall be prevented . . . or say, "It will probably rain tomorrow" in the absence of fairly solid meteorological evidence. Our probability-terms come to serve, therefore, not only to qualify assertions, promises and evaluations themselves, but also as an indication of the strength of the backing which we have for the assertion, evaluation, or whatever. It is the quality of the evidence or argument at the speaker's disposal which determines what sort of qualifier he is entitled to include in his statements: whether he ought to say, "This must be the case," "This may be the case," or "This cannot be the case"; whether to say "Certainly so-and-so," "Probably so-and-so," or "Possibly so-and-so."

Federal agencies use the term unwarrantable failure as a legal standard to judge the merits of individual and collective actions. When agencies apply this term, they hold individuals accountable for the quality of their actions and the adequacy of the evidence they present to justify their actions. When we call some actions warrantable, we accept that many actions will produce poor outcomes if conditions change, if unpredicted events occur, and if individuals legitimately cannot foresee the potential consequences of their acts. But these outcomes are not necessarily the product of unwarrantable action. When individuals are guilty of unwarrantable failure, they are guilty of knowingly, inexcusably, and unjustifiably failing to justify their actions *in writing* so that others could judge the adequacy of their reasoning.

The Problem of Timely and Adequate Documentation

To justify decisions about compliance, individuals must maintain a timely and adequate system of documentation.

The Mine Act specifies that mines must document conditions in the mines in writing. In a section entitled "Reports," the Mine Act specifies that

---

'support' and the like are the ones we use to mark the relation between statements cited as evidence and the possibilities whose relative credibilities are being examined" (p. 58). Toulmin recognized that changing conditions or incomplete information could mislead decision makers, but he distinguished between what he called an "improper claim to know something" and a claim that later turns out to be "mistaken." To attack a claim (as opposed to modifying a claim in light of new evidence), critics must also attack the argument leading up to it or the qualifications of the person making the claim.

Toulmin (1995) writes:

We [thus] distinguish between a claim that was improper at the time it was made from one which subsequently turned out to have been mistaken; and criticism directed against the claim as originally made must attack the backing of the claim or the qualifications of the man who made it—showing that in the event it proved mistaken may do nothing to establish that it was at the time an improper claim to make. (pp. 59–60)

"In addition to such records as are specifically required by this Act, every operator of a coal mine shall establish and maintain such records, make such reports, and provide such information, as the Secretary may reasonably require from time to time to enable him to perform his functions under this act" (Pub. L. No. 91-173, 30 CFR § 83-111 [b], 1969). Mines must conduct preshift examinations within three hours of each working shift in each working section of a mine (Pub. L. No. 91-173, 30 CFR § 75.304-2, 1990). If mine operators fail to conduct proper examinations and investigations, they can be cited for "significant and substantial" violations.[84] Unfortunately, what counts as a working section is not defined in the Act, and the term may be subject to multiple and sometimes conflicting interpretations.

During each preshift exam, examiners must test methane levels in each working section; examine seals, doors, roof, face, rib conditions, active roadways, travelways, belt conveyors, air flow, air volume, and velocity; and identify "such other hazards and violations of the mandatory health or safety standards as an authorized representative of the Secretary may from time to time require." As evidence of this exam, examiners must record the results of their examinations "with ink or indelible pencil in a book approved by the Secretary" (Pub. L. No. 91-133, 30 CFR § 75.303 (a), 1990).

Agencies are also held accountable for the quality of their documentation. When inspectors perceive danger, they must provide "adequate and timely" written notice of the conditions and the practices that would bring the mine into compliance.[85] The courts interpret this provision to mean that inspectors must state the problem "in such a way that the operator can clearly understand what it is that he has to do to achieve abatement."[86] This means that all parties involved in the decision must be able to recognize the hazards specified in the violation and the conditions that would alleviate (or "abate") the

---

[84]Section III of MSHA's Policy Manual III for Special Inspections reads: "Failure of a competent or qualified person to conduct adequate examinations or inspections can expose miners to unsafe acts or conditions and, under normal continued mining operations, may constitute a significant and substantial violation" (U. S. Department of Labor, 1996, p. 23).

[85] In IBMA72-27: Eastern Associated Coal Corp; IBMA 235, Dec. 27, 1972, the administrative court defined "adequate notice": "As a general proposition, when an alleged violation is sufficiently described to permit abatement, adequate notice of the condition is established. However, an operator is entitled to adequate and timely notice of this section of the Act or mandatory standard alleged to be violated in order to prepare a defense for a penalty assessment proceeding" (Cleveland and Turner, 1977, p. 7). In *The Compilation of Judicial Decisions that Have an Impact Coal Mine Inspection*, Cleveland and Turner (1977) interpret this decision for coal mine inspectors: "Notice must be stated in such a way that the operator can clearly understand what it is he has to do to achieve abatement and, specifically, what he has been charged with" (p. 7).

[86]Cleveland and Turner, 1977, p. 7.

potential risk. The alleged violation must be described in writing and the parties involved must have adequate notice so that they can prepare a defense (to mitigate penalties or have the citation removed). Good notes are essential, MSHA warns, "if facts are questioned."[87]

Every mine comes to a point where operators and management must make hard choices. In hazardous environments, individuals cannot guarantee that their actions will produce good consequences. But the law requires that mines make a good faith effort to document in writing the conditions that justify how and when they act to comply with standards.

The Problem of Unwarrantable Failure

The case of Secretary of Labor v. Consolidation Coal (Consol) demonstrates how courts can hold operators liable for the quality of their communication practices.

On March 19, 1992, a methane explosion at Consol's Blacksville No. 1 Mine in northern West Virginia killed four miners and injured two others. In Secretary of Labor v. Consolidation Coal Company, the court argued that Consol should be held liable for unwarrantable failure on the grounds that Consol officials knew about conditions but failed to communicate their findings to others. The court argued that "corporate balkanization" at the company had led to a situation where "officials in one division did not know what those in another division were doing." The court concluded that "confusion resulting from . . . inadequate communication" contributed to the disaster.[88]

The case demonstrates the importance of written communication in the determination of liability. The judge noted, for example, that management had assumed the risk of being cited for violations of the Mine Act when they failed to deliver appropriate supplies to miners prior to the explosion. But this failure alone did not constitute "misconduct, recklessness, or serious lack of care, amounting to more than ordinary negligence" and was therefore not accounted as unwarrantable failure.[89] Once MSHA inspectors had warned foremen, however, "Consol was on notice that greater efforts were necessary to address the cited condition." At that point, Consol was guilty of unwarrantable failure.[90]

---

[87]U. S. Department of Labor, MSHA, 1996, p. 24b.

[88]Secretary of Labor v. Consolidation Coal Company, 2000. The decision reads: "The confusion resulting from this inadequate communication and coordination was itself a contributing cause of the explosion. There was a serious lapse of judgment among Consol personnel in not ordering or ensuring that methane checks were made underneath the production shaft cap."

[89]Secretary of Labor v. Consolidation Coal Company, 2000.

[90]Secretary of Labor v. Consolidation Coal Company, 2000.

Written evidence ensures that miners and management do not willfully or inadvertently overlook the signs of significant and substantial danger. In rhetorical terms, unwarrantable failures cannot be supported by the available evidence. To insure that their decisions have ample warrant, individuals must document *in writing* the conversations, observations, and physical evidence on which they base their decisions.

Inspectors can cite operators for unwarrantable failures under Section 104 (d)(1) of the U.S. Mine Act. The term is not defined in the Mine Act, but the Secretary of Labor has interpreted the meaning to include "the failure of an operator to abate a violation he knew or should have known existed." The legal and legislative history of the term describes unwarrantable failure in terms of "indifference," "knew or should have known," "lack of due diligence," and "lack of reasonable care."[91] It does not include cases where an operator knew of a danger and took exceptional measures to bring the condition into compliance (Secretary of Labor v. Emery). Instead, the term refers to those cases where an operator knew of the condition but failed to provide adequate remedy.[92]

MSHA acknowledges that no single individual can be physically present to observe every aspect of mining. Ideally, observers can inspect every phase of the mining cycle, every shift in a multi-shift cycle, and every location in the mine, including escapeways, travelways, and "all other areas of the operation where conditions, procedures, practices, or methods could affect the health and safety of the worker."[93] Ideally, these inspections should take place when miners are working or—if miners are idle—only if conditions "are practically the same as they would be on working shifts."[94] In practice, however, observers may not have access to events that occur rarely or infrequently (such as blasting) in the mining cycle.[95]

MSHA holds inspectors accountable for the quality of their documentation even when they themselves cannot be present to observe all aspects of work. Inspectors must "evaluate enough conditions and practices and ask sufficient questions of miners to be reasonably assured that work is being safely conducted during the portions of the cycle that could not be observed."[96] If inspectors fail to show by a "preponderance of evidence" that a violation has occurred, an administrative appeals court can overturn inspectors' judg-

---

[91]Secretary of Labor v. Emery Mining Corporation, 1997.

[92]Secretary of Labor v. Consolidation Coal Co.

[93]U. S. Department of Labor, MSHA, 1996, p. 24b.

[94]U. S. Department of Labor, MSHA, 1996, p. 24b.

[95]U. S. Department of Labor, MSHA, 1996, p. 24b.

[96]U. S. Department of Labor, MSHA, 1996, p. 24b.

ments.[97] To protect themselves, inspectors must document "observations, conversations, and physical examinations of work conditions and practices."[98] The Mine Act acknowledges that this information may exist outside of texts in the conversations and undocumented observations that constitute the lived experience of risk. But inspectors are nonetheless accountable for documenting all forms of information relevant to safety.

In the wake of Pyro, labor unions criticized MSHA for failing to prevent the disaster. The union blamed inspectors for not documenting all of the areas of the mine. According to one handwritten memo, inspectors kept a "mental track" and did not actually document all areas of the mine.[99] Supervisors did not review inspectors' notes to verify that they had inspected the entire mine, and maps plotting the day-to-day progress of inspection were not kept up-to-date. The memo outlines MSHA's failures and concludes: "MSHA was aware of Pyro's disregard for the law. . . . Pyro's Failure to Comply resulted in the explosion."[100]

The notion of unwarrantable failure implies that we cannot distinguish in hindsight between the failure to discover warrants and the failure to document these warrants in writing. In either case, the actions that result are unwarrantable failures.

Agencies are also held accountable for the quality of their documentation when they write and publish new regulations. According agency personnel, the process of developing standards within large government agencies is "primarily a joint effort" between specialists and workers.[101] In this process, specialists draft regulations that go to public hearings where workers "are very active constituents."[102] In this ideal framework, regulatory agencies engage stakeholders and experts in an iterative process of problem definition, evaluation, and action that improves safety in the workplace.[103] But too much debate over the interpretation of new standards can paralyze the regulatory process and leave workers unprotected.

---

[97]Cleveland and Turner, 1977, p. 1.

[98]Cleveland and Turner, 1977, p. 1.

[99]According to MSHA, inspectors conducted three complete inspections at Pyro in FY89. The union's handwritten memo calls this "Bull" (Handwritten memo, 1990).

[100]Handwritten memo, 1990.

[101]M. Nichols (Personal interview, 1992).

[102]P. Silvey (Personal interview, 1992).

[103]Agencies differ across the board in their risk assessment procedures and in the degree to which they incorporate stakeholder input in risk management decision making. In 1997, the Presidential/ Congressional Commission on Risk Management in Risk Assessment and Management (1997) attempted to create a general framework that could standardize the recommended broad changes in federal regulatory development (see chap. 2).

THE NATURE OF TECHNICAL
DOCUMENTATION IN HAZARDOUS
ENVIRONMENTS

As this chapter demonstrates, agencies and individuals depend upon adequate and timely technical documentation to construct and enforce regulations in hazardous environments. This documentation builds a body of experience that agencies can draw on to help them improve the health and safety of workers.

The individuals who produce this documentation are not normally trained as writers, but they write frequently and regularly as part of their work, and their writing must demonstrate that they have adequate and sufficient evidence for their decisions. Unfortunately, few of these individuals have developed the rhetorical strategies they need to describe the relationships known and unknown, the theory and practice, the standards and phenomena they seek to regulate. As a result, administrative courts frequently overturn citations because agency writers have failed to provide adequate and timely warrants for their judgments.

But technical writers and rhetoricians have also not developed a workable, empirically based theory for understanding how the uncertain material conditions of hazardous environments affect the quality of technical documentation. We have not examined the rhetorical features that individuals employ when they describe their experiences or how this experience is transformed in writing. We have grounded our assumptions about technical documentation in systems where outcomes are predictable and certain. But we have not examined the nature of documentation in systems of profound material and institutional indeterminacy.

As we shall see in the following chapters, individuals and agencies have developed a wide range of rhetorical strategies to document events, conditions, and decisions in hazardous environments, but these documentation practices are neither systematic nor stable. As a result, written reports fail to capture knowledge that individuals need in order to understand the material conditions in which they work. The quality of documentation matters, I shall argue, because each document serves as a warrant that influences future policy and procedure throughout the industry. In chap. 2, I describe how these individual acts of documentation are linked in an ongoing cycle of texts and action within large regulatory agencies.

2

Moments of Transformation: The Cycle of Technical Documentation in Large Regulatory Industries

A serious injury is anything short of death, such as amputation of the hand or foot, broken arms, legs, ankles, knees, fingers, thumb or toes, skull, spine or pelvis or loss of the sight of an eye or any other injury where the person injured is admitted to the hospital for more than 24 hours.[1]

—Labor Research Department (1989)

The effect of the new arrangements for reporting accidents was to increase the number classified in the year as "major injuries" by about 30%. The increase in accidents compared with the previous year is 34.2%, but as there were fewer employees, the trend is not satisfactory.[2]

—Labor Research Department (1989)

The visible presence of risk in a mine, its everyday presence, local impact, and frequency contrast Beck's notion of a Risk Society—where risks are large-scale, invisible, unpredictable, and distributed widely among populations.[3] But the issues are similar. To prevent disaster, individuals must characterize hazards, determine the cause or causes of accidents, and communicate their findings with others. When technical documentation provides an inadequate picture of the events, conditions, and decisions that create disaster, agencies are paralyzed and writers cannot create policies and procedures to prevent disaster in the future.

---

[1]Labor Research Department, 1989, p. 5.
[2]Labor Research Department, 1989, p. 4.
[3]Beck, 1992.

Current research in risk decision making supports the assumption that the most successful risk assessments are characterized by broad-based public participation and a perception by the host community that proposed technical solution meets their specific needs.[4] Risk specialists no longer believe that expertise in risk is the sole province of technical experts who disseminate risk information to educate lay audiences about risk.[5] They have developed a variety of risk assessment frameworks to help them integrate stakeholder knowledge at all stages of the regulatory process,[6] and they have argued for increased public participation to overcome the public distrust of scientific risk assessments.[7] They have argued for a more interactive process that can involve stakeholders at many levels.[8] But researchers have not investigated how specific features of technical documentation might constrain or encourage stakeholder participation at each phase in the regulatory process. Nor have they examined how the process of risk assessment within regulatory agencies might interface with other activities like accident investigation and training.

In this chapter, I describe a rhetorical framework that can help us understand how technical documentation functions within large regulatory agencies. I have called this framework the Cycle of Technical Documentation in Large Regulatory Industries. The forms of technical documentation represented in this framework include a wide array of texts and genres: inspection reports, accident reports, statistical summaries, policy memos, instructions, and procedures. Although all genres are potentially available to writers throughout the Cycle, agencies and individuals have privileged certain forms of documentation at particular moments within the Cycle. These genres emerge from and shape rhetorical activity at particular sites.

This framework allows us to identify six critical moments of rhetorical transformation within the Cycle. At these moments, writers extract informa-

---

[4]See Kunreuther, Slovic, & MacGregor, 1996.

[5]The 1997 Presidential/Congressional Commission on Risk Management and Assessment was mandated by Congress to develop a more inclusive process of stakeholder involvement that would prevent the uncertainty and disagreement that had characterized previous attempts to define risks within individual agencies (p. 187).

[6]I describe these approaches in the Introduction. See, especially, Morgan, Fischhoff, Bostrom, & Atman, 2000; Morgan, Fischhoff, Bostrom, Lave, & Atman, 1992.

[7]See Shrader-Frechette, 1990.

[8]Specific studies arguing for increased stakeholder participation in specific technical domains (hazardous waste siting, nuclear power, toxic waste disposal, etc.) are too numerous to cite. Irwin's (1995) study, *Citizen Science*, remains a classic work in the public understanding of science. See also Irwin & Wynne, 1996. I am particularly indebted to my colleague Baruch Fischhoff for his continuing insights and support in this project. See Fischhoff, Watson, & Hope, 1984; Fischhoff, 1990; Fischhoff, 1991; Fischhoff, 1994a; Fischhoff, 1994b; Fischhoff, 1994c; Fischhoff, 1995.

tion that is presented in one modality (speech vs. gesture, written vs. oral, visual vs. verbal) and literally change the form so that the information can be re-represented for a different audience. These moments of transformation are largely invisible outside of agencies. But the results of these transformations affect future policy and procedure throughout the Cycle. Ideally, these moments help policymakers improve policy and protect workers. But unscrupulous writers can deliberately intervene in this process when they manipulate statistics or present an incomplete picture of events, conditions, and decisions in the workplace.

This chapter provides a general overview of the processes of transformation in large regulatory industries. Unfortunately, all of the rhetorical features that make documents persuasive are not necessarily commensurate. Even if we could recover all the source texts and oral narratives that produce a single document, several key features of these *Urtexts* might still be lost in the transformation: a speaker's viewpoint, the speaker's embodied sensory experience, and the non-verbal modalities like gesture that speakers employ when they talk about hazardous environments. In chapters 4 through 8, we analyze the processes of transformation in written documents and videotaped interviews. This analysis helps us make visible the rhetorical practices that are rendered invisible if we limit our analysis to written documentation alone.

Ultimately, this Cycle helps us see how the knowledge needed to manage hazards is a collaborative enterprise that must be constantly updated and recovered through the processes of transformation and documentation we describe in this chapter. When agencies document information, they attempt to stabilize information for the record. But the cumulative body of documentation in an industry is constantly in flux as individuals continue to create new knowledge of the environments they seek to regulate.

To set the stage for this discussion, I provide a brief overview of the 1997 Presidential/Congressional Commission Report on Risk Management and Assessment. The assumptions that govern the Commission report reflect an increasing awareness among risk specialists that increased stakeholder involvement can improve regulatory risk assessment. But the report also illustrates the tensions that arise when agencies fail to define a rhetorical framework for understanding how documents function in large regulatory industries.

THE NEED FOR A RHETORICAL FRAMEWORK

In 1990, Congress mandated a Presidential/Congressional Commission on Risk Assessment and Risk Management when agreement could not be reached during the drafting of the 1990 Clean Air Act Amendments about the best methods for establishing standards, describing risks, and evaluating outcomes. The mandate specified that the Commission would focus its investigation on "the policy implications and appropriate uses of risk assessment and risk management in regulatory programs under various Federal laws to prevent cancer and other chronic human health effects which may result from exposure to hazardous substances."[9] But the mandate also specified that the Commission might consider "the degree to which it is possible or desirable to develop a consistent standard of acceptable risk, among various Federal programs."[10]

The Commission ultimately interpreted its mandate broadly as a "clear need to modify the traditional approaches used to assess and reduce risks."[11] Building on previous models of risk assessment, the Commission sought to add "important new dimensions to the risk management process," actively engaging stakeholders in the process.[12]

In volume 1 of its final report, the Commission proposed a generalizable framework to help agencies integrate stakeholder values into the process of problem definition and risk management within large government agencies (Fig. 2.1). This framework depicts the process of risk management and assessment as an iterative process of problem definition and evaluation that allows stakeholders to serve as partners at any stage in the process. Although the Commission's mandate had originally been limited to issues of cancer and environmental toxins, the Commission hoped that this framework could be applied more generally across Federal agencies.

In its Final Report, the Commission acknowledged that stakeholders could contribute valuable knowledge at all stages of the process. The Commission

---

[9]Environmental Protection Agency, 1990.

[10]Environmental Protection Agency, 1990.

[11]Omenn, 1997. [Online] Available: http:www.riskworld.com/Nreports/1997/risk=rpt/htm/ epajanb.htm, p. i. The full text of Volume 1 of the final report of the Presidential/Congressional Commission on Risk Assessment and Risk Management, Framework for Environmental Health Risk Management, is available at RiskWorld.com in both HTML and PDF formats. The PDF version is a page-by-page replica of the printed report. The HTML version was posted at RiskWorld.com on January 29, 1997, and the PDF version was posted on February 5, 1997.

[12]Omenn, 1997, p. 1. The Commission defined the term stakeholders broadly, including those "who might be affected by the risk management decision," those who "have information and expertise that might be helpful," those who have "been involved in similar situations before," and "those who might be reasonably angered if they are not included." The notion of stakeholder thus includes community groups, local governments, public health agencies, businesses, religious groups, state and federal regulatory agencies, educational and research institutions, trade associations, and unions. But the Commission did not recognize that diverse stakeholders might be situated outside of traditional regulatory frameworks.

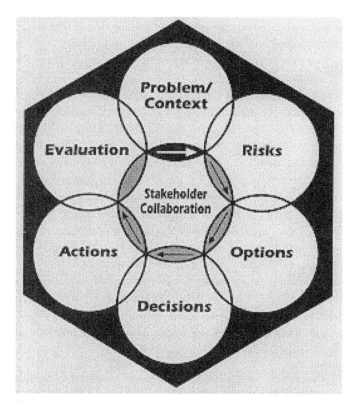

FIG. 2.1.   Risk management framework. Source: Presidential/Congressional
Commission on Risk Assessment and Risk Management (1997, p. 7).

maintained that "risk assessment was developed because Congress, regulators,
and the public require scientists to go beyond scientific observations . . . to
answer questions about what is safe."[13] Stakeholders could "bring to the table
important information, knowledge, expertise and insights for crafting workable solutions."[14] Commissioners recognized that stakeholders would be more
likely to accept risk management decisions if they were involved in shaping
policy. Stakeholder involvement was particularly important in risk decisions
because there were so many different views of the process.[15] Finally, they argued, risk assessments needed stakeholder knowledge because agencies could

---

[13]Presidential/Congressional Commission on Risk Assessment and Risk Management, vol.
2(1997), p. iii.

[14]Presidential/Congressional Commission on Risk Assessment and Risk Management, vol.
2(1997, p. 16.

[15]Presidential/Congressional Commission on Risk Assessment and Risk Management, vol.
2(1997), p. 39. The Commission argued:

not ethically replicate the events and conditions that produced disaster. Because scientists could not ethically set up the conditions to observe effects in humans, they were necessarily dependent upon stakeholders to provide information and data relevant to risk.

In volume 2, the Commission criticized both OSHA and NIOSH for their case-by-case approach to risk management. The Commission charged that hazards were not discovered and risks were not identified because workers were rarely involved in the day-to-day preparation or review of work plans.[16] The report concluded that many problems could have been averted if work plans had drawn on the "extensive experience and unique knowledge" of workers.[17] But the Commission framework did not define how workers' "unique knowledge" might be integrated rhetorically into the regulatory process.[18]

Criticism of the report illustrates the tensions that occur when agencies value stakeholder involvement but fail to develop rhetorical strategies that encourage stakeholder participation in the regulatory process.

When reviewers commented upon the first draft of the Commission report, they applauded the Commission's attempt to involve stakeholders, but they doubted that stakeholders could participate in the regulatory process without some sort of technical assistance. (These comments are summarized

---

Collaboration provides opportunities to bridge gaps in understanding, language, values, and perceptions. It facilitates an exchange of information and ideas that is essential for enabling all parties to make informed decisions about reducing risks. Collaboration does not require consensus, but it does require that all parties listen to, consider, and respect each other's opinions, ideas, and contributions. (p. 35)

[16]The Commission praised the Department of Energy for launching an "integrated" process of risk assessment that involved workers, "health physicists, industrial hygienists, safety engineers, and occupational medicine specialists who participate in the team along with managers, planners, and maintenance and operations supervisors" (Presidential/Congressional Commission on Risk Assessment and Risk Management, vol. 2(1997), p. 144). The Commission compared the DOE's new plan to previous processes within the agency that had produced "conflicting comments or work plans that were disconnected from actual conditions."

[17]Presidential/Congressional Commission on Risk Assessment and Risk Management, vol. 2(1997), p. 144.

[18]The Commission concluded: "Workers were rarely involved in either preparation or review of work plans, so problems that could have been averted because of the workers' extensive experience and unique knowledge of work conditions were not discovered and corrected until after a plan was released. As a result, most workplace deaths and serious injuries at DOE sites over the past five years can be attributed to inadequate hazard identification and control within the work planning process" (Presidential/Congressional Commission on Risk Assessment and Risk Management, vol. 2(1997), p. 144).

in the Commission's final report). Reviewers pointed to the institutional and disciplinary barriers that seemed to separate agencies and their constituents.[19] Others pointed to the Commission's failure to provide a "concrete scheme" for involving stakeholders or evidence that stakeholder participation would improve regulatory decision making.[20]

For scientists, the report's findings underscored the tensions between those who argued that increased research could produce more data and those who argued that data merely obscured the uncertainties in any attempt to quantify hazards in risky environments. One respondent praised the "utility" of the collaborative framework but argued that "this should not mean that risk assessment becomes a political process."[21] This reviewer concluded, "Risk assessment must be science based. Stakeholders can contribute by providing scientific information."[22]

In its final report, the Commission responded to reviewers' comments by adding "principles, guidance regarding implementation, and examples." But the Commission agreed with reviewers that stakeholders might need "technical and sometimes financial assistance to be effective participants in risk management decisions."[23] In the end, the Commission yielded to pressure from scientists and concluded that stakeholders "should not participate directly in the assessment itself" to avoid having risk assessments become "too politicized."[24]

There is a great deal of irony in the agency's failure to articulate the rhetorical strategies that might help stakeholders act as partners in the negotiations. Despite its own admonition that agencies had ignored "a decade of research [in communication] at leading universities and experience at all levels

---

[19]One respondent wrote, "No one denies that in many circumstances integration is absolutely critical to effective management. Most importantly, identifying how institutional barriers to integration can be overcome is more significant than acknowledging that integration should occur" (Presidential/Congressional Commission on Risk Management and Assessment, Vol. 2 (1997), p. 185).

[20]Presidential/Congressional Commission on Risk Assessment and Risk Management, vol. 2(1997), p. 186.

[21]Presidential/Congressional Commission on Risk Management and Assessment, vol. 2(1997), p. 185.

[22]Presidential/Congressional Commission on Risk Management and Assessment, vol. 2(1997), p. 185.

[23]Presidential/Congressional Commission on Risk Management and Assessment, vol. 2(1997), p. 185.

[24]Presidential/Congressional Commission on Risk Management and Assessment, vol. 2(1997), p. 185.

of government,"[25] the Commission ultimately failed to draw a connection between communication practices within agencies and the agencies' apparent failures to draw upon the "extensive experience and unique knowledge" of stakeholders.

The Commission viewed research in communication as a tool to "gain the confidence of stakeholders, incorporate their views and knowledge, and influence favorably the acceptability of risk assessment and risk management decisions."[26] They encouraged risk managers to listen to constituents to improve risk communication. But the Commission did not seem to perceive how existing rhetorical practices *within agencies* might already involve stakeholders in the process.

The Commission's framework depicts a cycle of problem definition within agencies, but this framework does not show the rhetorical interactions and documentation practices that shape technical documentation at each stage in the process. Second, the framework does not show how writers negotiate differences in documentation practices at different moments in the cycle. Finally, the framework does not enable us to see how the features of technical documentation might therefore reflect and structure activity throughout the cycle.

THE CYCLE OF TECHNICAL
DOCUMENTATION IN LARGE REGULATORY
INDUSTRIES

The Cycle of Technical Documentation in Large Regulatory Industries (Fig. 2.2) redraws the Commission framework in order to depict an ongoing cycle of documents and action within large regulatory agencies. This framework enables us to see how individual documents are the product of many individual moments of rhetorical negotiation throughout the regulatory process. In the process, local knowledge moves into the domain of science and engineering, where it is captured and transformed in writing.[27]

The framework is both temporal and spatial. Documentation, like knowledge, originates in specific material and institutional sites. It has temporal effects within a system, influencing future policy and procedures as the infor-

---

[25]Presidential/Congressional Commission on Risk Management and Assessment, vol. 2(1997), p. 43.

[26]Presidential/Congressional Commission on Risk Management and Assessment, vol. 2(1997), p. 39.

[27]For a complete list of documents involved in the construction of this framework, see Sauer, 1998, pp. 168–169.

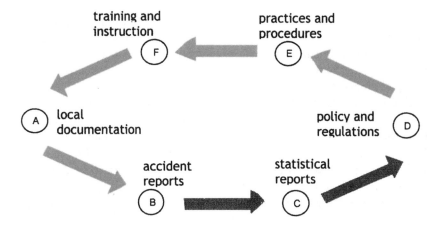

FIG. 2.2.   The cycle of technical documentation in large regulatory industries.

mation is reconstructed and re-represented for new audiences. The rhetorical properties of any single document are formed in collaborative interactions in response to specific problems in specific sites. As these rhetorical properties achieve stability over time, documents become templates for particular kinds of activities: inspections, accident investigations, risk calculation, policy-making, instructional planning, and instruction. Ultimately, the documentation structures activity, as when accident investigators seek only the information they need to complete a particular form or when workers follow instructions to the letter. Over time, this stability can create problems if documentation does not change to meet current needs.

The boxes in Fig. 2.2 represent the most visible forms of written documentation within large regulatory agencies: (A) local documentation; (B) accident reports; (C) statistical reports; (D) policy and regulations; (E) practices and procedures; and (F) training and instruction. These documents both reflect and structure activity at all phases of the process. When documents achieve formal status within institutions (as recognizable genres), these documents serve important functions within agencies and institutions. Documents create consistency; they provide an important history of the sites they represent. In some cases, however, the forms and activities have become so naturalized that writers may not perceive the degree to which formal documentation no longer suits the needs of changing material and institutional sites.

The Cycle in Fig. 2.2 begins and ends in *local documentation* in hazardous worksites (Document A in the Cycle)—although this starting point is clearly arbitrary. As we saw in chapter 1, the Mine Act requires individuals to document events, conditions, and decisions in writing. Mines must conduct daily preshift inspections, and they must record conditions in a daily log book. When accidents occur, investigators can return to this body of documentation to see whether management and workers had sufficient evidence to warrant judgments about risk.

Accident reports (Document B) provide a conduit from local sites to the agency's primary administrative offices. MSHA currently has 11 District offices that oversee the inspection and enforcement of standards at the local level. When accidents occur, investigators from the District Office join forces with investigators from the agency's administrative offices to investigate the accident. These individuals interview miners and document conditions at the site that might provide clues to the source (or sources) of the disaster. In their final investigation report, writers must transform the diverse and often conflicting local accounts of individuals into a single narrative that reflects the agency's technical perspective. Together with other accounts of conditions and practices, these reports produce the engineering history of an industry. (Chapter 4 describes this process in detail.)

Statistical reports (Document C) are situated within the agency headquarters. At this level, agency personnel categorize and abstract information in accident reports and industry production records to produce statistical summaries they can use to analyze trends in the industry. Agency personnel use this data to pinpoint sites and practices that require special attention. They can then determine where to focus their efforts to reduce and manage risk.

Policy and regulations (Document D) constitute the fourth set of documents in the cycle. As we saw in chapter 1, policymakers draw upon information from many different sources to create general standards that can be applied across diverse sites and institutions. These documents take many forms. Following an accident, agencies may issue a single policy memo to correct what they see as dangerous practices. Standards are written and rewritten to incorporate comments from many different stakeholders. Once the final standard has been agreed upon, these standards must be communicated in writing (promulgated) so that operators and institutions can begin the process of implementing standards in local sites.

Because agencies do not always specify how standards must be implemented, safety officials must transform regulations into *practices and procedures* (Document E) that can be applied across many different sites and conditions. At the local level, writers must develop ventilation plans, roof control plans, and general safety instructions that take into account differences in local conditions. In the mining industry, these plans must be approved at the District level before operators can proceed with their work.

Training and instruction (Document F) returns us to local sites where specific details of operation must be communicated to workers and management in new-miner training sessions, daily work orders, and annual refresher training. Workers need specific instruction in how to follow a roof control plan, how to install roof support, and how to maintain and monitor ventilation. Much of this instruction takes place on the job where novices learn as they work alongside experienced miners.

Ideally, investigations are thorough, data is complete and up-to-date, policy is wise and astute, instructions are well-written, training is persuasive, and individuals have adequate evidence to assess and manage conditions in local sites. But not all knowledge passes seamlessly into writing at critical moments in the cycle.

Hazardous environments are poorly understood and difficult to manage, even with the best technologies. Workers and management face production pressures that frequently invite disaster. Even under the best circumstances, compliance is difficult and complex. Within the Cycle, no single document can include all of the information that individuals would need to provide a 100% certain level of safety underground. Each document is ultimately the product of many rhetorical actions that may be so naturalized that writers no longer see these acts as rhetorical choices. If we attempt to build a theory of rhetorical practice based solely on our analysis of *written* documentation at any moment in the Cycle, we may miss the important rhetorical work that takes place at critical moments of transformation within the Cycle.

SIX CRITICAL MOMENTS
OF TRANSFORMATION

Writers play an important role within large regulatory agencies at all levels, although many of these individuals would not define themselves as writers. Regardless of their rhetorical training, these individuals have developed a range of rhetorical strategies to document, transcribe, abstract, summarize, and synthesize information at each moment in the Cycle. Once they submit their final reports, the rhetorical work of transformation is largely invisible outside of agencies. But the effects of these transformations influence risk decision making throughout the entire Cycle.

The Cycle allows us to identify six critical moments of transformation:

1. when oral testimony and embodied experience are captured in writing;
2. when information in accident reports is re-represented in statistical records;

3. when statistical accounts are re-represented as arguments for particular policies;
4. when policies and standards are transformed into procedures;
5. when procedures are re-represented in training; and
6. when training is re-represented to workers (Fig. 2.3).

The process of transformation that produces an accident report illustrates the kinds of transformations that take place throughout the Cycle. Following an accident, investigators draw on many sources to understand the accident: written documentation in local sites, scientific analysis and testing, and on-site observations before and after the accident. Following an accident, investigators must reconcile observers' testimony in order to produce a consistent narrative of events prior to the disaster. Because they stand in different positions—literally—in relation to the risks they describe, not all observers share the same viewpoint in relation to risk; not all eyewitnesses recognize the same hazards. Following an accident, investigators must make inferences about the location of the observers in relation to the risks they describe, the quality of the analysis, the accuracy of reports based on second-hand observation, and the relative precision of conflicting reports of events and conditions. Did observers smell smoke? Did they see flames? Did their ears pop? Did they hear noises on the surface or feel the rush of air following the explosion? Were investigators present at the mine? Did inspectors observe the conditions they describe? Is the speaker reliable? Was the speaker pressured by management to conceal events and conditions prior to the accident?

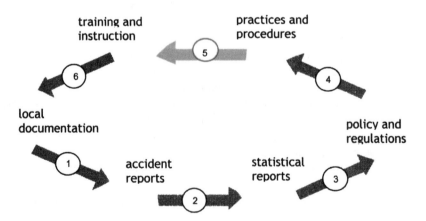

FIG. 2.3.   Six critical moments of rhetorical transformation.

To cite a mine for violations that precipitated the accident, investigators must fit miners' experiences and observations into categories of risk defined by the Mine Act. They must create a coherent narrative that captures the complexity of events, decisions, and conditions prior to the accident. And they must transform miners' embodied experiences of risk (underground and inside of the spaces they describe) into the language and viewpoint of engineers—above and outside of the experiences they describe. As we shall see in chapters 4–8, investigators draw important information from miners' accounts of local experiences, but they do not always systematically represent information encoded in gesture or a speaker's tone and body movements.

Within the Cycle, each document—recorded and remembered—eventually becomes part of the collective history of the industry. Individual accident reports are transformed into the engineering history of an industry. These engineering histories, in turn, become part of the official databases that agencies draw on to characterize risks across multiple sites. Databases must be retransformed to become policy, which must then be transformed into procedures, standards, and training.

Sometimes, the rules for these transformations are explicit; at other times, writers must make develop their own rules for categorizing data. Many agencies have explicit rules, for example, for categorizing data from accident reports and investigations. *The American National Standard Method of Recording Basic Facts Relating to the Nature and Occurrence of Work Injuries* provides one set of standards for categorizing workplace injuries.[28] This report explicitly acknowledges the degree to which the "quality and usefulness of any statistical data" is ultimately dependent upon "the quality of the original case reports upon which the analysis is based."[29] According to the report, ANSI standards are intended to be "straightforward and reasonably simple," but they must also "take into account the wide variations in the completeness and accuracy of the original accident reports from which the statistical data are derived."[30] The report concludes:

---

[28]American National Standards Institute, p. 7. ANSI standards provide standardized codes of practice, testing methods, and specifications for every product one can imagine—from childproof aspirin bottles to bricks and mortar. Nationally recognized specifications and standards like the ANSI standards or the Underwriters' Laboratory Standards allow writers and designers to specify objects and practices that meet safety criteria without having to construct a set of acceptable evaluative criteria of their own.

[29]ANSI, 1969, appendix A4.

[30]ANSI, 1969, p. 7.

If these reports are incomplete, biased, or otherwise deficient, the statistical tabulations will be equally deficient in meeting their intended purpose.

Effective analysis, therefore, must start with an effective case reporting procedure. Effective case reporting requires clear instructions to the persons who prepare the case reports indicating the kinds of information desired plus full cooperation by the reporters in obtaining and recording the information.[31]

My personal copy of the ANSI standards reveals the problem that agencies face when they attempt to incorporate any set of rules into their documentation practices. This copy has a sticky note attached from the analyst who gave me his copy of the standards. The note reads, "This one may no longer be an ANSI standard." Even so, it continues to govern agency practice.

Each moment of transformation has the potential to introduce new errors into the Cycle. MSHA's draft report, "Errors and Unexpected Data Encountered in MSHA's Address/Employment and Accident/Injury Databases and Miscellaneous Suggestions for Improving the Usability of Those Databases" reveals the number of errors that arise in the compilation of statistical data from accident reports and investigations. Some of these errors result from carelessness and imprecision, like the impossible seam heights reported in one database (the height of coal was reported to be 1 inch). Other problems reflect more directly the limitations of the accident narrative. The writers conclude that better documentation practices might improve the quality of the database, though they make no attempt to define what a "well-written narrative" might look like:

> Obviously, MSHA has little control over the narratives that are supplied by the mines, and insufficient resources to contact mines when poor narratives are submitted. Possibly, providing mines with guidelines and examples of well-written narratives (if this is not already being done) might help to improve the data received, although it is understood that many mines will report the minimum level of data that is necessary to comply with federal laws.[32]

A COLLABORATIVE NOTION OF EXPERTISE

In hazardous environments, the accumulated body of experience that constitutes local knowledge does not exist independently in a single document. Instead, this knowledge is distributed throughout the entire Cycle. Distinctions

---

[31]ANSI, 1969, appendix A4.
[32]U.S. Bureau of Mines, 1994, p. 3.

between expert knowledge (based in science), local knowledge (the property of interested citizens), and experience (tacit or craft knowledge) may thus obscure the degree to which writers draw upon many different types of expertise at specific moments within the Cycle.[33]

For my own purposes, I use the term "expert" inclusively, in the broadest meaning of the term. In this sense, experts have more knowledge, experience, and education than novices, but this definition does not presume that scientific expertise constitutes the sole criterion for expertise in a risky environment. Nor does it presume that any single individual is likely to have access to all aspects of expert knowledge. As a result, individuals must work collaboratively as teams, sharing information and communicating with colleagues and management in other sections of the system. Expert miners do not have the same knowledge base as expert statisticians, though both may draw upon expertise in another's field—visibly or invisibly—to assess and manage risk. In this sense, expertise is context dependent and highly situated; it reflects standards of disciplinary knowledge and education as well as highly local experience.

The notion of expertise in mining illustrates the collaborative nature of expertise in a hazardous environment. Ideally, individuals have a wide range of experience in all facets of mining. In reality, however, individuals vary in their ability, education, and experience. As a result, each individual observes different events and detects different types of problems. Individuals also differ with regard to their tolerance for and subjective judgment of significant and substantial violations. These differences may become exacerbated when, as in Great Britain, miners' salaries depend upon bonuses for increased production. As a general rule, however, experts make more reliable judgments than novices. They can identify problems in the environment and understand the complexities and consequences of their decisions. In simple terms, most people would rather rely on an expert than a novice in an emergency.

Experts, in the very general sense that I am using the term, represent problems in ways that facilitate problem solving.[34] They have acquired the tools to discriminate between relevant and non-relevant information in problem solv-

---

[33]As researchers in the social construction of science and the rhetoric of science have suggested, scientists use a variety of rhetorical strategies to construct knowledge claims and to warrant their findings and conclusions. Experts in science produce abstract and generalizable knowledge claims—what experts hold to be true and certain—within the realm of the laboratory or scientific community. Expertise as I am defining it, however, may occur in a variety of settings outside the laboratory (Latour, Woolgar, and Salk, 1986).

[34]See Larkin, & Simon, 1995; Bereiter and Scardamalia, 1989.

ing, and they use information efficiently to solve problems in the workplace.[35] They can communicate with others, work effectively in teams, and analyze situations from a perspective outside of their immediate environment. They understand the consequences of decisions and practices, and they have the awareness to recognize the meaning of changes in their environment.

Unfortunately, this broader notion of expertise becomes problematic when agencies attempt to define expertise as a set of competencies, skills, or standards based upon education and classroom performance. Because the length of training does not necessarily correlate with competency, British safety officials have created performance-based competency standards that measure expertise in terms of activity and communicative competence.[36] Novice miners must be able to describe the general procedures they will follow; the "basic theory, functions, operations, and operator maintenance" of the equipment they will be using; and the effects of mine gases and ventilation.[37] They must be able to explain trainees' personal responsibilities under the health and safety regulations and display the "potential" to work as members of a team. And they must show skills in "oral representation in the form of questions and answers during and/or on completing of each activity."[38]

But performance alone does not constitute expertise. Expert miners can identify hazards because they have experienced and observed hazards. They have analyzed the consequences. They can interpret standards and regulations, understand the systems in which they work, assess changes in the environment, and communicate these changes to colleagues and management.

Because mines differ in geology, equipment, and method of production, experienced miners (e.g., miners who have spent more time underground) will have a larger store of observation, analysis, and experience than novice miners—miners with little or no working experience underground. Both British and U.S. mines identify novices underground by their red caps—or mining helmets. Red caps must work under the supervision of an experienced miner (the time differs in U.S. and British mines). But the difference between expert and novice is not simply a matter of completing a 24-hour (in the United States) or 20-day (in Great Britain) mandatory training course.

---

[35]Bereiter and Scardamalia (1987) draw upon Glaser (1984) in order to describe differences in the ways that experts and novices attempt to solve problems. Their research suggests that experts can draw on familiar schemas and protocols. Experts are thus less likely than novices to apply problem-solving strategies to solve routine problems.

[36]RJB Mining (U.K.) (Draft report, Part I).

[37]RJB Mining (U.K.) (Draft report, section 2.1).

[38]RJB Mining (U.K.) (Draft report, section 4.1).

In craft industries like mining, workers must be able to apply generalized standards and scientific principles in highly local situations. This second form of competency—collectively known as pit sense—is more difficult to measure under normal conditions. Unlike activity in a factory, where units of performance are regular and predictable, hazardous environments require "site-specific geologic expertise" that is difficult, if not impossible, to measure and observe in classrooms and training sites.[39] Even if trainers could simulate conditions underground, they could not willfully and ethically expose novices to the kinds of hazards that might measure their true competence in difficult situations. As a result, trainers can teach miners performance skills, but they cannot demonstrate the site-specific geologic expertise that workers need to recognize hazards in a dynamic and unpredictable material environment.

This site-specific expertise is so local that miners with expertise in one area of a mine do not automatically transfer this experience to a new worksite. When accidents occur, MSHA categorizes three levels of experience related to fatalities in coal mines: total mining experience, experience at classification, and experience at the activity involved. Miners may have a long record of service in mining, but only a few years or months at a particular task.[40] During one six-month period, MSHA reported 24 fatal accidents in underground coal mines. Only two miners (of the 24) had less than six to seven years total mining experience. Eighteen miners had over 10 years' experience. But 19 of the 24 miners had less than eight years' experience at the activity involved in the accident.[41] In another six-month period, 21 miners died in underground coal accidents. Of the 13 miners whose experience was documented, only two had less than six years' total mining experience, but 12 had less than six years' experience at the activity.[42]

Even the most experienced individual (in all three senses above) does not recognize 100% of all hazards simultaneously. In one study, experienced miners could recognize only half of the detectable hazards in an en-

---

[39]Mark, Chase, & Iannacchione, 1991, p. 16e. In Utah, for example, management trained miners to drill test holes and record the results in simple logs to assist the company's geologist in identifying loose hanging rock that might predict a roof collapse. In Virginia and Kentucky, management trained miners to spot weak roof conditions that developed near longwall mines. In both cases, skills-based training did not help miners identify potential hazards, although researchers concluded that such training could be "quite helpful" as long as miners also had previous experience in the particular mine-sites (Mark, Chase, & Iannacchione, 1991).

[40]U.S. Department of Labor, Mine Safety and Health Administration, Metal and Nonmetal Mine Safety and Health, 1991a.

[41]U.S. Dept. of Labor. Mine Safety and Health Administration, 1991b.

[42]U.S. Dept. of Labor. Mine Safety and Health Administration, 1991b.

vironment prior to training. After they were trained in hazard recognition, their ability to recognize hazards increased to 91.3%.[43] At this rate, even the best miners would fail to recognize nearly 10% of the hazards in a mine—a critical problem in a coal mine, where even one undetected hazard could have fatal consequences. In a follow-up study, experienced miners trained by looking at three-dimensional slides scored higher than those trained with two-dimensional representations of risk, but only one in six miners achieved a target recognition rate of at least 90% in an underground walk-through of a mine.

Even if miners could recognize 100% of the risks in their section, no single individual has access to all aspects of the system simultaneously. Because all miners are dependent on the collective body of experience to detect and manage hazards, individuals must be able to communicate information with colleagues, management, and engineers in other parts of the system. The success of their ability to communicate may be apparent only in retrospect, when investigators determine whether particular individuals recognized hazards and acted appropriately (or inappropriately) to communicate problems to others.

One British training manual characterizes this collaborative expertise as the product of "clear thought, judgment, experience and the benefit of discussion with colleagues."[44] The manual concludes that "the best risk assessments" occur in groups of two or more experienced people: In these situations, each miner contributes his or her own "expertise" and "experience" to the collective knowledge of the group. This expertise includes knowledge of the scientific principles necessary to warrant decisions about safety as well as practical experience underground.

Risk specialists have created a variety of models to represent the collective body of expert knowledge. Morgan et al. (2000), for example, have created a mental models approach that integrates "diverse forms of expertise" in a single formulation.[45] Risk specialists can use this model to compare an individual stakeholder's knowledge with the idealized expert's knowledge.[46] The term expert in this model does not refer to individuals but to the collective body of expertise in the representation. The mental models approach demonstrates the collaborative nature of risk expertise, its distribution across many sites of knowledge production, and the heterogeneous nature of the knowledge included in the model. Figure 2.4 shows the basic structure of an expert model for influences associated with HIV and the resulting disease, AIDS.

---

[43]Mark, Chase, & Iannacchione, 1991, p. 9. See also Barrett & Kowalwalski, 1995.
[44]Herbert, 1989, p.
[45]Morgan, Fischhoff, Bostrom, & Atman, 2000.
[46]Morgan, Fischhoff, Bostrom, Lave, & Atman, 1992.

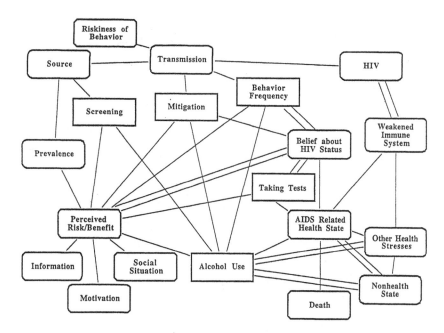

FIG. 2.4. Mental models approach: Influence diagram for risk produced by radon. Source: Bostrom, Fischhoff, and Morgan (1992). (In Morgan, Fischhoff, Bostrom, & Atman, 2001, p. 49). Reprinted with the permission of Cambridge University Press.

Rhetoricians do not ordinarily separate technical expertise from rhetorical practice because we believe that knowledge and its representation are insepa-rable. We also believe that rhetorical expertise can help us produce new un-derstanding in specific subject areas. In this sense, rhetorical knowledge is both epistemic and inventional. That is, when individuals represent knowl-edge in new rhetorical forms, they see the world differently through the lens of new representations; they produce new knowledge and new understanding in the transformation. Just as a graph or picture helps analysts see trends in otherwise unconstructed data, so new gestures and new documentation prac-tices also help individuals see new patterns in their work.[47]

The Cycle of Technical Documentation depicted in this chapter allows us to identify predictable moments when individuals have the opportunity to create new knowledge within large regulatory agencies: when investigators capture miners' narratives to determine the cause of the accident; when ana-

---

[47]This does not mean that technical expertise can be replaced by purely rhetorical expertise, of course. Nor does it assume that all knowledge is therefore either relative or wholly socially constructed.

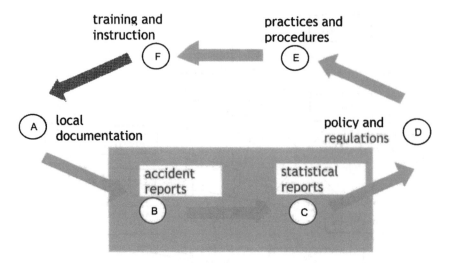

FIG. 2.5. Many writers limit the scope of their analysis to a limited set of documents.

lysts abstract statistics to pinpoint areas that need special attention; when policymakers interpret statistics in the light of new technologies and changing political economies; when trainers rethink standards from a miners' perspective; and when operators and management interpret standards and procedures to manage hazards in local sites. Although agencies and individuals can theoretically draw upon all of the collective knowledge captured in the documents represented in this framework, most writers generally restrict the scope of their analysis to the most familiar forms of documentation (Fig. 2.5). Writers are free to draw their arguments from many sources at any point in the cycle, combining and recombining information to meet the needs of new situations. But agency conventions and discourse practices may limit a writer's notion of what kinds of information should be included in a document. Each new document refocuses an agency's attention on particular hazards in the environment. If these documents represent accidents as local failures of training or practice, agencies may underestimate the magnitude of risk within the industry as a whole.[48] If agencies blame the industry more gener-

---

[48]According to one Bureau of Mines report, MSHA failed to change SCSR (self-contained-self-rescuer, an emergency breathing apparatus) training practices for the industry as a whole even after 27 men died in the Wilberg disaster, because the agency viewed these fatalities as isolated events (Vaught, Brnich, and Kellner, 1988).

ally for failing to protect the rights of miners, they may underestimate the role of individuals in preventing accidents. If writers perceive these choices solely as writing problems, they may fail to see how local choices affect the outcomes of deliberation throughout the agency. A writer's information design strategies are particularly important in regulatory agencies. Because the design and content of documents serve as models for future deliberation, highly local choices may inadvertently become a template for future deliberation.

The Cycle of Technical Documentation thus reinforces the commonsense notion that the collective body of knowledge represented in the Cycle must be constantly updated to remain "expert." By helping us identify the sites where knowledge is transformed and refigured, this framework allows us to help agencies and individuals examine how the rhetorical features of their interactions affect the negotiation and transformation of knowledge throughout the Cycle.[49]

THE RHETORICAL FORCE OF DOCUMENTS WITHIN THE CYCLE

The rhetorical processes of transformation matter because documents must persuade individuals to act to reduce and manage risk. All documents are not rhetorically equal. Some have more compelling effect on audiences. The rhetorical force may reflect highly local, but compelling evidence (e.g., a warning to a colleague to take care). Or it may result from the force of law—like the standards formulated in the Mine Act. In either case, texts that document information for the record will fail unless writers can persuade individuals to act to manage risk.[50] The "art" of rhetoric (its techne) is to discover what makes arguments persuasive to particular audiences.

Accidents have a particular rhetorical force within agencies. Because they draw media attention, they provide emotionally laden demonstrations of the failures of safe practice. A spike in accident rates can also serve as evidence

---

[49]Chapters 4 and 5 examine two specific features of this documentation: the writer's viewpoint and the representation of embodied knowledge in speech and gesture. Chapters 6–8 look at how these rhetorical features affect two specific moments of transformation—in accident reports and training.

[50]Sauer, 1993. This article is often cited as a justification of subjectivity and a critique of rationality. As Robyn Dawes (2001) reminds us, however, the opposite of rationality is irrationality. Thus, individuals may draw on rational arguments to make subjective judgments when there is a high degree of uncertainty. See also Dawes, 1988, for a discussion of rationality under conditions of profound uncertainty.

that something needs attention. By evoking fear in their audience (loss of a job, the potential loss of a limb, the sudden trauma of a roof fall), writers can increase the rhetorical force of their arguments.[51]

On the job, miners use very crude mathematical formulas to help them understand how simple changes in practice can affect the potential consequences of their actions. One British Coal training manual defines risk as the product of the hazard effect and the probability that conditions in the present will produce this effect:

$$\textbf{hazard effect} \times \textbf{probability} = \textbf{risk}[52]$$

In its simplest form, this formula expresses the direct relationship between risk and hazard: If you reduce the hazard effect—shielding miners, for example, or providing protective glasses—you can lower the risk for high probability events. If you reduce the probability of a large hazard like an explosion, events with potentially massive consequences (large effects) will result in a low risk. In practice, supervisors can implement particular precautions to reduce the effects of the hazard (by shielding miners, for example) or they can focus on precautions that reduce the probability that an accident will occur (by reducing the level of methane or removing sources of ignition).

When trainers focus on particular hazards, like roof falls or methane explosions, the increased attention should theoretically result in lower accident rates. But many factors influence local decisions about risk. When individuals believe that risk is high, they take precautions that actually reduce the potential for disaster in the long term. When individuals believe that risk is low, they may become complacent and inadvertently increase the likelihood of a bad outcome. Such a rational approach to risk ideally lowers the future potential for a bad outcome. If people's confidence is not warranted, however, their perceptions can create the potential for disaster (Fig. 2.6).

When writers downplay the significance of events and conditions— through ignorance, misjudgment, or deliberate manipulation, they not only disturb the long-term trends of statistics, but they affect the future health and safety of workers.

---

[51]Aristotle's notion of pathos locates the emotional response in the audience, not in the speaker. When speakers become emotional, they lose the credibility and trust (ethos) that is located in the speaker. See, Aristotle, 1991.

[52]Herbert, 1989.

Perception of High Risk Magnitude

→ Heightened Awareness

→ Heightened Action

→ Lower Probability of Bad Outcomes

Unwarranted Perception of Low Risk Magnitude

→ Complacency

→ Reduced Action

→ Higher Probability of Bad Outcomes

FIG. 2.6.   Perception influences the outcomes of risk decisions.

THE MATERIAL CONSEQUENCES OF MANIPULATING RISK ON PAPER

Risk analysts frequently adjust probabilities to account for new evidence or changing conditions. MSHA officials, for example, may question the credibility of accident rates based upon company estimates. Corporations known for honest reporting may have higher accident rates than companies suspected of under-reporting accidents. In this case, analysts might adjust probabilities to reflect subjective judgments about the company's credibility and its reporting methods. These adjustments can affect whether the agency decides to target a mine for special inspections or increased attention.

It is easy to cite examples of unscrupulous managers who manipulate risk calculations mathematically to serve their own interests. Following the privatization of British coal mines, one new mine manager claimed that he suspected the credibility of British Coal calculations and adjusted the accident rates for his mine accordingly. British miners countered that the company had manipulated accident rates on paper to support claims that privatization had not diminished British Coal's safety record.[53]

---

[53]The mine manager openly describes how risk managers reduced risk numerically by applying a new "yardstick" to previous records:

> The overall accident rate for the deep mine sector . . . cause that's the only one that I'm involved with—is very encouraging . . . uh we . . . we set off from British Coal in January and we used British Coal's full year of accidents uh for ninety three, ninety four because we couldn't put any credibility . . . uh . . . on the accidents we saw for the nine months up to December, so we used the full year . . . and the rate was 13.33—so that was the yardstick we were using, uh . . . we're currently standing at 9.74. [Anon. (Interview with mine manager, UK), 1995]

Workers can also underestimate the magnitude of risk, increasing the potential for disaster. They can easily become overconfident in their opinions, dismissing new data that contradicts old habits and beliefs. Rodrigue (1999) summarizes the results of research demonstrating how people adjust new data to fit their previously constructed frameworks:

> Related to this is a finding that people often make up their minds about an issue before being exposed to an adequate array of facts and arguments about it, often taking the position of a reference group they trust (Johnson 1993), and then become very confident in their opinions (Margolis 1996; Slovic 1991; Slovic, Fischhoff, & Lichtenstein 1982). Once the pattern gels one way or the other, new facts and arguments are fit into the framework in a way that further solidifies it, in order to avoid the cognitive dissonance of entertaining two mutually exclusive interpretations. That is, someone who decides that there is a significant risk in a situation or technology will dismiss data that counterindicate it as faulty or from a corrupt and self-interested source (Covello, Sandman, & Slovic 1991). Someone who has decided that there is no significant risk will equally dismiss data suggesting the risk is greater with similar denial mechanisms.[54]

When people artificially reduce risk—by manipulating the formulas by which they calculate risk or dismissing data that conflicts with preconceived notions of risk—they also increase the potential for disaster. Any manipulation that artificially increases or decreases the perception of risk can affect future estimates of risk and their outcomes throughout the entire Cycle.

Shifting the Focus of Institutional Attention

Institutions can make strategic decisions to reduce risk mathematically without affecting the number of accidents, injuries, and deaths. In coal mining, for example, accident rates are a function of production. The measure of production has a clear effect on the accident rate. Statisticians can calculate the accident rate as the number of accidents per ton, the number of deaths per 1,000 tons of coal, or the number of accidents per year. Higher production reduces these rates mathematically, but only if change in work practice does not induce a higher risk.[55] There are many factors that influence production and safety in mining. In the United States, longwall mining has increased productivity and thus decreased the number of injuries in proportion to the amount of production. In this case, the new method also removed miners

---

[54]Rodrigue, 1999, January 14.
[55]Leon, Davies, Salamon, and Davies, 1994.

from the site of greatest risk. But methods intended to reduce risk through increased production can increase the absolute numbers of miners injured if miners take shortcuts to increase production.

Table 2.1 shows how various combinations of risk factors distribute risk across different populations and turn our attention to different aspects of the problem. As Table 2.1 suggests, analysts can represent the potential threat to humans (as deaths, lost limbs, broken fingers, and torn ligaments). Or they can shift the focus of attention to calculate the probable effect upon institutions (production, profit, and investment) or environments (roof falls, explosions, and fires). By shifting the focus of attention from human to non-human factors, unscrupulous risk analysts can reduce the magnitude of risk mathematically without actually changing the number of deaths or injuries in the social world. When analysts focus on non-human elements, improved health and safety is not a necessary outcome of a mathematically reduced risk.[56]

Changes in data categories can also affect accident rates within an industry. Large regulatory industries must maintain consistent records and uniform models to avoid statistical variation and unreliable estimates. But outdated economic formulas can artificially increase the number of reportable incidents when the cost of damages increases through inflation. When agencies revise their formulas, they can artificially decrease the number of reportable events, decreasing the accident rate without changing the raw number of incidents.[57]

---

[56]Leon et al. (1994) note that "[e]ven in the most cynical mining context," these absolute numbers have practical significance:

Both the absolute numbers and the rates have practical significance. The absolute numbers give measures to the industry and to society in general of the burden of accidents. Experience elsewhere has shown that serious injuries and fatalities are associated with ever increasing costs. Thus, as the cost of medical treatment, compensation, interruption of production, payment of fines, etc. will increase in South Africa, the motivation to keep the absolute numbers of accidents as low as possible will increase. (pp. 55–56)

[57]Leon et al. (1994) demonstrate how a reliance on outdated data can affect policy decisions. The commission concludes:

It is now a matter for concern that there is undue reliance on data which is now nearly 40 years old, and that the results of the many thousands of dust measurements which have been made in recent years have not been analyzed and published in a form which makes it possible for experts in the field to describe the trends accurately, nor is it possible without access to the raw data to determine whether current sampling strategies and methods used to measure dust levels succeed in identifying the highest exposures accurately. (p. 55)

TABLE 2.1
Factors in Risk Analysis–Manipulating Risk

Factors	Per Number of Individuals in the Workforce	Per Number of Working Faces	Per Total Amount of Production	Per Unit of Time	Per Worksite
Number of Individuals	Deaths/ 100,000	Deaths/No. of faces	Deaths/ton	Deaths/ quarter	Deaths/ mine
Categories of Accidents	Accidents/ 100,000	Accidents/ no. of faces	Accidents/ ton	Accidents/ quarter	Accidents/ mine
Potential Accidents	Incidents/ 100,000	Incidents/ no. of faces	Incidents/ ton	Incidents/ quarter	Incidents/ mine
Events And Situations	Roof falls/ 100,000	Roof falls/ no. of faces	Roof falls/ ton	Roof falls/ quarter	Roof falls/ mine

Outdated document design can contribute to poor reporting practices. On one preliminary accident investigation form, one large section of the document asked for information that the agency no longer considered relevant. Another large section asked for information about mine ownership. Because of the poor document design, investigators were required to write a brief description of the accident in a one-inch by seven-inch block of text. Not surprisingly, these narratives provided inadequate and incomplete data about events and conditions that precipitated the accident.

The Consequences of Managing Risk on Paper

When a single type of accident like explosions produces a large percentage of relatively infrequent but costly disasters, analysts must determine whether or not to factor these statistics into the overall accident rate. The choice of which factors to include can have important consequences because individuals use these statistics to argue for changes in policy and practice.

When South Africa's Leon Commission calculated the risk of explosions, for example, they deliberately omitted explosion statistics from their accident data because, they argued, "explosions of this kind . . . tend to disturb the long term trend of statistics."[58] If they included data from explosions in long-term mine safety statistics, they feared they would "disturb" general accident

---

[58]Leon et al. (1994) acknowledged the uncertainties in mathematical calculations of risk. According to the report, South African coal mining statistics confirm that explosions of methane and/or coal dust play a major role in the fatality statistics in coal mining for 1993. The Commis-

rates and long-term trends at individual mines. If they ignored the effects of large explosions, they could legitimately conclude that the total number of fatalities in South African mines had decreased since 1955. But they would also be perceived as having manipulated statistics to reduce accident rates over the long term. In acknowledging their choice of method, the Commissioners recognized that each method could produce a different conclusion.

Because the Commission wanted to identify deficiencies in the mining industry, the Commission recognized that either method would have long-term consequences. By articulating the problem, Commissioners were able to present two different but equally true perspectives on conditions in the mining industry. If they had presented only one conclusion, they would have presented a valid but distorted picture of conditions underground.

THE RHETORICAL TRANSFORMATION OF EXPERIENCE

It is not possible to describe all of the rhetorical transformations that occur within the cycle in a single volume. The chapters that follow focus instead on two critical moments at the boundaries between agencies and the worlds they seek to regulate. My reasons are both methodological and theoretical. From a methodological standpoint, the rhetorical interactions that occur at these moments have been most heavily documented in agency depositions and accident investigation reports. Because draft revisions of documents within agencies are classified and highly protected documents, agencies cannot legally allow researchers to participate in the document revision process at other moments in the Cycle, although individual writers can speak in general terms about the kinds of transformations that occur at these moments.

My focus also reflects a deeper theoretical concern with the specific rhetorical transformations that take place at the rhetorical interface between agencies and the material sites they seek to regulate. Before we can speak about the problems of integrating stakeholder knowledge into the regulatory process, we need to understand the geographic and institutional uncertainty that affects judgments at critical moments in the cycle. Technical writers have investigated the rhetorical features of instruction, but they have not examined how workers manage this uncertainty and—more importantly—how this uncertainty affects their understanding of their work.

---

sion's report cites data from a study by H.R. Phillips that suggested that the percentage of miners killed in explosions had risen from 3.3% in the period 1955–1967 to 21.3% from 1981–93, despite a general decrease in the number of miners killed in coal mining accidents (p. 30).

I thus distinguish the rhetorical transformation of experience and its retransformation in instruction and training from the more visible rhetorical transformations that take place in writing at other moments in the Cycle. In making this distinction, I do not wish to suggest that writers within agencies might not also draw upon embodied experience to interpret and analyze information in written documents. As we have seen from social studies of science, much of the work of knowledge construction in science occurs as individuals negotiate and interpret data in social and institutional settings.[59] A study of the rhetorical transformations that take place at these moments might yield equally valuable insights into the processes of knowledge construction and risk decision making within large regulatory agencies. Unfortunately, such a study is beyond the scope of the present analysis.

In the chapters that follow, I focus largely on two particular moments of rhetorical transformation: accident investigations and training.

The Rhetorical Transformation of Experience in Accident Investigations

When accidents occur, labor unions, management, and agencies interview observers to determine the cause or causes of an accident. Before they can construct a coherent account of events and conditions prior to an accident, writers must transform this body of oral testimony into the language and conventions of an official report.

Accident investigators thus stand at a critical position as both interpreters and investigators of experience within the Cycle of Technical Documentation in Large Regulatory Industries. The results of their technical investigation will help the agency determine liability for the disaster. Their interpretations of experience also become part of the historical record that agencies will draw on to define problems, evaluate risk, and create regulations. The limits of any interpretation, like the limits of any technical investigation, propagate through the system, influencing policy and procedures throughout the cycle of risk management and assessment.

Following an accident, investigators interview all individuals who were involved in the accident—from workers at the site to management and engineers who made decisions that affected the outcome of the disaster. These interviews explore a wide range of technical questions concerning liability, responsibility, and practice. What kinds of decisions led to the disaster? What

---

[59]Tibbetts (1992) provides a good account of the problems of representing a reality "out there." See also Knorr, 1977.

failures in policy? How did training affect miners' responses in a crisis? Were standards inadequate? What policies can prevent disaster in the future? Investigators must also pay attention to verbal and non-verbal cues to determine whether witnesses are lying, contradicting others' testimony, or concealing incriminating evidence.

Investigators are not acting self-consciously as rhetoricians as they question witnesses, rephrase questions, and transform miners' narratives into the language of engineering. But they have developed a variety of rhetorical strategies to capture the non-verbal and visual elements of miners' testimony in writing—interpreting miners' gestures, using embodied knowledge as an index of the forces present in the mine, and helping miners recall critical details unarticulated in speech. (Chapter 9 discusses this process in detail.) In some cases, investigators indicate in the text that miners gestured or nodded in response to questions. In other cases, subtle traces in the text (such as when miners said, "I moved here . . . and then there") provide visible clues that miners are using their bodies to represent their embodied experience.

Following the interviews, investigators must make sense of 30 or more different interviews—each of which tells a different story of events and conditions preceding the accident. To reconcile differences, they must map each observer's position in relation to events and decisions in the disaster, reconstruct the time of events, and measure observers' rhetorical and physical distance from the events they describe. Ultimately, investigators must reconstruct the collective experience of risk as a single narrative from a single (agency) perspective. (Chapter 4 describes how one investigator reconciled these accounts.)

Accident investigations reverse the flow of information from agency to workers that we might have expected in conventional accounts of risk communication. In post-accident interviews, investigators can see events only in and through situated experiences of individuals who observe events and conditions in local sites. Investigators do not transmit information to lay audiences. Instead, the investigator's expertise provides a framework for analysis, a set of technical questions and issues that must be answered in the investigation, and conditions of logical probability (both scientific and commonsense) to test the validity of individual conclusions.

The Retransformation of Experience in Training

The Mine Act mandates training for new miners and annual refresher training for experienced workers. Management must also train workers on the job so that they can learn to operate specific types of machinery not specifically

covered in the Mine Act. In addition to this formal training, miners and foreman talk about their work. This informal discussion plays an important role
in helping miners maintain safety standards in their work.

The men and women who provide this training also stand at the rhetorical
interface between agencies and local experience within the Cycle of Technical
Documentation in Large Regulatory Industries, but trainers differ from investigators because they must reconstruct the local experience and observations
that have been extracted and transformed in the process of creating standards.

When trainers write instructions to help miners detect risk, maintain equipment, and follow safe practices under normal mining conditions, they must
help miners envision the physical conditions and moments of decision making
as miners would experience them in a crisis (inside the spaces they experience).
Trainers cannot ethically reproduce the conditions of a mine fire or explosion.
But they must represent these moments of crisis so that workers understand
how difficult it will be to remember instructions as they face a raging mine fire.
They must persuade miners that it is easy to imagine themselves walking cau-
tiously away from a fire, but it is harder to imagine crawling to safety as the aluminum structures melt around them in the heat of the fire.

Conventional written instructions can provide only a rough approximation of the kinds of knowledge that workers need to work safely in hazardous
environments. The amount of data needed to detect hazards is so great and so
complex that writers cannot construct precise and unambiguous rules of practice in local sites—particularly when problems like weakness or poor stability
are not immediately visible to the naked eye.[60] In theory, management could
provide miners with sonographic equipment, test monitors, and data-analysts
so that they could assess hazards and prevent problems. But few mines have
the economic incentives or personnel to assess hazards at the level of analysis
that would produce useful results.[61]

To overcome the problem of complexity, trainers must help miners visualize hazards in the environment from simple cues that may have little basis in

---

[60]Molinda & Mark, 1996, p. 2.

[61]In theory, all mines could create tunnels as safe and strong as highway tunnels; but workers
will still accept risks when they open the tunnels and build the supporting frameworks. Mines
could keep miners away from the face entirely, using robotic mining machines. When machines
break down, however, someone must be available to repair the machine, dig it out from a roof
fall, and conduct routine maintenance. Many mines deliberately choose to maintain an active
workforce rather than use robotic equipment because untrained and inexperienced workers
would be most likely to encounter danger in a crisis.

The economy of coal (what consumers will pay on the open market) governs safety choices
in a hazardous environment. Safety is expensive. Cheap coal from countries that do not have
the same level of protection for workers will undercut safety in unionized mines.

scientific theory.[62] Jumper wires and relay centers, for example, frequently indicate deeper problems with mine safety. Some miners watch the track and trolley system to see whether there is evidence of an arc, a poor bond, or a cable that is "jumpered out."[63] These simple rules of thumb teach miners to stay alert to potential hazards throughout the mine. The best training sessions allow experienced miners to describe how they recognize what one miner called the "tattle tale signs" of disaster. But these signs are never certain.

Training is difficult enough even when miners and management share the same assumptions about risk and safety. When mines push production or encourage miners to take shortcuts, trainers must help miners envision how production pressures and economic threats can influence miners' decisions about risk.

Within large mines, a small handful of safety managers may be responsible for envisioning the material consequences of as many as 250 individual events, situations, and decisions at any moment during three shifts per day. Ideally, writing takes the place of face-to-face communication. When all three shifts are running, safety managers cannot write instructions fast enough to account for changing conditions in multiple worksites.

Most of the writing in hazardous worksites takes place on-site—if it occurs at all. Too often, one Welsh safety manager complained, miners were already working before he had a chance to assess the potential hazards, check the facts, write a draft, pass it to management, and communicate with "the lads on the job." He knew "realistically" that he could not get around to speak with people and write an effective plan because he couldn't afford the time to conduct formal risk assessments. As a result, he depended upon workers to identify hazards before they started work. But often, miners could not detect hazards until they started to work on a specific task:

As much as I would like to see a world without hazard to workers, I recognize the limitations and tradeoffs that companies and workers accept in a hazardous environment. I am not advocating for lower standards or diminished enforcement of existing regulations. I too accept risks when I cross the street or drive my car. In theory, I would like the safest car possible (I imagine a tank), but I choose to drive my small compact car and buy insurance.

[62]The U.S. Bureau of Mines, for example, provides a simple classification system for operators who need to detect potential problems with the mine roof so that they can "begin to assess hazard potential or plan roof bolting schemes" to reduce the chance of a roof fall (Molinda & Mark, 1994, p. 17). The coal mine roof rating system provides a series of questions that help the user identify potentially weak rock formations in the mine strata. The guide does not replace the use of mechanical core samples to determine rock strength, but helps the geologist or engineer detect potential problems without testing so that engineers can plan appropriate roof support prior to mining. Molinda & Mark (1994) note that the guide allows engineers to conduct "quick assessment of roof rock strength for preliminary analysis with no testing" (p. 17).

[63]J. Oakes, (Interview with author), April, 1992.

You know realistically that you're not going to get round to speak with people. And I find in my job when they say, "write a scheme of work," your biggest time consumption is writing a draft, passing it to, say, management to make sure the facts are all in place, then passing it to the lads on the job. And all that takes time, just the feedback, and rewriting, and—the final copy's a good copy but it might a took well of the time ye cannot allow yourself to write schemes. . . .

The thing is the job is probably up and running before you draw up a scheme, and so you come along halfway through the [unclear] job and say, here's the rules now. And lads have been working.

And that could be poor organization on our part. You should have the rules wrote before the job's done. But I don't believe you can write a good scheme of work before the job's running.[64]

Even when mines have a written plan, supervisors and management may take calculated risks that will save time and increase production.[65] They know that the risks are remote, but the economic rewards are high. In the tight economic climate of British mines, for example, miners' wages are also tied to production levels because miners receive bonuses for increased production. Because miners' wages do not provide a sufficient level of income, some miners push production to increase their weekly salaries. But individuals at all levels also take risks simply to save time and effort. Thus, miners sometimes refuse to go to the surface to get proper lifting equipment because it takes too much time—even though this equipment can prevent painful back and neck injuries:

We tend to say, right, we've got a job coming on the weekend, and we need lifting equipment. Well, we'll manage. We'll manage with what we've got and a lot of the time the equipment is not adequate for what we need . . . for what we're doing. Well, if it's a particularly long job [we'll go to the surface], but if it's say, a shift job, you certainly wouldn't come to the surface to get lifting equipment, 'cause the shift would probably be over by then.[66]

Because there are so many reasons why individuals fail to follow written plans and procedures, trainers must find arguments that will persuade individ-

---

[64]Anon. (Interview with safety manager, UK), January, 1995.

[65]One Yorkshire miner described an imaginary conversation between the undermanager (concerned with safety) and supervisors who must choose between cutting coal or dusting the mine to prevent explosions:

Folks in the undermanager's chair obviously be saying, "Why can't you get some lads to stone dust?" But the supervisor on the night shift would probably say, "I've got nobody to do it . . . ye'll either gon' to cut coal or do you want me to stone dust" (Anon., Interview with miner, UK, January, 1995).

[66]Anon. (Interview with miner, UK), January, 1995.

uals to comply with standards in the face of pressure to take risks.[67] If trainers represent failures as local, discrete, and idiosyncratic, workers may fail to learn from the experience of others. If trainers do not represent these failures as part of a larger pattern of habits and behaviors, miners at other sites cannot learn from the mistakes of others. If trainers represent these events too abstractly, however, they will not be able to represent hazards as miners perceive them.

To encourage workers to follow safe practices, documents must persuade audiences that the consequences of their actions are probable and significant—and thus worthy of action—or low enough to be safely ignored. They must keep their messages concise so that tired workers will not tune out critical safety messages, but they must provide sufficient information so that workers will believe that they are not concealing or holding back critical information. They must display an aura of authority in order to persuade workers to follow safe practice and procedures. But they must also display humanness and responsiveness to overcome historical antagonism between labor and management—particularly if they are perceived as speaking for management or belonging to the same class or culture as management.[68]

THE UNCERTAINTY OF KNOWLEDGE IN LARGE REGULATORY INDUSTRIES

The Cycle of Technical Documentation in Large Regulatory Industries described in this chapter provides both a descriptive and inventional framework for investigating how this uncertain and dynamic material knowledge is cap-

---

[67]Following the Wilberg Mine fire, for example, MSHA did not question training practices at all operations when miners inside the fire failed to follow instructions for donning breathing masks (SCSRs) that might have saved their lives. In this case, the Wilberg miners' proficiency with SCSRs became just one of many issues that arose during the investigation at that site rather than coming to the forefront and causing a general questioning of miners' ability to use oxygen-breathing apparatus (Vaught, Brnich, & Kellner, 1988, pp. 3–4).

[68]Researchers in the public understanding of science have argued that miscommunication and misunderstanding may result when experts-in-science fail to draw upon the local knowledge and experience of lay audiences produced. As Irwin, Dale, & Smith (1996) suggest, official attempts at reassuring local populations about the low probability and high controllability of accidents are easily seen to be self-interested and untrustworthy. As a result, lay audiences may distrust risk communicators—especially when an audience's confidence has undermined by previous problems of misrepresentation and mistrust.

In describing rhetoric more generally as an art of design, Kaufer and Butler (1996) argue that speakers must develop strategies to balance the tension between predictiveness and responsiveness if they are to be perceived as both human and principled. Speakers who hold too firmly to principles without responding to audience concerns may be perceived as inflexible; but speakers who are too responsive may be perceived as unprincipled, obsequious or compliant.

tured and transformed in writing. As a descriptive framework, this Cycle provides the starting point for examining existing rhetorical practices as they are
produced and transformed within large regulatory agencies. It allows us to examine how agencies privilege particular forms of documentation at particular
moments; how these documents structure agency activity; and how writers
work to transform these documents for new audiences. The Cycle shows how
documents intended for one audience influence future policy and procedures—both visibly and invisibly—within agencies. The Cycle thus provides
the starting point for developing a more complete inventional map of documentation practices within large regulatory agencies.

If the Cycle allows us to recognize the rhetorical incompleteness of all documentation, it also allows us to see how the collective body of expert knowledge is continually captured and retransformed to produce new knowledge
within the system. Unfortunately, not all individuals have access to this collective body of knowledge—in part because agencies have created institutional boundaries and conventions of discourse that serve as barriers to understanding, but also because individuals do not have the rhetorical tools for
interpreting information conveyed in unfamiliar rhetorical forms. As I have
argued repeatedly throughout this volume, all accounts are inevitably incomplete, but this does not relieve writers of their obligation to increase their understanding of those rhetorical practices that have been rendered invisible
within a culture or institution.

Whatever strategies writers employ, writers must ultimately wrestle with
the uncertainty of knowledge within large regulatory industries. Hazardous
environments are dynamic and unstable. Personnel changes can affect the
quality of risk assessment. Individuals and agencies disagree about the methods of analysis and the factors that contribute to risk. As Aristotle recognized,
deliberation is necessarily uncertain because we do not argue about those
things that do not admit two possibilities.[69] We argue when we deliberate precisely because we admit that the future can be different from what we know in
the present.

This potential for a better future drives agencies to produce new technologies that will improve the health and safety of workers. It should also drive us
to rethink how technical and material uncertainty might challenge the assumptions that drive our rhetorical practice

Chapter 3 explores the sources of this uncertainty and its implications for
rethinking documentation practices in large regulatory industries.

---

[69]According to Aristotle (1993), "We debate about things that seem to be capable of admitting two possibilities; for no one debates things incapable of being different in past or future or
present, at least not if they suppose that to be the case; for there is nothing more [to say]" (p.
41).

3

Acknowledging Uncertainty: Rethinking Rhetoric in a Hazardous Environment

The uncertainty of risk analysis—the fact that the data do not provide unique answers—allows for a range of perceptions about the nature of chemical hazards and their effects on worker health. . . . Some place concerns about risk within a calculus of economic viability; others see risk from the point of view of those exposed to hazards. Each view has implications for responsibility and control.[1]

—Nelkin, (1984)

And we debate about things that seem to be capable of admitting two possibilities; for no one debates things incapable of being different in past or future or present, at least not if they suppose that to be the case; for there is nothing more [to say].[2]

—Aristotle, (1991)

From its beginnings in classical theory, rhetoric has been concerned with questions of probability, for, as Aristotle said, we do not argue about things that are certain.[3] Conceived as a theory of public discourse, rhetoric was defined as an inventional art—an art devoted to the study of a method to find out the available means of persuasion. In simple terms, rhetoric distinguished itself from analytic methods (dialectic) that promised to probe the truth of propositions through rigorous logical method. Instead, rhetoric provided rhe-

---

[1]Nelkin, 1984, p. iv.
[2]Aristotle, 1991, p. 41.
[3]Aristotle, 1991.

tors with a systematic method for discovering the psychological, ethical, and linguistic arguments (proofs) that would help rhetors deliberate, argue, and celebrate issues of public concern where purely logical methods might be inappropriate or misunderstood.

In traditional rhetorics, the uncertainty of any argument reflects the limits of the rhetor's knowledge base (the weight of the evidence) and the confidence with which a rhetor can argue a proposition.[4] In a postmodern rhetoric where neither signifier nor signified can lay claim to fixed and certain meaning, it becomes easy to argue that any attempt to talk about the material world will be fraught with linguistic and discursive uncertainty. In a society without foundational truths, uncertainties also arise when individual subjects conceive different and thus potentially conflicting views of the same situation.

Risk specialists have a more specific mathematical notion of uncertainty in risk management and assessment, but the problem of uncertainty raises similar questions about the methods we employ to measure the strength of an argument, the probability of our predictive models, the credibility of our observations, and the gaps in our knowledge. As Goodman argues, uncertainty "bedevils" any claim to a purely rational argument:

> The most fundamental of statistical questions—what is the strength of the evidence?—is related to the fundamental yet most uncertain of scientific question—how do we explain what we observe?
>
> This fundamental problem—how to interpret and learn from data in the face of gaps in our substantive knowledge—bedevils all technological approaches to the problem of quantitative reasoning.[5]

---

[4]The notion of *enthymeme* is particularly important for defining the difference between so-called logical syllogisms and rhetorical syllogisms. In a rhetorical syllogism, the argument is persuasive because it draws upon existing beliefs, attitudes, and values that reside in the audience or discourse community. Rhetorical syllogisms may be incomplete if rhetors omit premises that are so well accepted that they need no argument. Thus, the rhetorical (incomplete) syllogism "I always use Crisco; my mother did" relies on the belief that mothers know best. It appeals to the emotion (pathos) that resides in the audience. Similarly, "trust me; I'm a doctor" builds confidence through the missing rhetorical assumption that doctors are worthy of trust.

Aristotle's three types of proof (ethos, logos, and pathos) can also been seen as attempts to reduce uncertainty and strengthen the audience's confidence in a speaker's argument. On a more general level, the notion of rhetorical invention and the use of *topoi* provides a means for discovering proofs that can increase and audience's confidence in a speaker's knowledge. These rhetorical proofs do not, however, address the mathematical certainty of the argument. Classical enthymemes were frequently drawn from fictional histories or commonsense notions of the material world. When scientists attempt to calculate uncertainty, they are attempting to find empirical evidence that can help them find the margin of error in their mathematical models and statistical methods.

[5]Goodman, 1999.

As Morgan and Henrion (1992) demonstrate, uncertainty can arise from many different sources, including statistical variation, subjective judgment, linguistic imprecision, inherent randomness, disagreement, and approximation.[6] In policy analysis, the problem of defining uncertainty raises questions about how much knowledge we need to assess and manage risk. Ideally, policy analysts balance the need to simplify data (to reduce complexity and make the analysis more understandable) against the need to elaborate features of the analysis that affect the certainty of its conclusions and recommendations.[7] The process involves many iterations and continual testing to determine how much we need to know before we can adequately assess and manage risk. If this analysis is based upon too little detail, analysts may miss important aspects of the problem, omit significant assumptions, or draw unwarrantable conclusions.[8] Ultimately, however, every analysis includes significant assumptions that influence the outcomes of the deliberation.[9] Articulating those assumptions can help us understand the meaning of an assessment and its fit with current events and conditions. When similar problems re-occur, future readers can understand the grounds for our conclusions and recommendations.

Risk specialists cite the many benefits of articulating uncertainty in regulatory risk assessments. Risk assessments are only as strong or weak as their weakest component. When we identify these components, we can make better decisions.[10] Many risk problems have a "way of resurfacing" in new contexts. When we draw upon data in previous risk assessments, we can have "greater confidence that we are using the earlier work in an appropriate way."[11] We can understand how confidence in a model increases or decreases with new evidence.[12] Articulating uncertainty can actually provide knowledge that can help us evaluate the level of confidence in scientific models, evaluate competing claims, and draw better inferences.[13] On a practical level, the process of articulating uncertainty can help us determine whether the effort to gather additional information will improve decisions about risk and safety. If we articulate the uncertainties in our analysis, we can determine where to spend our efforts—gathering information or making a decision on

---

[6]Morgan & Henrion, 1992, p. 56.

[7]Morgan & Henrion, 1992, p. 38.

[8]Morgan & Henrion, 1992, p. 38.

[9]Morgan & Henrion, 1992, p. 38.

[10]Dowlatabadi, 1999.

[11]Morgan and Henrion, 1992, p. 3.

[12]Miller et al. (1997) write: "As new participants from non-scientific communities enter the debate, they may question previously unchallenged assumptions and methods, sometimes with considerable authority." See also Dombroski, 1991.

[13]Bradshaw & Borchers, 2000, p. 6.

the basis of available knowledge.[14] This process can help us develop more robust solutions even when we cannot resolve the uncertainty in the time available.[15] On a more theoretical level, the process of uncertainty can help us produce new knowledge of risk.

When policy makers identify areas of uncertainty, they also discover new areas for research and new questions for analysis.[16] When scientists articulate the uncertainty in their models and methods, they are forced to acknowledge the gaps in their knowledge frameworks, the limitations of their models, and the potential effects of this uncertainty in their judgments about risk and safety. They must examine how disciplinary frameworks, mathematical modeling techniques, and categories of analysis affect the degree of confidence they can attach to knowledge claims at specific moments of decision making within large regulatory industries. Like expert models, the process is highly reflective and epistemic. By drawing attention to what we do not know, the process can help us develop methods to increase our understanding of risk. This process can take place within scientific communities, but stakeholders can also contribute to the process.[17]

But the process can also create indecision and regulatory paralysis if agencies begin to focus their attention on gathering more and more information. The 1997 Presidential/Congressional Commission on Risk Assessment and Management warned that risk management might be "complicated by uncertainty and by the issue of how much information is enough to justify regulatory action."[18] The Commission report concludes: "The challenge for risk managers is to bring analysis to bear on the question of whether collecting additional data is likely to lead to better, more confident, or more widely accepted regulatory decision."[19]

---

[14]Dowlatabadi, 1999.
[15]Dowlatabadi (1999, p. 4) argues:

These decisions can be reflective, i.e., where to expend the marginal effort to gather information relevant to the problem at hand. They can also be to shape the solutions to the problem so that they are robust to uncertainties that may not be resolvable in the time available. In order to utilize fully these features of a formal integrated assessment, there needs to be systematic characterization of uncertainties in each domain, and in the interactions between the different forces represented in the framework.

[16]Miller et al., 1997, p. 5.
[17]Miller et al., 1997, p. 6.
[18]Presidential/Congressional Commission on Risk Management and Assessment, 1997, p. 91.
[19]Presidential/Congressional Commission, 1997, p. 91.

This chapter provides a brief overview of the most important sources of rhetorical uncertainty in technical documentation in the workplace: the dynamic uncertainty of hazardous environments; the variability and unreliability of human performance; the uncertainty of what one agency calls premium data; uncertainty in social structures and organizations; and the rhetorical incompleteness of any single viewpoint. As the examples in this chapter demonstrate, the material and institutional uncertainty of hazardous environments affects documentation practices at all levels. But documentation practices can also contribute to uncertainty, particularly when writers depend too heavily on agency conventions to structure recommendations and conclusions.

Risk specialists have developed sophisticated mathematical models to describe the methodological and scientific uncertainty in regulatory risk assessments. In this chapter, I focus on those aspects of uncertainty that are most affected by documentation practices in hazardous environments. To set the stage for this discussion, I describe the problem of uncertainty at the highest level of exigence: imminent danger.

UNCERTAINTY AT THE HIGHEST LEVEL OF EXIGENCE: IMMINENT DANGER

Sections 103–104 of the Mine Act give miners the highest level of protection under the law. Under Sections 103–104, inspectors must tour every mine at least four times a year.[20] If inspectors find evidence that mines have violated standards, they must issue a citation and estimate the time it will take for the mine to correct (abate) the problem (CFR 91-173, §103). If the danger is im-

---

[20]As we saw in chapter 1, the Mine Act defines imminent danger as the "existence of any condition or practice in a coal or other mine which could reasonably be expected to cause death or serious physical harm before such condition or practice can be abated" [Publ. L. 91-173, as amended by Public Law 95-164, 30 CFR § 75.3(j)]. Section 103 outlines the 4 duties of the inspector:

SEC. 103. (a) Authorized representatives of the Secretary or the Secretary of Health, Education, and Welfare shall make frequent inspections and investigations in coal or other mines each year for the purpose of (1) obtaining, utilizing, and disseminating information relating to health and safety conditions, the causes of accidents, and the causes of diseases and physical impairments originating in such mines, (2) gathering information with respect to mandatory health or safety standards, (3) determining whether an imminent danger exists, and (4) determining whether there is compliance with the mandatory health or safety standards or with any citation, order, or decision issued under this title or other requirements of this Act.20. . . . [30 CFR § 103 (a) 1977].

minent, inspectors can order the mine to close the affected area and bar all workers except those needed to correct the condition [CFR 91-173, §104 (b)]. Unfortunately, the definition of what counts as imminent danger is highly uncertain.

To cite a mine for imminent danger, inspectors must judge whether the harmful effects are likely to occur *before* workers can abate (or resolve) the potentially dangerous conditions. Inspectors must describe the problem in writing and state specifically what the mine must do to bring conditions into compliance. Any ambiguity in this citation can undercut the rhetorical force of an inspector's judgment, slow the decision process, and diminish the health and safety of workers throughout the entire cycle.

Although we might expect that imminent danger citations would persuade mines to act rapidly to reduce hazards, mine operators can appeal the citation if they believe the citation does not state specifically what the operator is charged with or what the operator must do to achieve compliance.[21] Under the Mine Act of 1969, the mine operator may apply to the Secretary of Labor to appeal any order of withdrawal (for imminent danger) or to challenge the time fixed for abatement of any violation. The mine operator may also apply for a public hearing on the amount of the penalty to be assessed for the violation. This process takes time and can delay action to reduce and manage hazards in local sites.

The Compilation of Judicial Decisions Which have an Impact on Coal Mine Inspections (Cleveland and Turner, 1977) documents the outcomes of these administrative appeals so that agencies can identify weaknesses in the interpretation and enforcement of standards. The following examples from the *Compilation* reveal the problems that inspectors face when they attempt to document evidence of imminent danger in writing.[22] As the following examples suggest, many of the conditions that inspectors confront are either undefined or open to interpretation; inspectors are nonetheless accountable for the quality of their evidence and the adequacy of their technical judgments.

Examples in the *Compilation* reveal uncertainty at many levels. Administrative courts have overturned citations because MSHA "failed to prove by a preponderance of evidence" that the mine failed to comply with standards. The courts have overturned other citations because judges believed there was "no evidence" to support the inspector's claim that there was a "significant and substantial" violation.[23] In all of these decisions, inspectors were faulted because they did not provide an adequate written account of their decisions.

---

[21]Cleveland & Turner, 1977, p. 7.
[22]Cleveland & Turner, 1977, p. 60.
[23]Cleveland & Turner, 1977, p. 60.

But the uncertainties of wording and interpretation also reflect the material, geographic, and institutional uncertainties of a hazardous environment.

First, everyday decisions about hazards draw upon embodied experience in hazardous environments (blowing dust off of a methane monitor), personal observation (watching a miner use an ax to test a roof), and on-site analysis of data (using instruments, test bores, and monitoring devices). Because these activities are highly local, subjective, idiosyncratic and dependent upon individual judgment, we might assume that no standard method of assessment might provide hard and fast rules for assessing hazards. Experienced inspectors can draw upon a body of experience and education to determine how and when to apply standards, but new conditions and technologies can change the rules and lead to high levels of uncertainty. With new technologies, we cannot predict outcomes because we may not know what the negative outcomes might be.

Second, individuals make judgments within complex and dynamic environments that do not allow them to have access to all of the information they need to make an informed decision. Each section, shift, team, and entryway isolates individual decision makers from information in other systems. But activities, events, and conditions in each section interact as a dynamic system that changes rapidly in time and space. As we saw in chapter 1, compliance with standards requires interpretation and judgment; acceptable (and even good) decisions in one section may produce dangerous conditions in other sections of the mine. Material conditions may change more rapidly than individuals can observe, analyze, and describe within large systems. We cannot predict outcomes because we do not have adequate information to predict future events with certainty.

Given the complexity and uncertainty of the geographic, material, and social environment in which they work, it is perhaps not surprising that inspectors produce incomplete and uncertain accounts of their experience.

THE DYNAMIC UNCERTAINTY
OF HAZARDOUS ENVIRONMENTS

Because mines are unstable (as the weight of the mountain shifts in response to pressure), mine managers must work to maintain ventilation, roof support, and drainage even when the mine is not working. The earliest mining laws required miners to maintain drainage and ventilation in order to maintain their claim to a mine.[36] Modern mines must also maintain haulageways and

---

[36]Agricola, 1556/1950. The earliest mine records document the effects of this high level of dynamic uncertainty. In the earliest scientific text on mining, Georgius Agricola noted that owners could lose their right to a mine "if anyone was able to prove by witnesses that the owners

roof support even when miners are not working because the stresses and forces within a mine are continually "working"—creating instability in the rock walls and the threat of a sudden collapse.

The Bilsthorpe disaster in Nottinghamshire, England, on August 18, 1993, illustrates how the dynamic instability of the material environment can affect the health and safety of miners. In this disaster, three miners were killed in an enormous roof fall. According to investigators, part of the mine roof sagged over a weekend when miners weren't working, causing stress fractures that created dangerous cracks and decreased roof stability.[37] Old waste from previous mining activity may also have destabilized the solid rock strata and contributed to the disaster. The report describes in detail the dynamic movement of the rock, the rapid changes in rock stability, and the number of distinct factors that intersected to produce the disaster. Following the first extraction of coal, the report notes, "the stresses [were] redistributed about the extracted coal and cause[d] a vertical stress increase on the solid rib and the occurrence of shear cracks located approximately 5 m into the solid coal ribside."[38] These stresses created a new equilibrium that destabilized the rock above the existing roof supports.

The Health and Safety Executive's report on the Bilsthorpe disaster stresses the unpredictability of rock behavior, the uncertainty of prediction, and the suddenness of the fall:[39]

> Common features of these falls were their unpredictability and the suddenness of collapse giving little or no chance of escape to those in the immediate vicinity. This point is highlighted because it may not be appreciated by the layman that there is no absolute clear understanding of strata (rock) stability criteria and the stresses associated with mine workings. Strata behaviour and its control is not an exact science and more work needs to be done to obtain a greater understanding and to be able to measure more accurately the effects on mine workings of coal pillars and gateside packs.[40]

---

had failed to send miners for three continuous shifts." Although these rules were revised by 1556, allowing owners to stop driving a tunnel, owners were still required to maintain the drains and repair ventilation holes in order to maintain their claim to the mine. Because an owner's claim depended on his ability and willingness to maintain a mine, inactivity could render meaningless a claim inscribed in paper (Agricola, 1556/1950, p. 62).

[37]Langdon, 1994.

[38]Langdon, 1994, p. 15.

[39]Miners disagreed that the fall was unpredictable. Chapter 5 presents alternative views of the problem.

[40]Langdon, 1994, p. 16.

The problems of uncertainty and instability are multiplied in a large mine. A typical mine may have seventeen different sections working at the same time to develop three parallel entries or tunnels that are driven into the coal bed. Each tunnel is connected by crosscuts that produce the effect of a large honeycomb or set of interlocking chambers.[41] In an active mine, stresses on the rock interact in three dimensions—changing the structure of the mine with each human cut, explosion, or rock fall.

My own experience shows how conditions underground can change rapidly, unpredictably, and without warning. In a large Yorkshire mine, a mining engineer was in the middle of an interview, explaining confidently that newly introduced roof bolting practices had reduced roof falls and improved roof support in British mines. In the videotape, he speaks with an educated British accent. Suddenly, a miner enters. The two switch to a thick Yorkshire accent. They tell me to turn off the videotape and stop taping. My camera records its sudden dislocation in space and time—a sweep of motion as I turn rapidly to face the newcomer. Then, as suddenly, the tape is blank. There are unexpected problems of roof control in the mine; conditions are dangerous; miners have tapped an underground lake. The engineer must decide—quickly—how to proceed.[42]

In such an environment, documents that construct a fixed and certain image of geographical space have only historical and limited value unless they are constantly updated. (Chapter 8 describes this problem in detail.) Con-

---

[41]Mallett, Vaught, and Peters (1992) describe a typical but imaginary mine:

Black Carbon Number One (pseud.) is an enormous mine, with access down one slope and eight shafts into two coal seams. These seams, separated horizontally by roughly 125 feet of strata, both average 48 to 54 inches in thickness. . . . The operation's layout, some 400 feet below the area's farmland, is extensive. Black Carbon uses the block system of mining, with 17 continuous mining sections being employed to set up three longwall units. Panel development at Black Carbon is done with "continuous mining" technology in which sets of three parallel "entries" (tunnels), each approximately 20 feet wide, are driven into the coal bed. These entries, whose centers are usually 80 feet apart, are connected by "crosscuts" (perpendicular tunnels), also about 20 feet wide, at intervals of 60 to 120 feet. Coal is produced during the process of driving these entries and crosscuts, and the points at which coal is actually being extracted are called "working faces."
    . . . For longwall development, two parallel sections will be driven simultaneously. These parallel sections, about 1000 feet apart, may be extended for several thousand feet, following surveyed projections, before reaching a property boundary or some other predetermined point at which they are to stop. (p. 33)

[42]Anon. (Interview with roof bolting engineer, U.K.), January, 1995.

ventional maps and written descriptions (as we understand them in the 20th century) provide a snapshot of a mine geography that has literally vanished.

THE VARIABILITY AND UNRELIABILITY OF HUMAN PERFORMANCE

Human variability can add to the sources of uncertainty in a dangerous and unpredictable environment. Davitt McAteer (1995) writes:

> By definition, the working environment changes all the time. As mined mate-rial is removed, an entirely new workplace is created in as little as 24 hours. Roof conditions vary from place to place.
>
> Explosive methane may be liberated. Mechanical systems may fail. A mine that is safe one day may be dangerous the next. A mine with a good safety rec-ord may confront new challenges daily. Personnel changes, too, can make a dramatic difference in the safety of a mine.[43]

At the Tower Colliery in southern Wales, Tyrone O'Sullivan described how personnel changes can disrupt the safety culture of a mine. O'Sullivan was (at the time of this writing) the manager of Tower Colliery, head of a miner-owned cooperative that bought the mine in 1995 when British Coal privatized British Mines. In the old days, O'Sullivan recounted, miners learned safety by working seven and a quarter hours daily as apprentices with an experienced miner. As they worked, experienced miners talked about football, sex, or women, but it was "all about learning." Experienced miners would tell apprentices: "This is the way to test the roof; dull means loose; pulls like a bell. Don't put steel on steel." O'Sullivan recalled: "You'd learn a lot from the men you were working with . . . They'd talk a lot, tell you about experiences. Sometimes it was like the man could see in the future. It drives you mad; it was natural; you could see danger."[44]

But talk was not enough. "We'd play games," O'Sullivan remembered. "The gob (loose rock) would break off in a straight line. We'd throw a stone and go back and fetch it. We didn't get hurt." During the games, young min-ers would get used to the system. "It worked," he said. "You could sound a roof and if it rang, O.K. . . . a dull sound [was] dangerous." In the old days, experi-enced miners were kept on at the mine, even when they could no longer cut

---

[43]McAteer, 1995, p. 6.
[44]O'Sullivan (Interview with author). January, 1995.

coal. They cleaned the belts, maintained haulageways, and looked after younger miners. Maintenance kept the mines clean and prevented minor accidents caused by slipping and falling or falling objects. And old miners kept the stories alive; they could see beyond immediate experience with an almost frustrating awareness of risk.

When British Coal introduced cutbacks in British mines to cut costs following the 1984 strike, they let go the older miners. Mines in the north of Wales no longer spent critical production time on maintenance. And safety suffered, according to one safety officer in another British mine, particularly after privatization. Because Tower miners could control their own wages and production, they agreed to take one day per week for maintenance. "The safety manager [at Tower]'s got it easy," one British manager sighed. "They've got such clean mines (touch wood)."[45]

When miners worked as apprentices, master miners could adjust their learning styles to the needs of their apprentices and the specific conditions. Young miners could learn a variety of methods by working closely with different miners. When British Coal introduced standardized training schemes, however, these new procedures could not take into account individual differences in knowledge and experience. One miner claimed that safety standards were created by safety managers who were "totally separate" from the safety culture at the mines.

Miners like O'Sullivan lament the loss of the old safety culture: "Fifty years ago, apprentice miners did all jobs of mining under the supervision of an experienced miner. You learned the conditions and how to respond. . . . New training programs take a 15½ year old boy just out of school and attempt to turn him into a full-time face worker in 80 days. . . . Now, all you've got is talk."

MSHA statistics seem to confirm that personnel differences affect the level of safety. In the United States, inexperienced miners have the highest accident records.[46] But experience as a miner is only one factor. Miners experienced on one face (section of the mine) can become inexperienced when they move to a new face or new conditions. Because miners cannot be present in all aspects of mining, telling stories, playing games, and learning from other miners become ways of transmitting experience without the costs.

---

[45]Anon. (Interview with safety manager, UK), January, 1995.

[46]There are various other explanations for these statistics. In some cases, the victim's experience has little relation to the accident. In the J and T mine disaster, miners entered the mine and were crushed because miners on a previous shift had failed to follow the roof control plan. In another case, a woman cutting wood on her farm was killed by flying debris from a mine. She had no experience at all. See U.S. Department of Labor, 1991c.

Risk specialists argue that effective risk communication must build upon the knowledge and experience of lay audiences. But what if those audiences have profoundly different experience? What if they express that experience in play and narrative?

As we shall see in the chapters that follow, conventional written documentation may fail to capture an individual's "embodied sensory experience." Conventional forms of workplace discourse like instructions and procedures can render invisible the diverse viewpoints of observers situated literally and physically in a different relation to risk. Written documents may also fail to capture knowledge embodied in speech and gesture. Before we can speak with confidence about what audiences know, we must find ways to fill the gaps in our own rhetorical practice.

THE UNCERTAINTY OF PREMIUM DATA

Categories of analysis serve important functions in regulatory risk assessments. But these categories create uncertainty if we do not have an adequate framework for understanding what purposes they serve and how they were created.

Categories of analysis enable us to organize and make sense of the collective body of knowledge in hazardous environments. They enable analysts to generalize and rationalize the messy details of uncertain and unpredictable environments, to determine the frequency of particular types of accidents, and to determine the level of risk in particular environments. Because they are necessarily reductive, categories also erase small but insignificant differences and nuanced meanings. Poorly constructed categories can render invisible the human tragedy in mining accidents—when, for example, MSHA categorizes the death of a miner as a haulage or electrical accident. When categories are too general or imprecise, they can conceal details critical to effective risk management and assessment.

The following examples show how imprecise classifications can affect the health and safety of workers. All three accidents were classified as a major injury in a mine. In the first accident, a 33-year-old development worker slipped in water and fractured his left fibula.[47] In the second accident, a 33-year-old faceman fell while he was cleaning the top of the powered support. In the third, a 35-year-old faceworker fractured his left ankle when he lost his balance while changing his trousers at the end of a shift.

---

[47]Herbert, 1989.

Such classifications may be useful in defining the level of injury. (All three suffered the same level of injury.) The categories can help us determine the frequency of accidents or lost work-time in a mine. But these categories provide little information about the cause or causes of the accident. If we wish to protect the health and safety of workers, we need more information than the data provide. To determine a solution, we need to understand the events and conditions that produced the accident.

In categorizing all three accidents as major injuries, written reports may conceal the hazards to workers from poor mining conditions (in the first accident); unsafe practices (in the second); and faulty protective clothing (in the third). These categories fail to indicate whether management took action to warn workers about potential hazards or whether conditions in the mines, like sloppy maintenance, might have contributed to the falls. And although we may laugh about the likelihood of injury from putting on trousers, female miners in the United States have described the hazards of company-issued protective boots, hardhats, gloves, and overalls sized one size fits all. For small miners, extra-large hardhats can obscure vision; large gloves encumber hand movement; large boots cause slips and falls in wet, rough, or slippery conditions; and baggy overalls can encumber movement.

Outdated categories can also create uncertainty. When agencies fail to update categories, reports may not provide the data investigators need to assess hazards in emerging technologies. One data analyst at MSHA, for example, attempted to collect data on longwall mining from accident reports and investigations. These reports did not specifically mention the method of mining. But the analyst was able to infer when mines employed longwall mining from other details in the report. His data necessarily contained high levels of uncertainty, but the numeric results were reported and repeated in other documents as if they were collected directly from accounts of longwall mining.

One Bureau of Mines report (1994) shows how an agency's outdated categories can affect mine safety.[48] According to the report, the agency's classification system for Underground Work Location (ULOC) and Underground Mining Method (UMETH) did not provide categories to differentiate whether accidents occurred in the entry, beltline, or haulageway. As a result, agencies could not determine the specific underground work location with the current categories because all accidents were classified in a "generic continuous category."[49] The report warns: "Both of these fields track information that is very important for health and safety research. . . . These latter mining methods are currently 'hot' safety topics in the mining industry."[50]

---

[48]U.S. Bureau of Mines, Human Factors Group, 1994.

[49]U.S. Bureau of Mines, 1994, §11.

[50]U.S. Bureau of Mines, 1994, §11.

If agencies update categories too often, however, analysts will not have consistent data or sufficiently long risk histories to make informed judgments about risk. When new technologies change, they may also change the database for recording toxic exposures and long-term risks. New diesel emissions standards, for example, may change the level of harmful gasses in the mine atmosphere, but workers with five to 10 years' exposure under old conditions may not suffer symptoms for 20 years. If mines collect data under the new standards, they may fail to link toxic exposures to illnesses with long incubation periods.[51]

When analysts classify and abstract data from accident narratives, they participate as rhetorical agents in the rhetorical transformation of experience within large regulatory agencies. Because these rhetorical transformations affect the adequacy of documentation, they always have the potential to create uncertainty. The ANSI standards for recording data that we described in chapter 1 note that the most effective documents are produced when observers first "describe events as completely as possible in [their] own words, then classify data and organize information into categories."[52] In theory, this process "insures uniformity in the treatment of data and avoids the introduction of analysis errors resulting from different interpretations of the classification procedure."[53] In practice, however, classification systems and categories of analysis can become rigid protocols for activity if investigators focus on filling in blanks in standard accident reporting forms. Thus, while we might expect that scientific inquiry would inform policy, policy embodied in documentation also shapes scientific inquiry.

MSHA's 1988 *Coal Accident Analysis and Problem Identification Instruction Guide* shows how a system of classification designed to collect premium data can become an artificial template that structures not only the activity of the inspection but also the conclusions that inspectors can draw from their data.[54] The *Instruction Guide* provides a set of checklists, tally sheets, and "planned inquiry sheets" to help inspectors categorize accidents.[55] The classification system in the Physical Barrier Analysis Matrix (Fig. 3.1) seems to support the conclusion that physical barriers are either "not possible or practical" in most of the cases reported on the form—a conclusion that might exempt management from responsibility for protecting workers. Because data from such an analysis directly affects policy and procedures, classification systems can have

---

[51]Anonymous. U.S. Bureau of Mines personal comment.

[52]American National Standards Institute, 1962, p. 34.

[53]ANSI, 1962, p. 35.

[54]U.S. Department of Labor, Mine Safety and Health Administration, 1988b.

[55]U.S. Department of Labor, Mine Safety and Health Administration, 1988b, p. 17.

PHYSICAL BARRIER ANALYSIS MATRIX

	1	2	3	4
	Barrier on the Energy Source	Barrier between Energy Source and the Person	Barrier on the Person	Time/Space Barrier Separating the Energy & the Person
A Was Not Possible and Practical	1, 2, 4, 6, 7, 8, 9, 10, 11, 12, 13, 14, 15, 16, 17, 18, 20, 21, 22, 23, 25, 26, 27, 28, 29, 30, 31, 32, 33, 34, 36, 37, 38, 39, 40, 41, 42, 43, 44, 46, 47, 48, 49, 50	1, 2, 3, 4, 5, 6, 8, 9, 10, 14, 15, 16, 17, 18, 19, 20, 21, 22, 23, 24, 26, 27, 28, 29, 30, 33, 34, 35, 37, 38, 40, 41, 42, 43, 44, 45, 46, 48, 49	1, 3, 4, 5, 6, 7, 8, 9, 10, 12, 13, 14, 15, 17, 18, 19, 21, 22, 24, 25, 26, 27, 28, 30, 31, 32, 34, 35, 37, 38, 40, 41, 42, 43, 44, 45, 48, 50	1, 2, 3, 4, 5, 6, 7, 9, 10, 11, 12, 13, 14, 15, 16, 18, 19, 20, 21, 22, 23, 24, 25, 26, 27, 28, 29, 30, 31, 32, 33, 34, 35, 36, 37, 38, 39, 41, 42, 43, 44, 45, 46, 47, 48, 49, 50
B Was Possible and Practical but Not Provided	3, 19, 24, 35, 45	13, 31, 50		17
C Was Provided but Not Used		7, 11, 12, 25, 36, 39, 47	2, 11, 16, 20, 23, 29, 33, 36, 39, 46, 47, 49	8, 40
D Was Provided and Used but Failed	5	32		

FIG. 3.1.   The Physical Barrier Analysis Matrix. This document seems constructed to support the conclusion that physical barriers are "not possible or practical" in most of the cases reported on the form. Source: U.S. Department of Labor, Mine Safety and Health Administration (1988a, p. 63).

a powerful effect on the health and safety of workers—particularly in the anti-regulatory atmosphere of the Reagan Administration.

Because the quality of writing varies widely in an industry where writers have little training in rhetorical theory, the quality of narratives also varies. The title of the U.S. Bureau of Mine's draft report on "Errors and 'Unexpected data' " speaks for itself. ("Unexpected data" refers to data that is either physically impossible or extremely unlikely, like the 75 reported examples where the seam height was either less than 10 inches or greater than 200 inches.[56]) Accident narratives contain many typographic and spelling errors. Writers abbreviate key words to fit the accident narrative into the space allo-

---

[56]U.S. Bureau of Mines, 1994, p. 2.

cated on accident-reporting forms.[57] Without a consistent vocabulary, analysts have a difficult time classifying information in the report.

To improve consistency in reporting and classifying data, agencies may construct their own list of recommended terminology.[58] In the most extreme case, these terms can actually influence a writer's vocabulary and structure the prose narrative. Accident reports thus become tautological documents that literally imitate the agency's already preconceived conclusions of how accidents occur. A sample mine training bulletin (Fig. 3.2) shows trainers how to use the language of the agency's accident analysis to structure accident-report narratives.[59] Writers who lack confidence in their writing can imitate the passage.

Ideally, writers are free to analyze and report their observations and conclusions in ways that can help agencies rethink policy to improve safety. In practice, however, documentation practices not only structure the activity of reporting but they also shape the agency's analysis and conclusions.[60]

UNCERTAINTY IN SOCIAL STRUCTURE AND ORGANIZATION

As researchers in the sociology of science (SSK) have argued, written documents conceal the compromises, negotiations, and alternative representa-

---

[57]U.S. Bureau of Mines, 1994, p. 3.

[58]U.S. Department of the Interior, Bureau of Mines, 1993, p. 55.

[59]U.S. Dept. of the Interior, Bureau of Mines, 1993, p. 55.

[60]In the most extreme example, MSHA (1987) proposes a "mad model" of writing derived from a Mad Magazine game that the authors describe for those not familiar with the reference. To play the game, readers are given a story line with parts missing and they choose from columns of words to fill in the missing parts. They provide the following example:

When the _____ falls _____ will _____ in the _____. (p. 61)

The authors provide a list of words to fill in gaps in the sentence. The authors note: "As you can see, what is filled in the blanks can be as logical or illogical as one wishes to make it. The choice of one inappropriate word can turn the otherwise meaningful statement into sheer nonsense" (p. 61). Using the "mad model," inspectors can create a problem list or summary that they can use to make recommendations to the mine operator.

Because it is so explicit, this model demonstrates the effects of classification systems on future mine safety and health as well as the ideological bias in the process. The writers conclude: "[The process] permits us to isolate the causes of problems and accidents so that the operator can better utilize efforts to eliminate causes having the greatest impact" (p. 87). If these recommendations are followed, they claim, "our recommendations will not only be more likely to eliminate the cause of the problems, but applying the correct solutions will be much less costly for the operator" (p. 65).

Mine Training Bulletin

Hand(s) on Rotating Steels or on Drill Head

Observation

A bolter at times will position a drill steel and keep a hand on the steel either when starting to drill a hole or when removing the rotating steel from the hole.

Also, roof bolters will lay their free arm and hand (the one not used to operate the controls) across the drill head as it is being raised to the top during drilling.

Common Hazards

Falls of loose rock are the leading cause of roof bolting accidents. In several of the accidents, workers were injured when roof rock fell either inby the ATRS or through the ATRS ring. The most likely time for pieces of rock to fall from the roof is during the drilling process.

When a miner's hand is around the steel, the hand is out from under canopy protection. It is directly below the top area which is being disturbed by the rotating steel. Also, a rotating steel can cause lacerations to the hand or "catch the glove" and wrap the hand and fingers around the steel. And, a miner's free arm may be hit by pieces of falling roof rock if the arm is extended towards the steel, and out from under canopy protection.

Injury Analysis

Falls of loose roof rock and hands on rotating steels accounted for a large percentage of the drilling accidents at this mine. Min-

ers hands were sprained while grasping a rotating steel and also by falls of roof rock as they were holding onto the drill steel.

Example: Roof bolter operator was trying to remove a drill steel from the hole with hand on the steel, and started the rotation which caused a hand sprain. Example: The miner's hand was caught on a rotating steel as worker grabbed the steel when steel was coming out of the hole. Hand and thumb injured.

Example: Roof bolter was placing the drill steel in the drill head and closed the steel guide on finger, resulting in a fracture. Example: The miner was installing bolts in a high top area of 10 to 11 feet. With hand on the steel to guide it up into the hole, worker was struck on the forearm by a piece of rock from the roof.

A Safe Approach

The company should adopt a "hands-off" safety procedure for drilling holes. This can be developed as a safe work procedure in the form of a job aid.

Roof bolters also should be made aware of the danger from falling pieces of roof rock. It should be emphasized that bolters should not place their free arm on the moving drill mast during drilling. Instead, they should keep this arm under the canopy's protection.

FIG. 3.2.   Sample mine training bulletin. This document shows trainers how to structure accident report narratives. Source: U.S. Department of the Interior, Bureau of Mines (1993, p. 55).

tions of data in rough drafts, management reviews, and brainstorming sessions.[61] This research suggests that individuals and social groups draw upon complex and largely unarticulated "chains of unquestioned assumptions" (Woolgar, 1992) in order to construct the meaning of data within organizations.[62]

In calling attention to the performative role of texts within historically specific social situations, social studies of science have helped rhetoricians situate rhetorical practices within networks of continuing social and institutional activity beyond the framework of written texts.[63] These studies underscore the radical uncertainty of knowledge claims, the incommensurability in machine performance, the ambiguity of standards, and the tension between objectivity and relativism in the construction of scientific fact. Many of these arguments are not new, but social scientists have extended these arguments empirically in ethnographic studies that document the presence of social interactions beyond the framework of textual practices—in the institutions, laboratories, and public forums where knowledge is constructed. Researchers in the public understanding of science have extended this work to argue for a more reflexive approach to science and technology studies that draws on the local knowledge and experience of lay audiences.

---

[61]Latour, Woolgar, & Salk; 1986; Myers, 1990.

[62]Woolgar (1992) writes:

> Reference to the current temperature, the shape of the trace and the direction it is taking, all pass as adequate at particular instants in the interaction. This adequacy is achieved through participants' reliance upon chains of unquestioned assumptions. . . . What features of the organization of the interaction make possible the accomplishment of descriptive adequacy? (p. 140)

[63]Pickering's analysis provides an excellent overview of the issues in and major contributors to the debate. In a chapter called "Living in the Material World," Pickering (1995) outlines the major controversies in the philosophy and sociology of scientific knowledge. He himself dismisses relativist concerns about the failure of correspondence theories of representation He agrees that what counts as knowledge is a function of the "specific material-conceptual-disciplinary-social-etc. space in which knowledge production is situated" (p. 185). But this "space" and its "contingent practices" do not entirely determine knowledge (p. 885). Pickering thus shifts the focus of attention from problems of correspondence to problems of performance. Rather than arguing that scientific "facts" represent or fail to correspond to "the world as it is," Pickering argues that "[s]cientific knowledge [eventually] . . . converges on a mirroring relation to nature" (p. 186). Pinch and Bijker (1984) have demonstrated how the meaning of facts and artifacts are constructed in social interactions. Their analysis of the social construction of the bicycle demonstrates thus extends research the notion of social construction to include science knowledge and the knowledge embodied in technical artifacts. Sismondo (1993) reminds us that there are also many theoretical frameworks that constitute what we understand to be the social construction of reality.

Hazardous environments challenge the dichotomy between lay and expert knowledge. In these environments, individuals make judgments within social systems that attempt to mediate the idiosyncratic judgments of individuals against a set of standard practices, social behaviors, local knowledge, and conventions of discourse. What counts as local knowledge is also the product of negotiation. At each moment of transformation in the Cycle, what counts as an adequate warrant must be renegotiated and reconstructed in new modalities for new audiences. The rhetorical processes of transformation thus produce a hybrid knowledge that draws upon the collective expertise of many individuals throughout the system. It is no longer possible to speak of a discrete form of local knowledge.

MSHA's (1996) guidelines for inspectors demonstrates the multiple levels of interpretation and negotiation that take place within institutions before inspectors can actually document the violation or write a citation.[64]

To achieve uniformity, consistency, and fairness in the application of standards, the Mine Act orders the Secretary of Labor to develop guidelines for inspection and investigation under the law.[65] According to the guidelines, inspectors must consider all aspects of a mine's history beyond the limits of his own personal knowledge, including violations and citations from previous inspections; documentation and certification of training programs; mine ventilation plans; escape and evacuation plans; legal identity reports; injury, illness and accident reports; information concerning independent contractors at the mine; petitions for modification granted for the mine; and "any other significant information concerning health and safety at the mine to be inspected."[66]

Before they inspect the mine, inspectors must conduct a pre-inspection conference with mine officials and miners' representatives to discuss the scope and purpose of this investigation. Following an inspection, they must meet again with management and miners again in a post-inspection conference.

---

[64]U.S. Department of Labor, Mine Safety and Health Administration, 1996, pp. 21–22.

[65]The Mine Act (1977/2000) reads:

The Secretary shall develop guidelines for additional inspections of mines based on criteria including, but not limited to, the hazards found in mines subject to this Act, and his experience under this Act and other health and safety laws. For the purpose of making any inspection or investigation under this Act, the Secretary, or the Secretary of Health, Education, and Welfare, with respect to fulfilling his responsibilities under this Act, or any authorized representative of the Secretary or the Secretary of Health, Education, and Welfare, shall have a right of entry to, upon, or through any coal or other mine.

[66]U.S. Department of Labor, Mine Safety and Health Administration, 1996, p. 21.

Finally, inspectors must find ways to be both omniscient and omnipresent in a mine, despite the fact that no individual can observe and analyze all aspects of mining underground simultaneously. MSHA instructs inspectors to observe each phase of the mining cycle that occurs during each regular inspection of the mine, though the agency acknowledges that inspectors cannot be present to observe all aspects of mining during the walk-around. If mines operate on more than one shift, inspectors must observe mining practices on all three shifts. In large mines, however, inspectors cannot be physically present for all aspects of mining simultaneously. MSHA's (1996) policy manual for inspectors thus cautions:

> While inspectors are not required to observe the complete mining cycle at every working place, they must evaluate enough conditions and practices and ask sufficient questions of miners to be reasonably assured that work is being safely conducted during the portions of the cycle that could not be observed.[67]

In the post-inspection conference, inspectors must defend their decisions against the objection of management, who frequently have a financial incentive to overturn inspectors' decisions. At the conference level, assessors and conference officers often feel that they make judgments on too little information and incomplete inspection statements.[68] Operators want inspectors to write inspection reports at the mine in order to increase efficiency, but inspectors fear that the process will turn into haggling between the operator and inspector.[69] As a result, inspectors write reports after they have left the mine and must rely on notes and their own memories to complete their assessment.

Writing plays an important role in the process. MSHA (1996) requires inspectors to keep "clear, concise, and factual notes."[70] During their "walk-around" of the site, they must document all "observations, conversations and physical examinations of work conditions and practices" on the appropriate MSHA Field Notes (a record-keeping system). "Clear factual notes"[71] are essential, the manual advises, in both the post-inspection conference and final report.

---

[67]U.S. Department of Labor, Mine Safety and Health Administration, 1996, p. 23.

[68]*Oversight hearings on the Coal Mine Health and Safety Act of 1969 (Excluding Title IV)*, 1977, p. 480.

[69]*Oversight hearings on the Coal Mine Health and Safety Act of 1969 (Excluding Title IV)*, 1977, p. 477.

[70]U.S. Department of Labor, Mine Safety and Health Administration, 1996, p. 23.

[71]U.S. Department of Labor, Mine Safety and Health Administration, 1996, p. 23.

Faced with complexity, some inspectors choose the safest path, imitating previous reports and constructing formulaic reports that reflect the agency's terminology and format. If they use the agency's formulas to structure the inspection, they can simplify the process of transforming notes and observations into writing. But the process can endanger miners' lives. In a 1985 memo presented as evidence in 1987 *Congressional Oversight Hearings of the Mine Safety and Health Administration,* Joseph Lamonica tells William L. Querry that 10 fatalities have occurred in less than five years at mines operated by Messrs. George, Clyde, and Duane Bennett. Seven of the 10 fatalities were caused by roof falls; two of these fatalities occurred within a seven-month period in the same mine. "This level of fatalities is unacceptable," he concludes. "In each Report of investigation, the investigator states, in effect, that the fatality was caused by management's failure to take certain safety precautions."[72]

The hearings present seven accident reports as evidence, each of which follows the same formula. Each blames management for allowing workers to work under "known loose roof that had not been properly supported."[73] Yet no one acted to reduce and manage risk. If the conditions were known, why were they not reported? Who knew? Did workers have sufficient supplies? Did they know the proper procedure? In each case, writers document events for the record without taking action to prevent similar occurrences in the future. The accident-reporting structure becomes an activity in itself. The activity of reporting becomes a formulaic act that has no rhetorical force beyond the agency's requirements for documentation.

There are many reasons why writers follow safe formulas. As we shall see in chapter 9, the complexities of institutional authority in an Appalachian mine create uncertainty about the locus of authority and responsibility for decisions that affect safety. The 1987 oversight hearings show how questions of conflict of interest clouded many aspects of the investigation. These hearings describe how inspectors were taken off the job after they had charged mines with major violations.[74] Inspectors were transferred to new assignments after

---

[72]*Oversight of the Mine Safety and Health Administration, Hearings before the Committee on Labor and Human Resources . . . on examining activities of the Mine Safety and Health Administration,* 1987, p. 243.

[73]*Oversight of the Mine Safety and Health Administration, Hearings before the Committee on Labor and Human Resources . . . on examining activities of the Mine Safety and Health Administration,* 1987, p. 247.

[74]*Oversight of the Mine Safety and Health Administration. Hearings before the Committee on Labor and Human Resources . . . on examining activities of the Mine Safety and Health Administration,* 1987, p. 220.

companies complained to their superiors. With the larger coal companies, one inspector commented: "I have been told, 'This guy, he does a good job, so give him every break you can.' "[75]

In these environments, social and institutional authority is itself a source of uncertainty that affects documentation practices as well as decisions about safety.

THE RHETORICAL INCOMPLETENESS
OF A SINGLE VIEWPOINT

Rhetorical theory was not developed to help individuals document the dynamic uncertainty of a hazardous environment.

In classical theory, Aristotle's notion of artistic and inartistic proofs (technic and atechnic proofs) assumes that rhetors have the artistic means to resolve the ambiguous and contradictory meanings in the evidence given to rhetors. Ciceronian and (later) Renaissance notions of the good man trained in the art of speaking well assume that a rhetor's character (ethos) and integrity form the ethical foundations of a coherent and unified argument.

Rhetors could, of course, embrace multiple viewpoints—pro and con—in exercises designed to sharpen the wit or argue against an opponent's construction of evidence. In the sophistic tradition, rhetors did not necessarily maintain a consistent viewpoint across different or unrelated arguments—like lawyers who serve multiple clients.

The notion of a rhetorical self also implies that individuals can maintain more than one viewpoint by suiting their character and discourse to the rhetorical situation.[76] Rhetors can "talk different realities" in order to manage discord within social interactions.[77] To orient listeners to these (assumed) roles and characters, individuals must "annotate speech" with "elaborating and contextualizing clauses and phrases that provide explicit definitions of the content, a definite sense of role and character, and classically rational arguments designed to persuade the hearer that the speaker's symbolic reality is true or correct."[78] Speakers may also manage these multiple roles by changing dialect or style—aurally marking differences in stance or position by chang-

---

[75]*Oversight of the Mine Safety and Health Administration. Hearings before the Committee on Labor and Human Resources . . . on examining activities of the Mine Safety and Health Administration,* 1987, p. 226.

[76]Greenblatt, 1984.

[77]O'Keefe, 1988.

[78]O'Keefe, 1988, p. 202.

ing from standard English to non-standard dialects.[79] But few theorists allow speakers to hold multiple—and potentially conflicting—viewpoints simultaneously. When speakers assume multiple personae to argue contradictory and probabilistic truth, they open themselves to charges of moral relativism, opportunism, and manipulation.

Pedagogical practices in rhetoric and a long tradition of logical argument in current traditional rhetoric also reinforce a unitary notion of argument. To demonstrate proficiencies in argument, student writers must construct and defend a single thesis with a coherent narrative structure and logical and consistent proof. Although classroom teachers value the diverse and often contradictory viewpoints expressed by individual subjects, their classroom practices may reinforce linear argument structures that inadvertently exclude multiple and competing viewpoints. Process models of critical thinking, constructivist theory, and critical pedagogies thus exist in a fragile tension with the demands of a consistent and linear narrative.

In practice, of course, no speaker can ever achieve dramatic consistency and internal coherence, particularly with naturalistic as opposed to scripted speeches. Kaufer and Butler (1996) describe the limits of predictiveness and extreme consistency in rhetorical argument. Positions that are "predictable in the ideal case" claim to be complete and consistent.[80] But every rhetorical argument is ultimately incomplete. "Every rhetor must do his or her best with a position that is in some respects incomplete (causing indecision) and latently or overtly contradictory (causing no consistent decision to result when the whole point of holding a position was to converge on a consistent outcome decision)."[81] Even when a rhetor constructs a consistent argument that reconciles contradictory positions in the past, they argue, rhetors must inevitably confront new contradictions as they enter "the current here and now of audience and opponent with energized curiosity about the uncertain."[82]

Because hazardous environments are by definition unpredictable and uncertain, decision makers have no scripts. They can preplan actions and prepare a plan, but their plans must provide them with the ability to adapt to unpredictable and rapidly changing conditions. Plans provide minimum standards, but individuals must use their own judgment when they encounter unplanned conditions. In these environments, individuals must be able to conceive multiple, and potentially conflicting, outcomes without the sense of contradiction or ambivalence that characterizes classical unity and coher-

---

[79]Johnstone & Bean, 1997.
[80]Kaufer & Butler, 1996, p. 97.
[81]Kaufer & Butler, 1996, p. 98.
[82]Kaufer and Butler, 1996, p. 98.

ence. As Kaufer and Butler (1996) conclude, predictiveness without responsiveness leads ultimately to unpredictability to the extent that writers fail to
document important assumptions and conditions. Predictiveness in this sense
leads to bureaucratic behaviors so well rehearsed that actors fail to take into
account the local and human features of the rhetorical situation.[83]

As I argue throughout this book, hazardous environments are by definition
profoundly uncertain. Seasonal changes in pressure and temperature change
the flow of gasses and may release naturally explosive pockets of methane
known. Human activity produces a different kind of change with less predictable results. Personnel changes can affect the expected certainty of outcomes
predicated on notions of reliability and standards of human performance. Inactivity can also produce dangerous changes—if rock begins to settle or gasses
accumulate in an unworked section of the mine. In these environments, all
representations are admittedly reconstructions. All documentation inevitably describes a temporally outdated and geographically inaccurate space. The
problem of representational correspondence becomes moot in real-time attempts to document rapidly changing events and conditions underground.
But individuals are still held accountable for the adequacy of their warrants
and the confidence they attach to evidence in support of their claims. To improve the confidence and consistency of data, agencies have developed conventions of discourse and models for imitation. But these practices can increase uncertainty if they exclude information important to the analysis.
Even when we recognize the most immediate hazard to human health and
safety (imminent danger), warrants can be overturned if writers lack sufficient evidence or if writers cannot provide adequate documentation to support their judgments in writing.

For rhetoricians, the profound uncertainty of hazardous environments ultimately challenges the conventions of technical writing in the workplace
and causes us to rethink the fundamental assumptions that guide rhetorical
practice. By identifying "areas of uncertainty," social scientists can help scientists discover new areas for research and new questions for analysis.[84] When
rhetoricians articulate the uncertainty in scientific models and methods, they
too help scientists discover new questions for analysis. What kinds of war-

---

[83]Kaufer and Butler (1996) argue that "predictiveness without responsiveness leads to dog-
matic behavior and, ironically, unpredictability, insofar as important unstated conditions and
implications are bound to remain implicit and undocumented; predictiveness without humanness leads to robotic and bureacraticized behavior, leads to people so well rehearsed going into
the rhetorical situation that they don't dare set the script aside and engage the audience as a fellow human" (p. 82).

[84]Jasanoff, S. (2001). Personal communication with the author. Ithaca, NY. April, 2001.
See also Miller et al., 1997

rants do writers employ to assess and manage risk? Can individuals move outside of their situated positions rhetorically so that they can produce a better understanding of the events and conditions that produce disaster? What other modalities (besides writing) do individuals employ to describe their understanding of hazardous environments? How do these modalities help speakers negotiate and construct new knowledge of risk—collaboratively and individually—in workplace interactions?

The following chapters focus on the rhetorical transformation and re-transformation of experience at the boundaries between agencies and the sites they seek to regulate. In exploring these transformations, we begin to discover the full range of rhetorical practices that constitute the Rhetoric of Risk (Fig. 3.3).

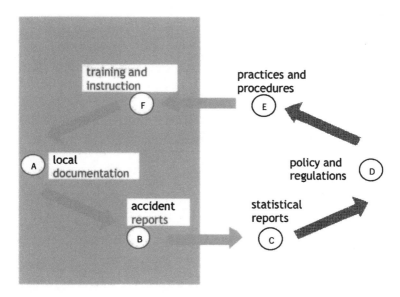

FIG. 3.3.   The Rhetoric of Risk focuses on the rhetorical transformation and re-transformation of experience at the boundaries between agencies and experience.

II

MOMENTS OF TRANSFORMATION

4

Reconstructing Experience: The Rhetorical Interface Between Agencies and Experience

More than 20 percent of the 2,685 reportable incidents in this study could not be analyzed because report narratives did not contain sufficient detail to identify a specific job task.[1]

—Klishis, Althouse, Layne, & Lies, (1992)

They were dragging compressors all over the mine and not concerned with the fact that they had the equivalent of a portable flame thrower. . . . That compressor has umteen gallons of automatic transmission fluid which burns when the compressor fails and safety devices fail.[2]

—Anon. (Interview with MSHA official), April, 1992

In a large accident, fatalities rarely occur because of a single event. Large accidents occur because unpredictable events and uncertain outcomes interact in time and space to produce disaster.[3] In the Wilberg fire in Orangeville, Emery County, Utah (December 19, 1984), for example, miners were 7600 feet—more than one mile—inside the mine. To escape, they needed oxygen (from self-contained breathing devices called SCSRs); they had to find exitways not blocked by previous roof falls or explosions; and they had to de-

---

[1]Klishis, Althouse, Layne, & Lies, 1992, p. 78.

[2]Anon. (Interview with MSHA official), April, 1992

[3]Studies in the history of large-scale technological disasters have described how the interaction of unpredictable events can produce disaster (Perrow, 1984; Trento, 1987). Petroski (1985) shows how engineers learn from error, but the process requires that engineers document their work and communicate problems.

termine a safe way out of a maze of interconnected passageways, burning ventilation curtains, and smoke-filled entryways.[4]

The Wilberg mine fire was one of the worst disasters since the passage of the Mine Act. Twenty-seven miners died in the Wilberg fire, mostly from smoke inhalation. A single miner, Kenny Blake, escaped. MSHA ultimately identified nine contributing factors in the disaster. Because of the magnitude of the fire, there were rumors of arson. The FBI was involved in the original investigation, and MSHA hired an independent expert on mine fires, John Nagy, to develop independent conclusions based upon events in the disaster.[5] Based upon Nagy's reconstruction of events in the hours preceding the fire, MSHA located the technical cause of the fire in an overheated compressor that had operated continuously for 69 hours because miners had bypassed the over-temperature safety switch that might have prevented the compressor from overheating.[6]

But fires do not necessarily kill miners. There are some 20 compressor fires each year. Not all of these fires produce disaster.[7] As Wilberg demonstrates, large-scale technological disasters often result from "tightly coupled" but unrelated events.[8] In these situations, no single event can be said to be the sole cause the disaster, yet removing any single event might have interrupted the chain of events that produced the fatal outcome.

Twenty-seven miners died in the Wilberg fire because management had extra men underground in order to increase production. The fire spread rapidly because the compressor where the fire originated was not permanently housed in a fireproof material. The heat was particularly intense because miners had inadvertently used mineral oil, turning the blazing compressor into a "flame thrower," according to one agency official.[9] Rescuers could not turn on the water pumps to fight the fire because workers in the main office had turned off the main power to prevent the spread of what they thought might be an electrical fire.

Poor training also affected how miners responded to the fire. Blake, the only miner who escaped the section, used his SCSR to travel a difficult dog-

---

[4]Miners were trapped behind the fire because MSHA had approved a two-entry mining system (instead of the usual three-entry system). One of the two entries had been blocked by a previous roof fall, leaving miners with no alternative means of escape (Huntley et. al, 1992).

[5]Huntley et al. (1992) note: "Eyewitness accounts of the quickness with which the fire spread in the early stages seemed to support arson as a possible source" (p. 25). The FBI entered the investigation in September, 1985.

[6]Huntley et al., 1992, p. 88.

[7]D. McAteer, Personal interview with the author, 1992.

[8]Perrow (1984) demonstrates how many different "tightly coupled" events must interact to produce a disaster. See also Petroski, 1985; Trento, 1987.

[9]Anon. (Interview with MSHA official), April, 1992.

leg route contaminated with gases from the fire. Thirteen victims were later found along this route. None of these miners had used their SCSRs correctly, and they did not know the location of exits. In the final *Report of Investigation* (Huntley et al., 1992), MSHA cited management for failing to train miners to use SCSRs, for failing to provide an accurate and up-to-date map, and for failing to house the compressor in a fireproof structure (among other violations).[10]

Given the number of events that intersected to produce the disaster, it's not surprising that writers struggled to produce a coherent narrative of the accident.[11] Because agencies draw upon data derived from accident reports and investigations to justify new regulations, investigators must produce a "precise identification of the problem."[12] To cite mines for violations, they must fit events to the categories of violations in the Mine Act. But writers were also aware that unraveling every link in the chain could lead them to too many disparate conclusions. "It's legitimate to say you can't put all eggs back together," one labor lawyer complained.[13] Whose viewpoint would count? How far back should investigators go? How could they satisfy the agency's need to identify the technical cause of the accident, yet still provide sufficient information to help regulators prevent similar accidents in the future?

Fortunately, MSHA investigators left a detailed record of their investigation.

MSHA's final *Report of Investigation* was published in 1992, eight years after the disaster.[14] The report is a massive, heterogeneous document that includes 92 pages of narrative and 25 appendices. The documents in these appendices include victim data sheets, mine maps, selected photographs taken during the investigation, evaluation and test reports on various pieces of

---

[10]The union contended that the report omitted the history of violations at the mine and ignored MSHA's responsibility for approving a two-entry mining system that left miners inby the fire with no path to safety because a previous roof fall had blocked the one remaining escapeway (Anon. (1992). Interview with former inspector, U.S. April, 1992).

[11]Anon. (Interview with miner, U.S.), April, 1992.

[12]Querum (1977), speaking on behalf of the National Coal Association, Bituminous Coal Operators' Association and American Mining Congress, argued that the need for "precise" data placed an unfair burden on operators. *Oversight hearings on the Coal Mine Health and Safety Act of 1969 (Excluding Title IV)*, 1977, p. 328.

[13]McAteer, D. (1992). Personal interview with the author. Shephardstown, West Virginia. April, 1992. At the time, McAteer was a labor lawyer for the Council on Southern Mountains.

[14]On December 23, 1984, rescue personnel determined that the mine was too hot to continue rescue and recovery operations. Because of the risk of explosion, MSHA withdrew all personnel, deenergized power supplies, and sealed the mine permanently on January 7, 1985 (Huntley et al., 1992, pp. 22–23). The mine was reopened for the underground portion of the investigation on November 2, 1985.

equipment likely to have contributed to the disaster, a detailed account of the mine recovery operations, and copies of the citations and orders issued as a result of the investigation. Some reports document MSHA's findings during the investigation and recovery of bodies; others report test results on various pieces of equipment found during the investigation.

These documents allow the public to evaluate MSHA's conclusions based upon the evidence presented in the investigation.[15] In this sense, they allow us to see the rhetorical transformations that occur as local information moves into the domain of science. At the same time, the variety of viewpoints represented in the documents creates a complex and heterogeneous representation that exists outside of—and in fragile tension with—the agency's technical findings and conclusions.

While it would be impossible to examine all of the viewpoints[16] that writers represent in all 25 appendices, two documents in particular help us analyze the rhetorical processes by which local knowledge moves into the domain of science and public policy. Appendix F contains Blake's hand written account of his escape, written immediately after the accident. Appendix H contains Nagy's independent analysis of miners' testimony following the accident. Blake's narrative enables us to view the uncertain and dynamic geography of a mine from the viewpoint of a miner focused on his own survival. His viewpoint is limited to his own experience, but his narrative reveals how he used sensory cues in the environment to save his life. Nagy's report reveals the rhetorical strategies that one fire consultant employed as he attempted to reconcile conflicting accounts of the disaster in order to reconstruct a coherent and consistent narrative of events and conditions that precipitated the disaster.

Together, these two documents help us understand the processes by which agencies reconcile and reconstruct the diverse and sometimes conflicting viewpoints of individual observers at one critical moment of transformation. Blake's

---

[15]Other documents appear outside of the report.

[16]Viewpoint differs from standpoint, although the two terms share metaphoric origins in a notion of situated knowledge. In the sense that argument theorists use the term, standpoint is the outcome of reasons in support of the standpoint. But it does not control or produce the argument in the same way that a physical viewpoint controls a reader's visual and sensory interpretation of events and conditions. Van Eemeren, Grootendorst, & Henkemans (1996) argue that "verbal utterances" are "standpoints or reasons when they serve to express a certain position in a (potential) difference of opinion or to defend a certain position in a context of a (potential) controversy. Standpoint reflects an opinion that results from analytic and logical argument without reference to geographic and institutional space and time. Haraway (1991) argues that theories of knowledge production must take into account the situated viewpoints of individuals For a complete summary of the notion of standpoint, see Tannen, 1993, pp 14–56.

For an early bibliographic essay on issues relating to gender in professional communication, see Allen, 1991. Brasseur (1993) provides a review of feminist issues and feminist methodologies in professional communication.

report (Appendix F) documents his first-hand experience from a viewpoint inside the mine. Nagy's report (Appendix H) allows us to see the process by which agency writers consciously transform the testimony of observers who disagree about the precise timing and location of events at the time of the accident into a single agency perspective. But it also reveals the frustrations he feels at the inadequacy of any single conclusion. In the end, Nagy's report raises questions about the rhetorical effectiveness of the agency's narrow viewpoint as a guide to future policy within large regulatory agencies.

The following discussion provides a brief overview of the physical structure of a mine to help readers understand how individuals are situated in geographic space and time. This discussion helps us understand the literal and figurative structure of situated knowledge in a hazardous worksite.

THE STRUCTURE OF SITUATED KNOWLEDGE IN HAZARDOUS WORKSITES

Coal miners experience hazards from viewpoint inside of the spaces they describe. Agencies and management view these same experiences from a viewpoint outside of and above the same environment. Miners pass through ventilation curtains and move inside of haulageways that appear as subtle markings on a two-dimensional map of the ventilation system.

In these systems, what an individual can see from his or her virtual location within the system is physically constrained by the natural and man-made geography of the site. Because the physical geography limits the viewpoint (the physical vantage point) of any individual within the system, knowledge of the whole system is always dependent upon the collective knowledge of many individuals whose individual viewpoints must be coordinated and reconciled so that writers can make sense of the whole. But these viewpoints are not always commensurate, particularly if individuals orient themselves in relation to sensory cues like smoke and heat. As a result, writers like Nagy must work to reconcile conflicting viewpoints, testing individual observations against what is known and what is possible in the physical and institutional spaces they inhabit.

The Literal and Figurative Structure of Work in Hazardous Worksites

Literary notions like standpoint, situated knowledge, and viewpoint take on literal meaning within the institutional and geographic structure of a mine.

In its literal sense, the term *viewpoint* describes the scope and range of what observers can see from their virtual positions in geographic space.[17] In a coal mine, an individual's geographic viewpoint frequently overlaps his or her institutional viewpoint. Thus, managers and engineers in offices on the surface are distanced geographically and institutionally from the sites they manage. They stand in different relationship to risk, and they are situated in jobs that place them close or far from hazards. Miners farthest underground bear the greatest risk. In English mines, pit baths (where miners shower before leaving the mine) mark the institutional and geographic boundary between underground labor and the cleaner intellectual work that occurs on the surface.

In a coal mine, the division of labor between surface and underground has its roots in the geography and geology of the mine. The underground structure of a mine necessarily follows the direction of the seam (or vein of coal). The height of the seam can vary from 27 inches (in so-called low-seam coal) to towering caverns of 10 feet. The working area of a mine is called the face. To travel horizontally, miners ride in battery and diesel-operated locomotive-like devices. To travel vertically to the deepest sections of the mine, they ride in elevator-like structures called cages. Deep mines are the most dangerous because faceworkers are farther from the surface in the event of a fire or explosion. In a deep mine, the roof or ceiling of the mine is also under greater pressure from the mass of mountain about the mine.[18]

Although the history of the political and social relations between surface and underground is beyond the limits of this discussion, no discussion of surface and underground is ever free from the political and economic antagonism between labor, management, and local and Federal government. Despite the apparent separation of the two worlds, the underground world exists in fragile tension with the world of the surface. To control and manage risk, managers need local information as well as a bird's eye view of the entire system. Landowners on the surface are also affected by events underground. If mines collapse (either deliberately or by accident), structures on the surface can collapse into large sinkholes. The resulting subsidence on the surface can lower the entire landscape by as much as 10 or 12 feet. But mines must also

---

[17]Jackendoff (1996) describes how cultures differ in the ways that speakers orient themselves in geographic space. These orientations define a viewpoint that must be re-negotiated when speakers confront new systems of geographic and spatial meaning. See also Levelt, 1996; O'Keefe, 1996.

[18]South African gold mines are some of the deepest mines in the world. In these mines, roof control is particularly difficult. The depth of a mine is only one reason for the number of rock falls in South African mines.

pay attention to surface features like lakes and rivers. If mines cut too close to a body of water on the surface, the entire mine can be inundated.

Nor is it easy to represent these two worlds simultaneously on paper. To map the geographic and institutional architecture of a mine, cartographers must represent two or more physically incommensurate viewpoints simultaneously in the same visual field. A cutaway map can show how passageways are related in vertical space, but not the location of passageways and exits at each level. A bird's eye view can show the structure of individual sections and the location of exits, but not the depth or inclination of the passageways or the height of a coal seam. Even with improvements in three-dimensional imaging, most mine maps represent only a partial and extremely limited view of the whole system. Because many mines do not identify locations underground, miners must learn to visualize their location in relation to objects and features in their environment. As a result, the relationship between a miner's understanding of space and a cartographer's bird's eye view from the surface is often highly uncertain.

The Vocabulary of Situated Viewpoints

Miners have developed a rich vocabulary to describe their locations within the geographic and institutional architecture of a hazardous environment. Miners describe their positions "inby" and "outby" fixed reference points in geographic space. Miners inby a mine fire must visualize alternative escape routes and must communicate their position to others outby the fire.

The terms "inby" and "outby" (these terms rhyme with "pie" and drive copyeditors mad) describe the situated position from which miners observe and experience hazards. The term "inby" refers to a miner's location between (we might say "inside of") a known location (such as a mining machine, numbered entryway, conveyor belt, or compressor) and the working face. The term "outby" refers to a miner's location between (or "outside of") a known location and the mine exit. When miners work inby the compressor, they stand between the compressor and the face. When they work outby the compressor, they work between the compressor and the mine exit. Miners inby a fire are trapped and must move through the fire to escape. Miners outby the fire stand in a much safer position relative to the fire.[19]

---

[19]It is possible to be both inby one physical location and outby another simultaneously. Miners outby the fire were still inby other locations in the mine. I have borrowed these terms to describe the rhetorical viewpoints that agencies and individuals assume in relation to the risks they represent.

The terms "inby" and "outby" describe an individual's physical location within the mine, but they also serve as a measure of an individual's position in relation to hazard. Thus, miners may be 50 feet inby the fire or 500 feet inby the fire. Individuals closest to the fire are at risk from the fire itself; those far away are not as likely to be injured directly, but they may be trapped by smoke and gases. Investigators can judge the quality of an individual's representation by knowing what the individuals should see or experience at each position inby or outby the fire. Thus, individuals close to the fire may feel heat and see flame. These sensory experiences can be used to locate miners' viewpoints following a disaster.[20]

The map in Fig. 4.1 shows the areas inby and outby the fire at the Wilberg mine. As this map shows, 13 miners were trapped inby the fire because the working face of the mine is the furthest extension away from the exit.[21]

The Rhetorical Uncertainty of Documentation in Hazardous Worksites

The physical structure of a coal mine provides a metaphor for the rhetorical uncertainty of documentation in a hazardous environment.

Because no single individual has access to all aspects of mining simultaneously, individual documents will represent different aspects of experience. These differences reflect the viewpoint of individual subjects within and outside of the sites they seek to describe and evaluate. These documents are designed to support the agency's factual findings, but any individual document must be tested against other documents within the same geographic and institutional space. As a result, any single account cannot make the same claim to fact that we might ascribe to knowledge claims in science.

Instead, the adequacy of a document—the extent to which any single document might be said to be adequate, accurate, complete, or sufficient—must necessarily reflect the limitations of the observer's viewpoint. This viewpoint can easily be obscured by local conditions. It can be blocked by structural features of the environment, or literally distanced by the physical separation of observers from the sites they seek to manage. Inside risky environments, machines, dust, and darkness may block an observer's line of vision. Protective

---

[20]Such estimates may reflect tautological reasoning: We assume that miners see particular events because they are located in particular sites. We then draw conclusions about events and conditions based upon their presumed location.

[21]This map was based upon Nagy's estimate of many factors in the fire. Although Nagy recognized that this was merely an estimate, the results of his investigation (like the accounts of previous disasters that he himself drew upon to make these estimates) will provide support for future documentation within the Cycle.

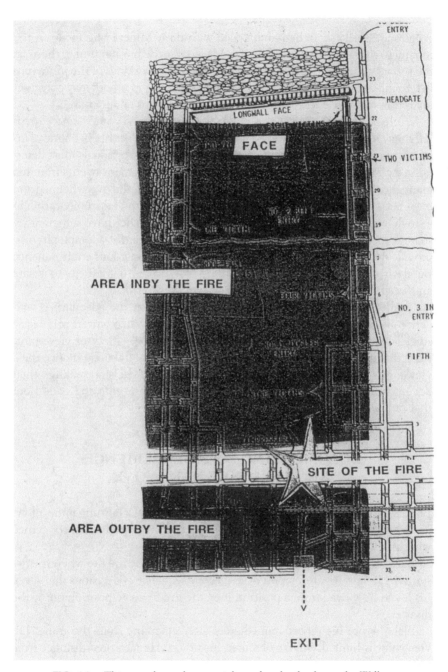

FIG. 4.1. This map shows the areas inby and outby the fire at the Wilberg mine.

goggles are easily scratched and fogged with dust. Miners who focus on the roof may trip on poorly maintained roadways. At the most extreme, those on the surface (above ground) cannot literally see the workings of the mine (underground and beneath them). As a result, engineers in offices may construct maps that bear little resemblance to actual conditions underground.

While the complexity of this collective body of observation and experience may improve the quality of risk decision making, the sheer volume of information creates difficult rhetorical problems when individuals must determine a course of action. If individuals attempt to document events from too many incommensurate viewpoints, they can easily become overwhelmed and overcautious in the face of risk. Individuals who limit their documentation to a single viewpoint may face less anxiety, but they will not have sufficient resources to operate in conditions of profound uncertainty. Complexity can provide decision makers with the fullest account of hazardous environments. But decision makers must have the tools to sort through complexity to isolate and weigh how each individual factor contributes to risk.

Ideally, writers make strategic choices based upon the adequacy of evidence and their confidence in the writer's method. When writers privilege a single viewpoint, they may inadvertently silence the variety of viewpoints that constitute the collective history of an institution. If writers silence these viewpoints too soon in the processes of analysis and decision making, their documentation may not help others understand the events and conditions that produce disaster.

DOCUMENTING LOCAL EXPERIENCE: KENNY BLAKE'S NARRATIVE

Blake's handwritten account of his narrow escape from a burning mine allows us to see the Wilberg fire from the viewpoint of a single individual concerned with his own survival.

Blake, the sole survivor of the Wilberg fire, was inby the fire when the fire started. His handwritten report describes the sequence of decisions that saved his life as he moved from a position inby the fire to a safe position outby the disaster.

Blake wrote his report immediately after emerging from the mine. His viewpoint is limited. Smoke obscured his vision. He measured distance from moment to moment as he headed into the air flow and observed patterns of smoke "coming around a corner." Stunned and shocked by the events he describes, he does not provide a coherent account of the origins of the fire or the technical problems that contributed to the deaths of 27 miners.

In the following excerpt, Blake describes how he sensed physical changes in the environment inby the fire and responded accordingly. As a technical description of a burning mine, the narrative is confused and disorderly. It reveals little about his escape route or the location of specific stoppings (ventilation controls). But the narrative recalls a series of highly local and, in retrospect, effective decisions as he assessed conditions in the burning mine. Each "So" (in bold face, added) marks a new decision that initiates new acts that literally saved his life. The spelling and capitalization are Blake's:

> **Blake:** .. then I lost tom [sic] and could not find him,
>
> **So** I wandered around trying to find them but I couldn't even find—the Belt line again,
>
> **So** I turned around and headed the Other Way again looking for more Stopping doors, I can't remember but I think I found another Stopping door it was not hot,
>
> **So** I went through, then I headed with the air, I could feel it then. I walked about 200 to 300 feet when it Started to Get hot and I could hear the top falling and fire burning,
>
> **So** I turned around and headed into the air Flow and walked for about 10 min when I noticed the air on the Floor was clear.
>
> **So** I laid on the Ground and Saw the smoke coming around a corner
>
> **So** I walked about 200 to 300 ft before I found another Stopping door it was cool.[22]

Blake's account enables us to view the uncertain and dynamic geography of a burning mine from the viewpoint of a miner focused on his own survival. In this narrative, Blake senses changes in the physical environment that help him find clear air. Trainers can draw upon this narrative to help miners learn from Blake's experience without endangering their own lives. But Blake's viewpoint in this passage is too limited in scope to help investigators understand the full range of institutional decisions, practices, and conditions that precipitated the disaster.

In the final *Report of Investigation* (Huntley et al., 1992), MSHA transforms Blake's embodied experience into statements of material fact that serve as evidence for MSHA's conclusions. Blake's escape is described from a third-person viewpoint outby the disaster:

> After traveling about 250 feet, Blake eventually found a mandoor in a stopping and went through the door into No. 5 entry, 1st North. The air in this entry, which was the main return air course for 1st North, was smoky but cool. Blake

---

[22]Blake, 1985, pp. 1–2.

traveled with the return air for a short distance when he heard the fire roaring and the top falling. He turned around and traveled against the current farther into the mine. He noticed that the smoke was clearing near his feet, and got down on the floor and looked up the entry, where he saw the smoke coming around a corner. He walked past the corner and encountered clear air. Blake went through a door in an overcast and into No. 4 entry at crosscut 40.[23]

In the agency's final representation, no single narrative can fully represent the truth of the experience. The agency's account must weigh this first-hand account against other representations of the disaster. Fortunately for rhetoricians, Nagy's analysis explicitly documents the processes by which this transformation occurs.

CONSTRUCTING THE AGENCY'S PERSPECTIVE: OUTBY THE DISASTER

When the fire started, each miner viewed events from a different location relative to the source of ignition. To reconstruct a coherent account of the disaster, MSHA investigators needed to reconcile the observations of many different individuals who observed the fire from many different locations within the mine.

Blake viewed events as he experienced them during his escape. Twenty-seven victims died inby the fire and thus could not testify to the events and conditions that caused their deaths. Thirteen other miners viewed events and conditions from the surface or from a location outby the fire. Their accounts would help investigators pinpoint the origin of the fire—and thus the potential source of ignition (in an overheated compressor)—but these miners could not provide the first-hand information necessary to determine positively the source of ignition.

Nagy recognized that his narrative must achieve a unified and coherent perspective—a viewpoint not possible for any single observer underground or above ground, but possible *rhetorically* if he could assume "the proper perspective of the actions."[24]

---

[23]Huntley et al., 1992, p. 11.
[24]Nagy (1987) writes:

The order and time in which the events occurred is of importance to get a proper perspective of the actions. No one person consistently and accurately noted the precise time for a series of actions. However, fair correlation may be obtained by assigning reasonable "reconstructed" times for events, such as stopping the belt in No. 5 Right, shutting off the electric power in the mine, and the initial use of the telephone. (p. H-7, Italics added)

But whose perspective counts? How wide should his viewpoint extend?

Nagy limited his focus to the cause and location of the fire in its early stages. His report "is not an exhaustive account of the fire, but rather focuses on evidence indicating the cause and location of the fire in its early development."[25] He analyzes evidence from examination of the fire area in January, 1987 (three years after the fire), and he uses studies of fire literature to confirm his conclusions.[26] As a result, he does not address the issue of MSHA's approval of a two-entry mining system, and he does not examine the history of violations at the mine or events and conditions more than one hour prior to the fire. Thus, even though Nagy notes that "the compressor was reported to be operating for at least 2-1/2 days prior to the fire without attention," Nagy does not question why the compressor was left "without attention" for nearly 69 hours prior to the phone call that alerted miners to the fire.[27] Nor does he explain why miners and management failed to detect problems with the compressor in the two days before the accident.

Since there were no witnesses to the fire, Nagy determined the "most-likely source" from "indirect evidence."[28] Because of the controversy surrounding the fire, Nagy carefully documented his methods of analysis and data collection. As a scientist, he derived many of his conclusions from logical deduction and analysis. But he also depended heavily on previous accident reports and his own experience to confirm his construction of events in the accident.[29]

The point here is not to question Nagy's conclusions as findings of fact, but rather to analyze the process by which writers within agencies map the location of individuals, reconstruct the timing of events, and draw conclusions about the technical cause of the accident. Hired as an independent consult-

---

[25]This institutional focus may reflect his long career as an employee of the Bureau of Mines as an expert in explosions and fires. The final report of investigation explains that "MSHA engaged John Nagy as a consultant to perform an independent study and analysis of the Wilberg fire. Mr. Nagy is a renowned mine expert, having spent his entire 42-year career, most of it with the Bureau of Mines, researching and investigating mine fires and explosions" (Huntley et al., 1992, p. 29).

[26]Nagy, 1987, p. H-2.

[27]Nagy, 1987, p. H-23.

[28]Nagy, 1987, p. H-7.

[29]When Nagy uses the results of previous accident investigations to support his conclusions in the Wilberg fire, he complains that these reports provide inadequate information about the events and conditions that precipitated these disasters. Ironically, Nagy's own report tells us little more about why individuals who were "dragging compressors all over the mine not concerned with the fact that they had the equivalent of a portable flame thrower" (in the words of one investigator) did not perceive the magnitude of the risk. (Anon., Personal interviews with the author, 1992).

ant to locate the site of the ignition, Nagy's conclusions suggest that he recognized the limitations of his own findings even as he lamented the inadequacy of previous accident reports and investigations.

Mapping Embodied Positions

Miners view events from embodied positions that are situated, literally and physically, in three-dimensional geographic space and time. Experienced miners maintain a sense of their bodies' relation to the surface so that they can escape in a mine fire or explosion. But spatial relationships can be deceiving underground. "Down" can mean "toward the face"—down into the mine, even when the geographic landscape literally slopes up. As a result, miners frequently locate themselves in relation to permanent equipment, overhanging rocks, drill bits, electrical cables, low clearances, and moving conveyer belts.[30]

To work safely, miners must also pay close attention to the location of colleagues and management.[31] The grim cartoon figure Three-Fingered Joe reminds miners to pay attention to their own fingers, toes, hands, and bodies when they work with machines that can easily take off a finger. When British apprentices described how to stay safe, they cited two rules: know your body position and communicate with your buddy.

When accidents occur, investigators must transform these embodied locations into coordinate positions that can be located in relation to the numbered passages (crosscuts) on a two-dimensional map of the mine. Nagy can identify these numbered passageways because he knows the map locations of telephones, belts, and spillage sites. But he must rely on miners' memories to locate their embodied positions.

Nagy tells how he determined each miner's location according to miners' activity at the site (e.g., coal dumping) and by their location in relation to numbered passageways (e.g., "from No. 5 Right"). Nagy thus couples two ways of defining a miner's location in one hybrid event. Italics in the following passage indicate how miners' activities are melded seamlessly into the language of engineering:

Data are given in Table 1 for the events occurring at Crosscut 34, where *the belt from No. 5 Right dumped* on to the Main North belt. Two men, Price and Tidwell, were at that location *shoveling [sic] spillage.* After Silsbury *made his two*

---

[30]In the following passage, Strayer (1987) locates himself in relation to the feeder (a piece of equipment) and his colleagues ("them guys") working ahead of him: "And I started down towards the feeder to help them guys start belting off . . . And as I was going down there . . . they started splitting a stump" (Strayer, 1987, MSHA video).

[31]Anon. (British mine manager), January, 1995.

telephone calls, O'Neill, who was at Crosscut 25, came to Crosscut 34, talked to Price and Tidwell, and then continued inby to Crosscut 43 and thence into No. 4 entry.[32] (Italics added)

The rhetorical transformation of embodied experience into the language of engineering is even more striking when Nagy reveals that the numbered crosscuts were not actually marked underground. Miners differentiated these crosscuts by the presence (or absence) of various pieces of equipment. Nagy (1987) explains: "Except for the equipment (compressor-power center), there were no marked distinguishing characteristics in the mine to indicate which crosscut was which."[33] (As we shall see in chap. 8, miners' memories about their location are highly uncertain, and the transformation is not a simple process.)

Not surprisingly, miners at different locations disagreed about many of the events in the disaster. Nagy argues that these disagreements actually reinforce the rhetorical credibility of their testimony since agreement might suggest that miners had agreed among themselves on a single narrative.

Reconstructing Time

The timing of events in a disaster is critical in determining the cause of the accident. Nagy explains that "the order and time in which the events occurred is important to get a proper perspective on the actions."[34]

But miners disagreed about the precise time of events as they perceived them at different locations. Salisbury claimed he made a phone call to the surface at 9:10 PM. Two miners underground (Bloomquist and Nelson) and two on the surface (L. Cox and Leavitt) corroborated his testimony but disagreed about the precise time of the phone call.

Since no one noted the "precise time" of these events during the disaster, Nagy explains that he achieved a "reasonable correlation" of events by assigning "reasonable reconstructed times for events" such as the stopping of the belt or the first phone call to Mounteer:

No one person consistently and accurately noted the precise time for a series of actions. However, fair correlation may be obtained by assigning reasonable reconstructed times for events, such as the stopping of the belt in No. 5 right, shutting off the electric power in the mine, and the initial use of the telephone. (p. 7)

---

[32]Nagy, 1987, Italics added, p. H-7.
[33]Nagy, 1987, p. H-15.
[34]Nagy, 1987, p. H-7.

TABLE 4.1
Reconstructed Time of Events in No. 4 Entry

"Reconstructed" Time	Reported Time	Person	Reference	Event
9:05 p.m.	9:00 p.m.	Salisbury	(10)	Smoke at overcast
9:10 p.m.	9:00 p.m.	Salisbury	(11)	First phone call to Mounteer
9:15 p.m.	9:10 p.m.	Salisbury	(12)	Second ook [sic], fire at Crosscut 36
9:20 p.m.	9:15 p.m.	Salisbury	(13)	Second phone call
. . . .				
9:50 p.m.	—	Taylor	(19)	Fire on east rib, smoke at Crosscut
9:40 p.m.	—	Boyle	(18)	Saw Blake at Crosscut 39 or 40
9:45 p.m.	—	Boyle	(20)	Took Blake outside.

Source: Nagy (1985): Wilberg Investigation, Appendix H, Table 2, pp. 9–10.
Note: Numbers in parentheses refer to the page number in the appropriate miners' testimony.

Since all but one of the observers died in the fire, Nagy catalogues events as miners outby the fire perceived them. Nagy organizes these events in a table (Table 4.1).

The location and order of events in the narrative is critical because it helps investigators answer important questions about the site of origin and the speed of the fire as it spread throughout the mine. Nagy can eliminate a belt fire as the source of ignition, for example, because three miners agreed that the belts continued to run after miners observed smoke.[35] The equipment in the No. 5 Right section could not have caused the fire, he reasoned, because miners reported that the fire was "external to the 5 Right section."[36] (Eventually, Nagy bases his determination of the cause of the fire on the speed of the fire and the density of the smoke.)

Both the reported times and the reconstructed times are highly uncertain. To reconcile differences, Nagy must evaluate the rhetorical distance between the observers' locations and the events they describe.

Categorizing Rhetorical Positions

To reconcile conflicting accounts, Nagy evaluates the location of each individual's experience in relation to the disaster. R. A. Cox, for example, "could only feel the heat"; he did not observe the fire directly.[37] L. S. Cox assumed an observer's viewpoint. He located the smoke "inby Crosscut 36" and flames

[35]Nagy, 1987, p. H-16.
[36]Nagy, 1987, p. H-18.
[37]Cox, R. A., 1985. In Nagy, 1987, p. H-37.

"outby" the crosscut.[38] Nagy can use these first-hand sensory experiences to locate events within a mine.

But miners also report secondhand information that they heard from others closer to the events. Thus, Lyons reports what "the guys that was there before us said."[39] These experiences enable Nagy to categorize the observer's rhetorical position as first-hand, second-hand, or merely "reported."

Nagy distinguishes between events, reported events, reconstructed events, secondhand information, references, and statements. Reported events enter the chronology of reconstructed events when they have been confirmed or verified as credible observations. Statements must be tested to confirm their reliability.

Thus, Nagy discounts Lyons' "secondhand information" because Lyons "states" that "other people told him" the fire moved inby 50 feet before he arrived on the scene.[40] Instead, Nagy bases his conclusion on Nelson's firsthand observation that the fire did not move because Lyon's "secondhand information is not confirmed by the statement of other miners"[41] and E. R. Nelson "specifically states" that the fire did not change position during this time.[42]

In the final narrative, first-hand experience counts more than secondhand narrative, even when the secondhand observer reports information from miners who presumably had temporal advantage as observers during the crisis. When miners disagree, Nagy can test the rhetorical certainty of their observation by measuring the degree of distance between the observers' viewpoint and the events they describe.[43]

Reconstructing the Collective Experience of Risk

Ideally, we could train miners to be so alert to signs of hazard in the environment that they would perceive hazards immediately. In reality, however, hu-

---

[38]Cox, L. S., 1985. In Nagy, 1987, p. H-37.

[39]Lyons, 1985. In Nagy, 1987, p. H-37.

[40]Nagy (1987) notes, for example, that "E. R. Nelson specifically states the fire did not change position during a 1/2 hour period. D. G. Lyons states that other people told him the fire moved inby 50 feet prior to the time he got there" (p. 37). He thus concludes: "[T]he foregoing data and observations support the conclusion that the Wilberg fire did not move to any marked degree inby following its ignition" (p. H-38).

[41]Nagy, 1987, p. H-37.

[42]Nagy, 1987, p. H-37.

[43]Originally, Nagy (1987) had planned to assign a "probability of ignition" to each factor "because of lack of eyewitness observations." However, as the study progressed, "the accumulation of evidence that the fire started at the compressor became so overwhelming and so conclusive." He ultimately decided that "an assignment of a probability of ignition to other sources would be misleading" (p. 19).

mans shift their viewpoints, both deliberately and inadvertently, to focus on different features of the work environment. Miners focused on production may become so intent on their work that they fail to recognize signs of hazard. Miners focused on the safety of colleagues may fail to notice hazards around them. In one MSHA training video, Dave Garry explains how his buddy Charley ("Jaz") was so intent on the job that he failed to pay attention to the condition of the roof support. Garry is amazed at his buddy's concentration. According to Garry, the noise was "really loud," but Jaz never heard it:

> So on the ride out, uh, I asked him, I said, Jaz, What, didn't you, you know, what took you so long, didn't you, you know, see anything happening. . . . You know . . . and he said that he was so intent on doing the job that he was doing that he really wasn't paying attention to the post situation and the roof.
>
> He said, you know, the noise of the roof breaking up was really loud, and he never heard it, I mean, he said, he just all of a sudden he was in a dust filled situation and in the dark.[44]

In theory, both miners should tell the same narrative. Instead, Garry observes the falling roof because he first hears a noise and turns his attention to the situation. Jaz focuses on his job until he sees nothing but dust and dark. Neither miner's narrative provides a complete representation of the accident because these miners do not have access to conditions beyond human view within the roof strata. Garry focused on geographic conditions and thus lost his focus on production. Jaz focused on production and failed to notice the signs of a falling roof.

When we compare viewpoints, however, we can begin to speculate about why miners were injured. Each individual's viewpoint is relative to a whole and perfect but inaccessible representation of a mine. As we attempt to reconcile these viewpoints at each location within the institutional and geographic architecture of a mine, we can begin to construct the collective experience of a disaster.

The eight tables that accompany Nagy's report reconstruct the collective experience of all 13 miners who were present during the early hours of the disaster. Tables 1–3 construct three separate chronologies of events at three temporally synchronous but geographically different locations underground and on the surface.[45] These events intersect at three critical moments in the

---

[44]D. Garry, 1987, MSHA video.

[45]Table 1 (Nagy, 1982) reconstructs the time of events at Crosscut 34—the presumed initial site of the fire. Table 2 reconstructs the time of events at the No. 4 agency as observers rescued Blake and attempted to fight the fire. Table 3 reconstructs the time of events on the surface. Table 4 lists the locations that miners reported as they described the initial location of the fire in the number 4 entry.

disaster: (a) when miners underground telephone colleagues and management on the surface; (b) when controllers on the surface restore power to run the pumps; and (c) when Blake emerges from the mouth of the mine and calls Turner at home.[46]

At each location, miners observe and remember different events and conditions. At Crosscut 34, for example, Tidwell smelled smoke; Blake reported that the shearer was not operating; Tidwell noticed flames inby the mandoor; Price looked at the belt drive; and Cox noticed that the overcast was collapsed. Miners who arrived at the scene in the early stages of the fire focused on the safety of colleagues and thus paid proportionately less attention to the location of the fire. Individuals on the surface were temporally and physically distant from events underground. These individuals described how they communicated with miners underground—answering phones and controlling power to stop the spread of the fire.[47]

Nagy's report thus represents the accumulated experience of all 13 miners who testified in the official investigation. Based upon this accumulated experience, Nagy "places" the initial location inby Crosscut 34 and outby Crosscut 37.[48]

Constructing Collective Agreement

In the end, the collective disagreement of the 13 miners who observed the fire becomes evidence of collective agreement. Nagy acknowledges that any agreement is quite remarkable, given the "excitement of the fire and the knowledge that more than 25 buddies were trapped."[49] Although Nagy himself has no way to substantiate his belief that the miners did or did not speak with one another, he concludes that disagreement "lends credence to their statements given several months after the event" since it suggests that miners did not collude with one another in their testimony.[50]

Nagy's report gradually erases all signs of disagreement, however, as he generalizes the "several statements" of the 13 observers into a single collec-

---

[46]Nagy, 1987, p. H-11.

[47]Not surprisingly, these miners disagree about the precise location of the initial stage of the fire. This disagreement is visible in the map that accompanies Nagy's report. This map marked 13 possible sites of origin based upon the 13 different accounts of the disaster.

[48]Nagy, 1987, p. H-13.

[49]Nagy (1987) writes: "Considering the excitement of the fire, the knowledge that more than 25 buddies were trapped, and that 1/2 to 1 hour had transpired between the start of the fire and their arrival, it is remarkable that the agreement among the several statements is as good as it is" (p. 15).

[50]Nagy, 1987, p. 15.

tive conclusion in which these men "state" the location of the fire. In the process, the individual voices are first summarized, then subsumed into a single technical conclusion that places the site of ignition and locates the cause of the fire:

> The statements of the thirteen persons who came early into No. 4 entry do not establish the location of the initial fire precisely.
>
> The several statements are summarized in Table 4.
>
> The 13 men . . . did not agree precisely on the exact location of the fire, even though general agreement was had.
>
> Tidwell, Price, and O'Neill . . . clearly state the fire was inby the 34 Crosscut and could be seen when looking in that direction through the mandoor in the overcast.
>
> These men, who were in the No. 4 entry shortly after the fire started, state the fire was between 37 and 34 Crosscuts.[51]

But the location of the fire does not identify the practices and decisions that caused the fire. Nor does it suggest how miners might prevent similar disasters in the future. To arrive at these larger conclusions, Nagy must compare his data with data from previous accident investigations.

Drawing on Previous Experience

Nagy stands at one critical moment of transformation within the Cycle of Technical Documentation in Large Regulatory Industries. The results of his report will help the agency determine liability in the Wilberg fire, but his conclusions also become part of the historical record that future investigators will use to draw conclusions about future accidents. If previous accounts are inadequate, Nagy recognizes, his own account will reflect those inadequacies.[52] In the ongoing cycle of risk management and assessment within large regulatory agencies, the rhetorical force of Nagy's conclusions will also influence future accounts of risk. Uncertainties in previous accounts of risk propagate through the system, influencing policy and procedures throughout the entire Cycle.

---

[51]Nagy, 1987, pp. 13–14.

[52]Researchers at the Bureau of Mines recognize that investigation reports do not provide the details that agencies need to understand the cause of the disaster. One study concluded that "More than 20 percent of the 2,685 reportable incidents in this study could not be analyzed because report narratives did not contain sufficient detail to identify a specific job task" (Klishis et al., 1992, p. 78).

In a section entitled "Related Facets of Mine Fires," Nagy explicitly uses the results of previous accident investigations to determine the cause of the Wilberg fire. He cites directly from three previous reports—all nearly to 20 years prior to Wilberg—to "illustrate the serious effect of high air velocity on fire development and control."[53]

Nagy acknowledges that these reports are inadequate for his purposes. He wants detailed observations and technical principles he can use to draw conclusions about the Wilberg fire, but "most reports do not give specific information on [such issues as] the effect of air velocity on flame spread." These reports take for granted that these effects are "common knowledge" and thus provide little help with the current case.[54]

But where does this inadequacy come from?

In Nagy's report, we see the tensions between the agency's need to determine the technical cause of the disaster and the larger problem of assessing the practices and policies that permitted technical problems to occur in the first place. At first, Nagy seems to take a narrow view of the accident, focusing on the technical cause of the fire and making logical guesses about the likely (and in some cases) necessary relations between facts and their likely consequences. But his conclusions—surprisingly—do not reflect the purely technical focus one might expect. Hired to locate the site of the ignition, Nagy could have concluded his reports with the paragraph that precedes his conclusions:

---

[53]Nagy, 1987, p. 34. The quotations he excerpts from each of these reports talk about the spread of the fire. But these reports make no reference to the source of the fire or its relation to the compressor at Wilberg. Nagy cites the following:

Report of coal mine fire, Dawson Mine, Harrison County, West Virginia, July 28, 1966: . . . . The fire spread rapidly toward the belt and track haulageway portals (underline sic, p. 35)

Report of coal mine fire, Pursglove No. 15 Mine, Monogalia County, West Virginia, October 22, 1965: While water from a 2-inch line was being sprayed on the fire at the porta-feeder and adjacent (outby) track haulage where the fire was spreading. (p. 35)

Report of coal mine fire, Joanne Mine, Marion County, West Virginia. May 29, 1865: Figures . . . show the fire spread downward and not upwind. (p. 36)

[54]There is another reason, of course, why previous reports do not provide the comparative technical information he needs: Fire acts generally in predictable ways, but the local circumstances of the fire at Wilberg were unique. Miners had accidentally used mineral oil in the compressor. As a result, the fire burned with "like the equivalent of a portable flame thrower," according to the speaker in the quotation at the head of this chapter.

The foregoing data and observations support the conclusion that the fire in the Wilberg mine did not move to any marked degree inby following its ignition. This further substantiates that the fire was ignited at the compressor location.[55]

But this result does not provide an adequate explanation of the practices and production pressures that produced the accident. In the end, Nagy seems to argue for a larger conception of the cause of an accident in "defective materials, poor installations, or practices that should not be tolerated in coal mines."[56]

Evaluating the Outcome

In the final section of his report, Nagy explains that he "feels it appropriate to quote from a Bureau of Mines' 1954 Miners Circular called 'Explosions and Fires in Bituminous Coal Mines,' " written some 35 years before the accident. He also cites an abstract of a Bureau of Mines paper he co-authored with Ward Stahl in 1967 entitled "Stop Playing with Fire."[57]

The first quotation implies that electrical fires are most often caused by "defective materials, poor installations, or practices that should not be tolerated in coal mines." Mines frequently make "makeshift repairs to prevent a temporary loss of production," but these practices have "resulted disasterously" [sic] in many instances. The passage concludes: "Prompt and satisfactory repairs to electrical machinery and accessories may entail temporary loss of production but, over a period of time, are justifiable from both the operator's and worker's standpoint."[58]

The second quotation presents a series of commonplaces familiar to all involved with mine fires and explosions. I have separated sentences in the following citation to demonstrate the commonsense virtue of his conclusions:

- Experience has shown the value of preventive measures.
- Stop ignition and the fire is eliminated.
- Many ignitions will be eliminated by preventive maintenance, use of adequate size and quality of cables, proper grounding, fuses and circuit breakers.
- Prevent an incipient fire from becoming a disaster by being prepared.

---

[55]Nagy, 1987, p. H-38.

[56]U.S. Dept. of Labor. Bureau of Mines. (1954). "Explosions and Fires in Bituminous Coal Mines (Circular No. 50). March, 1954. In Nagy, 1987, p. H-38.

[57]Nagy, 1987, p. H-38.

[58]Nagy, 1987, p. H-38.

- Equipment should be adequate, properly maintained, and strategically located.
- The mining crews should know their responsibilities and be well-trained.
- Fighting a fire is a team effort.
- The best fire-control system has little value if communications fail or if the equipment is crusty dirty and corroded from neglect.
- When a fire occurs underground, the immediate problem is the safety of the men working downwind. Under all circumstances, these men should be notified promptly and operations conducted for their safe removal.[59]

All of these commonplaces apply to mine fires in general, yet Nagy fails to identify the specific practices, materials, repairs, or preventive measures that might have caused the Wilberg fire. At first glance, one suspects that Nagy could have written his conclusions without reference to Wilberg at all. Nagy himself makes no attempt to connect these quotations to Wilberg either by framing the quotations with an introduction or by explaining the relevance of the quotation to the particular case.

In fact, however, poor practices, makeshift repairs, faulty installation, and defective materials had produced the Wilberg disaster. In the days before the fire, someone had deliberately bypassed the safety switch on the compressor, causing the compressor to overheat. Someone had accidentally used mineral oil in the compressor, causing the fire to burn with "the equivalent of a portable flame thrower" according to one safety official who commented on the disaster. The compressor was not housed in fireproof housing, and no one had recorded its location in the log books. As Nagy's quotations demonstrate, miners focused on production failed to perceive the signs of disaster. In locating the site of ignition, his report does not answer questions about why miners focused on production missed the signs of imminent danger.

One can only speculate about Nagy's purpose in adding such general conclusions to his report. If previous accident investigations were inadequate, as Nagy claims, was Nagy trying to find support in these documents for his suspicion that the real cause of the fire would be found in the decisions and makeshift practices that occurred in the days before the fire? Was he trying to push MSHA to investigate these factors, or was he trying to create a list of warnings, as if to say, "These things keep recurring? Why can't we stop them?"

Buried in appendix H in an overwhelmingly detailed accident investigation report, couched in such general terms that they seem to apply generally

---

[59]Nagy, 1987, p. H-39.

to all mine disasters, Nagy's conclusions weakly suggest the limitations of the agency's limited perspective even as he opens up the possibility of an enlarged viewpoint that would integrate the diverse perspectives of workers and management.[60]

THE POTENTIAL FOR MULTIPLE
VIEWPOINTS

Large accidents like the Wilberg fire produce large effects within the Cycle of Technical Documentation in Large Regulatory Industries. Any uncertainties in the narrative have a large effect on policy and procedure throughout the cycle. One author of the final report described the accident as an "awakening" within the agency because the agency had not previously given "due consideration" to the conditions that produced the accident.[61]

Because of the possibility that arson had caused the fire, the Wilberg report is much more explicit about documenting the processes of transformation than most similar investigation reports. Most reports provide only a chronological narrative that supports the agency's technical conclusions. As Klishis suggests in the quotation at the head of this chapter, these reports are frustratingly inadequate as sources of information about the history of an accident. More than 20% of the incidents Klishis studied could not be analyzed "because report narratives did not contain sufficient detail to identify a specific job task" involved in the accident.[62] Because of the controversy surrounding the origins of the fire, MSHA investigators were particularly careful to document the rhetorical processes and technical analyses they used to determine their conclusions.

This chapter describes how individuals capture experience in writing at one particular moment of transformation within the Cycle of Technical Doc-

---

[60]It is also impossible to determine whether the rhetorical uncertainty of Nagy's conclusions results from his own lack of skill in writing persuasive recommendations or whether the vagueness exists in the texts he draws upon to warrant his conclusions. In either case, the quotations he cites lack the rhetorical force to persuade readers that the problem warrants our attention.

[61]Anon. (Interview with agency personnel), 1992. Following the fire, MSHA issued a series of policy memos that identified specific problems that contributed to the disaster. These memos noted that aluminum stoppings had melted in the heat of the fire; that the two-entry mining system had trapped miners in the fire; and that another exit was blocked by a previous roof. They criticized training programs because miners had failed to don self-contained rescue systems in the heat of the fire.

[62]When Nagy (1987) attempts to use previous reports to confirm his conclusions, he too finds them frustratingly inadequate.

umentation in Large Regulatory Industries, but it does not prescribe how individuals should act to reduce and manage hazards. Nagy's reconstruction of experience shuts down the narrative complexity that might be possible if we could include all of 13 miners' perspectives in a single document. In the chapters that follow, we see the rhetorical complexity that is possible when individuals do not limit their focus of their documentation to a single "proper perspective."

5

Learning From Experience: Enlarging the Agency's Perspective in Training and Instruction

Within large regulatory industries, training reverses the process by which experience is transformed into a single collective voice. But training must also convey the lessons learned through experience. These two purposes may be at odds when miners attempt to apply generalized rule-based procedures in ethically and geographically complex situations. Miners need rules derived from experience, but they also need training that will help them understand how the everyday boredom of their job and the stresses they face in a crisis will affect their ability to judge hazards.

For new miners, the problem is particularly acute. Because they are inexperienced, novices may focus intently on the job at hand and fail to perceive how simple miscalculations in cutting coal can weaken structural supports and precipitate a roof fall. Or they may misjudge the magnitude of risk because they lack experience to perceive the consequences of poor practice. Paradoxically, novices may use their own lack of experience to justify poor practice. Mallet et al. (1992) explain: "The more times miners spend under unsupported roof, the more likely they will be to doubt that such activity is dangerous, and the greater is their chance of being harmed."[1] But experi-

---

[1]Mallett et al., 1992, p. 33. Souder (1988) suggests that three factors can reduce mine fatalities: "First . . . fatalities can be substantially reduced by improving the general quality of mine management and first-line supervision. Second, miner training should be more concentrated on improving miner pattern recognition skills and perceptual information processing abilities. The ability of miners to perceive ambiguous cues of pending accidents, e.g., poor roof conditions, is a key to accident prevention. Third, training mine supervisors to emphasize and reinforce alertness skills among their subordinates" (p. 9).

enced miners may also violate the most basic rules of safe practice if they are exhausted, overwhelmed by rapidly changing conditions, or too focused on production to pay attention to safety.

This chapter begins by looking at the problem of anomalous behavior in well-trained miners. Anomalous behaviors occur when experienced and/or well-trained individuals commit seeming irrational, life-threatening acts. These behaviors raise questions about the nature of work and about the kinds of documents (including visual and verbal representations) that can help individuals recognize the consequences of these seemingly irrational acts. Why do experienced workers violate simple rules of safe practice? How can agencies help miners learn from experience without endangering their lives in the process? What can these behaviors teach rhetoricians about documentation practices in hazardous environments?

To answer these questions, this chapter describes how agencies and individuals transform experience to learn from the past and prevent accidents in the future. The first section examines how the agency transforms miners' experience in 12 training videos. These transformations are not entirely successful because miners resist the efforts of MSHA interviewers to reduce their actions to a single rule-based formula—what the agency called the "proper perspective" on the event. In the second section, a British miner critiques these MSHA training videos. His comments reveal the number of alternative viewpoints that are silenced in the agency's narrow reconstruction of events. The third section demonstrates how even the most limited documents can prompt miners to recall a wide range of specific experiences. The final section reveals the number of perspectives that one miner can manage when he attempts to understand a complex technology. This miner speaks as an active rhetorical agent who can assume a wide range of institutional viewpoints beyond the limits of his own experience.

As this chapter argues, anomalous behaviors can be viewed as problems of focus and attention—the consequence of the uncertainty and complexity of work in hazardous environments. In each case, workers focused on a single task placed themselves in danger because they failed to turn their attention to other aspects of their work.[2]

This discussion challenges traditional notions of workplace practice as a linear sequence of hierarchical procedures focused on a single task or procedure. It raises questions about the rhetorical skills we need to manage multiple viewpoints—achieving a balance between too many conflicting view-

---

[2]As Klishis et al. (1992) suggest in the quotation at the beginning of chap. 4, accident reports provide a limited account of the specific tasks that miners were engaged in at the time of the accident.

points and the dangerously narrow focus that reflects the agency's notion of a single "proper perspective."

As we saw in the previous chapter, the agency's focus on a single viewpoint can shut down the kinds of flexible thinking and planning that enable miners to work safely in an uncertain and dynamic environment. But this chapter also raises the possibility that even the most limited documents (called FATALGRAMs) can function as inventional tools that produce new knowledge if workers are encouraged to speculate about the cause or causes of an accident.

The examples in this chapter are not exhaustive, but the training methods described in this chapter are typical of the training that miners receive under the standards of the Mine Act. These examples are illustrative, not definitive. The following discussion is not intended to represent a full-scale rhetorical analysis of training. But it suggests that individuals can overcome the limitations of the agency's narrow perspective if they have the rhetorical understanding and training to manage multiple perspectives. (If this were not so, writers within agencies could not represent events and conditions beyond the walls of their office.)

This analysis has important implications for rhetorical theory and suggests that rhetoricians must reexamine how individuals and agencies might manage these diverse perspectives when they work to construct an understanding of hazards in their work. Chapters 5–9 explore this idea in detail.

THE PROBLEM OF ANOMALOUS BEHAVIOR: RETHINKING INSTRUCTION AS HIERARCHICAL PROCEDURE

Hazardous environments place enormous stress on workers engaged in difficult and potentially dangerous tasks. In addition to the dangers of operating machinery, the environment itself can kill, maim, or injure individuals working safely according to established rules of practice. Individuals can also be distracted by the environment so that they inadvertently place themselves at risk. To follow instructions, individuals must keep their attention focused on the task; to work safely, they must also turn their attention to signs of danger all around them.

In these environments, human error frequently results from exhaustion and workload.[3] When employees are overloaded beyond their capacity, they make errors in judgment. But the converse is also true. When individuals are

_____

[3]Souder, 1988.

understimulated, they commit errors as a result of boredom and inattention. These anomalous errors are difficult to manage precisely because they involve individual judgment and behavior.[4]

Individuals vary in their capacity for handling overload and in their flexibility in adjusting to change. We know that some experienced drivers adjust to fatigue and bad road conditions by increasing their concentration and focus. But fatigue can dull drivers' response time and diminish their capacity for judgment.[5] Souder (1988) describes a cycle in which increased stimuli create fatigue, decreasing an individual's capacity for judgment and increasing risk. Seemingly trivial failures can push the cycle toward a fatal accident, as stress increases and individuals tire under pressure.

Souder (1988) cites examples of "behavioral accidents" in which normally safety-conscious employees acted "irrationally."[6] He notes that many individuals who engage in anomalous behaviors are frequently experienced and well-trained personnel. These examples seem irrational because the behaviors seem so deliberately to violate the assumptions that govern ordinary instruction and safe practice:

1. Three experienced divers and lifesaving instructors ignored normal safety procedures to go on a night dive, incompletely equipped, in an unexplored underwater cave. All three drowned.

2. An experienced 45-year old supervisor assisting a work crew suddenly turned and walked into the path of the crew's bulldozer and was crushed before anyone could stop him.

3. An experienced mine employee knowingly walked under a bad roof, pointing out various roof flaws to his companion. He was crushed by a sudden fall of roof.[7]

Bureau of Mines researchers have also interviewed miners to determine why miners engage in unsafe practices like walking under unsupported roof. The results of this research suggest that these behaviors are not irrational in the sense that miners were making risk judgments without reason. This research suggests two different explanations for miners' apparently anomalous behavior.

---

[4]Activity theorists have examined how workers restructure their environments to simplify work tasks that have a regular pattern or set of procedures (Hutchins, 1996). The present analysis extends these studies by looking at how individuals attempt to regularize discursive and embodied activity in unpredictable contexts.

[5]Souder, 1988, p. 8.

[6]Souder, 1988, p. 2.

[7]Souder, 1988, p. 2.

Peters and Randolph (1991) suggest, first, that miners acted rationally, but made incorrect judgments about their ability to assess roof conditions by visual inspection. They argue that individuals who engaged in these anomalous behaviors understood what they were doing but did not believe that they were intentionally violating safe practice.[8] Their data suggest that most miners who walked under unsupported roof first looked at the roof to assess conditions. Miners who observed colleagues reported that they believed that most of these miners were aware that they were under unsupported roof. This research also suggests, however, that "a significant proportion of people who work at the face *unintentionally* go under unsupported roof."[9]

My own interviews with miners confirm that many miners violate safe practice unintentionally. One miner described how her colleague was injured when he went only a short distance under unsupported roof. (She indicates this short distance with her gestures.) This miner claimed that her colleague was so involved with other aspects of production that he failed to notice his position inside the last row of supported roof:

> I don't think he intentionally went out from under bolts. They were turning the cross cut, it was the first cut out of it, and when they had mined the face, they got it off center to the right, and when they pulled back to the left it meant that the row of bolts going off that rib were almost in a half circle. They were going in a straight line up that rib and when he's turned that, he went to set over to get the right side, he cut the left side in, which was the curvy side, which cut the other side. He just stepped out there to throw the cable. He was probably that far from the last roof bolt, that's it.[10]

Of course, not every incident of anomalous behavior leads to an accident. Accidents result from a combination of human error and external conditions. Drivers who swerve to avoid ice may skid into an oncoming car, but only if the car is present. Not every miner who walks under unsupported roof unintentionally is injured; not every roof collapses. But miners must learn to pay

---

[8]According to Peters & Randolph (1991), several miners had gone beyond unsupported roof because there were no warning devices posted. Others noted that the area had been rockdusted, leading them to believe that the area was also bolted. Peters & Randolph (1991) conclude: "Several of those who said that they were walking somewhere when they discovered that they had gone under unsupported roof noted that no warning devices had been posted at the last row of bolts. In a few cases, miners noted that the area they were walking in had been rock dusted, leading them to believe that it was bolted" (p. 28). See also Peters & Wiehagen, 1988.

[9]Peters & Randolph, 1991, p. 28. (Italics added)

[10]Anon. (Interview with miner, U.S.), Osage, WV, February, 1999.

attention to conditions that require heightened alertness before problems turn into crises.

When trainers represent these behaviors as anomalous or irrational, they may fail to teach miners how habitual behaviors, reinforced through repetition in the workplace, often take the place of reasoned judgments about risk. We expect miners to become better—more experienced—as they perform more and more iterations of an action. But we may not perceive how these iterations become embodied habits that miners can perform mechanically without thinking.

In my own interviews with miners, expert roof bolters described two radically different versions of the amount of attention they paid to their work. One miner claimed, "You don't think about it at all." Another agreed: "She's not thinking about anythin' [sic], [she's] pretty much doin' it closed mind when you're drillin' that hole." It's "something you do subconsciously." "You're listening to that sound, you're not thinking about it. You just do it." But a third miner, a vocal advocate for safety, challenged this perception: Mining required her full attention, even if the process was tedious. This miner did not want to be under a roof with a colleague who was not thinking about her work.[11]

To manage risk in a crisis, individuals must maintain their focus on normal hazards, to prevent even greater losses, even as they adjust their focus to handle rapidly changing conditions. Experience is a poor teacher if it does not demonstrate the consequences of inattention and lack of focus. As chap. 6 demonstrates, individuals may also violate simple rules of safe practice in order to obtain sensory cues from the environment. These miners may injure their hands when they place their hands on drill bits in order to sense the workings of the drill in the rock.

Fortunately, individuals can draw upon the collective experience of others without risking their own lives in the process, if training helps them retransform the embodied experience lost in previous transformations within the Cycle.

CONSTRUCTING THE PROPER PERSPECTIVE: MSHA RECONSTRUCTS MINERS' EXPERIENCE

MSHA recognizes that experience is a poor teacher because it creates a false sense of risk magnitude, especially for new miners, who may not immediately perceive hazards of going inby the last row of supported (and thus safer) roof.

---

[11]Anon. (Interview with miner, U.S.), Osage, WV, February, 1999.

If miners assess risk based solely upon their own limited experiences, they might actually take more risks. Mallet et al. (1992) conclude:

> It is very likely that if they walked under unsupported roof a few times, nothing would fall on them. This would, unfortunately, tend to make them even less afraid of going inby supports.[12]

To counteract the paradoxical effects of experience, MSHA developed 12 training videos that they hoped would instill "fear of unsupported roof" in all new miners before they had a chance to experience tragedy.[13] These videos were intended to help novices experience the full range of viewpoints they were likely to occupy as observers, colleagues, and victims in an accident. The agency deliberately varied the focus in each interview, but the format is similar.[14] In each video, an agency interviewer prompts miners to describe their own experience in their own words. Then the interviewer reframes events from a single viewpoint, what the agency calls the "proper perspective."[15] Prompted by the interviewer, miners then confess how they endangered themselves and others because they failed to follow a simple rule, even though they risked their own lives to rescue a buddy.[16]

Dave Morone's video documents how Morone recovered the body of a friend killed because he went inby roof supports.[17] In this video, Morone describes the effects of witnessing his friend's death. Dave Garry's video tells how Garry attempted to rescue a colleague injured in a roof fall.[18] Garry observed events from a position outby the accident. As the smoke and dust cleared, he could see the mining machine but he couldn't see "anything" so he started calling for the operator of the machine, "calling his name, and we called several times." When the operator answered, "it sounded like he was far away . . . his voice was real small."[19]

Larry Strayer describes how he lost his leg while working with a friend.[20] Strayer's narrative re-presents experience as he experienced it from a position inby the disaster. In this excerpt, Strayer remembers how it felt to lie on the ground as the roof began to collapse over his head:

---

[12]Mallet et al., 1992, p. 33.

[13]Mallet et al. (1992) note: "It is especially important to instill fear of unsupported roof in all new miners before they even have an opportunity to go under it" (p. 33).

[14]U.S. Dept. of the Interior, Bureau of Mines, 1987, p. 4.

[15]U.S. Dept. of the Interior, Bureau of Mines, 1987, p. 4.

[16]Morone, 1987 (MSHA video).

[17]Morone, 1987 (MSHA video).

[18]Garry, 1987 (MSHA video).

[19]Garry, 1987 (MSHA video).

[20]Strayer, 1987.

Strayer: Yea, was, we wasn't in real far, we just like maybe two or three feet like I said I really didn't even get time I just turned around come down. But I can remember laying there and looking up at the bolts hoping that the bolts above my head was gonna hold the rest of the roof up there.

Interviewer: So you almost got back under the bolts before it fell.

Strayer: Yea, like I said, you know, all the guys thinks they're fast in a mine and they can get out of the road in time but for as fast as that roof come like I said, I just turned around and I didn't even get chance to jump or nothing, the whole place was down.[21]

The rhetorical situation in which these miners speak is very different from the private and anonymous interviews I conducted with U.S. and British miners. In the MSHA videos, miners testify in public about their own experiences in order to help other miners prevent accidents in the future. The MSHA training videos are not overtly scripted, but they have a deliberate rhetorical purpose—a "fear message."[22] MSHA's *Instructor's Guide* describes their form and content:

> The videotapes all follow the same format. First the miner tells about the event and then an interviewer asks some specific questions so that important points will be covered. While all three of the tapes illustrate the dangers of unsupported top, each presents the information with a different focus.[23]

The video thus sets up a dichotomy between a miner's natural reaction and the proper action which must be "drilled into miners" through instruction and training.[24] In one video, Garry admits that his actions were unreasonable—that is, without thought. "In an emergency situation like that," he confesses, "a lot of times you just do."[25] In the second half, the interviewer pushes Garry to think how he might have behaved differently "if you had thought to do that." Prompted by the interviewer to reframe his actions, Garry contrasts his "natural reaction to go and try to save your buddy" with the agency's presumably correct and reasoned response.[26]

Strayer describes how "inattention" caused his accident. Strayer confesses that he and his colleague Barry were so "involved in what we were doing"

---

[21]Strayer, 1987 (MSHA video).

[22]Mallet et al., 1992, p. 33.

[23]U.S. Dept. of the Interior, Bureau of Mines, 1987, p. 4.

[24]U.S. Dept. of the Interior, Bureau of Mines, 1987, p. 1.

[25]Garry, 1987 (MSHA video).

[26]The rhetorical situation in which miners confess their failures in a public forum contrasts the private and anonymous interviews I conducted with U.S. and British miners. Although miners' narratives are not scripted, they cannot be counted as naturalistic testimony.

that they failed to see that they had moved inby the bolts and thus underneath unsupported roof:[27]

> **Strayer:** I guess whenever me and Barry was pulling that roof down we got pretty involved in what we was doing there. If we would have just look like we was inby the bolts there a few feet. If would uh you know just stopped and thought about what we was doing there . . . Before—you know—we went ahead and did it, we'd probably be all right today.[28]

When the usual verbal warnings, safety talks, and even disciplinary actions appeared to carry little force in preventing miners from violating safe practice, the creators of the MSHA videos decided to incorporate a "fear message" to persuade audiences to "modify their attitudes and intentions" as well as their practices.[29] Mallett et al. (1992) explain:

> Those who make use of fear messages hope that employees will perceive the recommended behavior as leading to a reduction of the threat and that they will begin following the recommended actions. . . . In the case of roof fall accidents, this would not be hard to do. The consequences of being hit by a roof fall are often harsh. They are the most frequent cause of accidental death in coal mines. If the person survives, it often takes them an extended period to recover, and they may be permanently disabled.[30]

If documents instill too much fear, however, miners may not be able to respond rationally in a crisis. Although MSHA writers wanted to reduce their message to a single rule, they recognized that the situations that miners describe are technically and ethically uncertain. Miners must fear going under unsupported roof, but they must also manage fear in a crisis to make effective risk judgments. Mines are not simple problems that allow miners to maintain a single focus of attention. As we have argued repeatedly, documents that

---

[27]Strayer's case is a bit more complicated. In this video, Strayer describes how he and a colleague assessed difficult roof conditions before deciding not to try to pry down loose pieces of rock in the mine roof. Although we might expect that the miners' decision not to bar down loose rock might be the cause of the accident, the video does not judge the quality of this decision or its effect on the outcome. Instead, the interviewer focuses on Strayer's failure to stay under supported roof.

[28]Strayer, 1987 (MSHA video).

[29]U.S. Dept. of Labor, Bureau of Mines, 1987, p. 1. Mallet et al. (1992) note that the authors wanted to go beyond the "usual techniques for discouraging miners from going inby supports" that "appear to be limited to those verbal warnings about the danger of going under unsupported roof made during safety talks, and in some cases, threats concerning disciplinary actions that might be taken if an employee engages in such behavior" (p. 33).

[30]Mallett et al., 1992, p. 32.

construct events from a single viewpoint may not adequately represent the complex chain of events and conditions that interact to produce disaster.

MSHA focused on a simple rule in these videos because it provides miners with a clear guide to action. If following a single rule counts as rational behavior, behaviors that violate these rules are anomalous or irrational. Unfortunately, this simple answer does not identify the practices and decisions that caused miners to violate safe practice. Nor does the agency's limited viewpoint suggest how miners might prevent similar accidents in the future, since the videos do not describe the technical and institutional causes that precipitated the accident. To arrive at these larger conclusions, the *Instructor's Guide* advises, trainers must help miners "relate the discussion to their own setting."[31] The guide encourages trainers to talk about the conditions that prompt normally safety-conscious individuals to step out of character and knowingly commit unsafe acts: Did miners commit errors of judgment because of "carelessness, inattention, thoughtlessness, poor habits, machoism [*sic*], bandwagon effects, and thrill-seeking"? Did the institutional setting influence miners' judgment? Can equipment or work procedures be changed to prevent that situation in the future?[32]

Even as the agency interviewers work to limit the number of perspectives represented in the video, its own instructional material recognizes that the agency's limited perspective must be enlarged and retransformed in the context of training to help miners learn from experience. The guide thus leads us back to a world where there are multiple and often diverse answers to these questions.

THE BRITISH MINER'S PERSPECTIVE: FREE TO SPECULATE, MINERS CAN ARTICULATE GAPS IN THE AGENCY'S CONSTRUCTION OF EXPERIENCE

If documents like the MSHA training videos attempt to limit the discussion to a single focus, documents can also open up discussion when individuals are free to acknowledge the gaps in the agency's narrow construction of events.

To evaluate MSHA's training materials, I videotaped British coal miners as they were viewing the MSHA training videos. The miners themselves commented freely as they watched the video. These sessions were not struc-

---

[31]U.S. Dept. of the Interior, Bureau of Mines, p. 4.
[32]U.S. Dept. of the Interior, Bureau of Mines, p. 4.

tured as formal interviews. I asked questions only to help them elaborate their answers.

One British miner viewing the MSHA videos complained that the videos focused on the victim's actions and not management's responsibility both before and after the accident. His analysis was highly critical of management's failures in the accident and equally critical of MSHA's attempt to blame the victim. In the Garry video, he noted that Garry's actions had little to do with the outcomes of the accident since he arrived on the scene after the roof fall.[33] This miner wondered about the locus of responsibility and about management's role in helping miners stay safe: Why, for example, did MSHA focus on Garry rather than the foreman? Why did the agency focus on the rescue rather than the roof fall? Why did it focus on individual decision making rather than institutional policy and procedures? Who is responsible for setting posts and insuring that miners follow safe procedures? Why were there no barriers to prevent miners from walking inby safe support?

Like MSHA's interviewers, the British miner viewed events from a rhetorical position outside and above the experiences and observations of the miners he observed. Because he is not American, he can comment upon the events at a safe distance from the institutions and persons he observes. This miner was blackballed in England as a result of his own union activities during the 1984 strike. He is currently unemployed, though he is active in the labor movement and travels widely.

The British miner's comments reflect his ideological and economic position as a strong advocate for labor. But his comments provide a more nuanced view of these videos than we might expect. This miner raises questions about human factors (the role of management and the kinds of planning decisions that preceded the accident); equipment failures (the unexpected and awkward technical problems caused by the angle of the boom); and general conditions in the mine.

As this miner recognizes, falls of roof are a general condition of mines; but mine operators have a variety of options they can apply to create safer condi-

---

[33]Garry (1987) describes how he approached the scene with his foreman (Italics added):

Garry: *'Bout the time I ran into him, we were coming up*—when it cave in, it rolled way back, it came back past, through the intersection and the dust was just so thick you couldn't see anything . . . So the foreman asked, he said, you know, is everybody out. And . . . uh . . . we counted heads real quick and we were missing one. And we pretty much figured it had to be the operator— . . . At that point then I hoped that he'd stayed in the cab, you know, and didn't get out and try to run it . . . so . . . foreman sent one of the buggy operators up to the phone to call dispatcher and let him know that we'd had a fallen miner and . . . uh . . . *as the smoke was clearing the foreman and myself we started walking up in towards the cut* . . . to see what we could see how bad the conditions were . . .

tions for miners. His questions show how he can assume a range of viewpoints as he describes the technical and institutional problems that miners like Garry and Strayer must manage in their work. I have quoted a lengthy segment to illustrate the number of viewpoints (in brackets) he is able to assume in his assessment of the problem:

> **Miner:**   What I'd like to know is . . .Who draws up the method of support? Who's there to tell him that he mustn't work in front of unsupported ground?
>   [viewpoint of supervisor or safety manager]
> What's to prevent him physically from doing it?
>   [viewpoint of safety engineer]
> Falls of road in unsupported roadways is actually a major cause of accidents in Britain too—whatever method of support you use. But it's a matter of fact that whether you're using roof bolting or traditional arch supports, you must break down the loose ground in front of before you can set them . . . and what we need to do is to make sure that they—that you—find a method of support in advance of setting the permanent support—through temporary means of timber and try to do everything from this side of the support . . . um . . .
>   [viewpoint of trainer]
> But what I'd like to know from this fellow is . . . Did he sue the company? Where's the company guy saying, "In hindsight, we should have—we should have protected our worker? That we should have taken better care of you. . . ."
>   [viewpoint of the labor advocate]
> **Interviewer:**   How could they protect them?
> **Miner:**   By drawing up a method of support that could prevent him from being in that situation. . . .
> You could have temporary supports that you run right to the face of the heading of the development. You can have extra long supports that if you're talking about arch girders, for example, which temporarily pin the front while you set the support, the permanent support. There's similar methods you could use with rock bolting. From the permanent support use a temporary support. And make sure that all—all—unsupported ground becomes supported. And do any work that has to be done at that side. You do it from this side either by—
>   [viewpoint of the rock-bolting engineer]
> They was using a pinch bar, we call a rock bar, it depends on the length of it, how long was it, was it long enough to get from this side to that side?
>   [viewpoint of the foreman or trainer]
> Part of the reason they couldn't get it down with the miner—I suppose he's talking about the Joy continuous miner, was because the angle of the boom was too acute to reach it. And that's OK, that depends upon how well you did the job on the prior cut.[34]
>   [viewpoint of foreman or safety manager]

---

[34]British miner, Personal interview with the author, August, 1994.

FATALGRAMS IN U.S. TRAINING:
ENCOURAGED TO SPECULATE,
MINERS RETRANSFORM THE AGENCY'S
LIMITED PERSPECTIVE

Social constructivists have argued that the way in which documents are introduced into the interaction is more significant than their material character (e.g., whether the document was a chart rather than a speech utterance).[35] This argument reinforces the commonsense notion that the teacher is more important than the text, particularly in the context of training where a teacher's experience can augment even the most limited document. But this argument may overlook important differences in the material character of documentation and its function within particular rhetorical situations.

Trainers draw on a variety of documents and use these documents in different ways. Both the nature of the documentation and the trainer's questions determine the character of miners' responses.

To test this hypothesis, I observed an 8-hour MSHA refresher training session. In this session, trainers varied the format in each one-hour module.[36] In the first module, the trainer asked miners questions based upon a set of printed hand-outs that described fires and explosions. Despite the trainer's stated desire to "present the material in a way that will encourage[s] some participation by you," miners remained silent during the discussion or offered single words like a chorus as the trainer fished for answers.[37]

> **Trainer:**   what is fire? [SILENCE] We see it every day. It's something you take for granted. [pause] Rapid oxidation through a chemical process. [pause] Combustion. Process. It burns. What's a Class A type? [SILENCE] Class A would be . . . What type of material? [SILENCE] Wood? Paper? Class B.
>
> **Miners:**   Flammable
>
> **Trainers:**   Such as . . . Biggest . . . Diesel fuel? Class C?
>
> **Miners:**   Electrical.
>
> **Trainer:**   Does electricity itself burn?
>
> **Miner:**   No.[38]

---

[35]Woolgar, 1992.

[36]Trainer. (Canterbury Coal Refresher Training, Interview with trainer), April, 1997. I am grateful to Mark Radomsky, of the Pennsylvania State University, for his help in this project.

[37]Trainer, (Canterbury Coal Refresher Training, Transcript of training session), April, 1997.

[38]Anon. (Canterbury Coal refresher training, Transcripts of training session), April, 1997.

In another module, the trainer presented accident data based on recent events at the mine. Although the data was theoretically anonymous, miners quickly identified their colleagues; they knew the details of the case and attributed cause of the accident to particular flaws in the miner's character, behavior, or attitude. Thus, they laughed at "Bob" [not his real name] who walked into a ladder. When the instructor pointed out that the ladder was in the wrong place, they laughed again. Bob always walked into ladders. Miners had heard the details; they were not interested in hearing this information again.[39] When they focused on particulars, they did not generalize to a more general notion of safety that they could apply to their own work.

The most successful module from the perspective of both trainers and miners involved what may be the oddest set of documents, MSHA's so-called FATALGRAMs.

When miners die, MSHA sends FATALGRAMs, Fatal Illustrations, and (full color) Fatal Slides to all operators and interested third parties. The FATALGRAMs provide a brief narrative of the accident and its causes as well as recommendations for preventing similar disasters in the future. Each FATALGRAM contains a cartoon-like black-and-white drawing of the victim or victims at the moment of the disaster based upon a brief Abstract of Investigation constructed in the early phases of the accident investigation. Every six months, MSHA republishes these FATALGRAMs as Abstracts with Fatal Illustrations (Fatal Abstracts).[40] Both the FATALGRAMs and Fatal Abstracts construct what agency personnel describe as a snapshot or picture of the accident at the moment of disaster.

There are two types of FATALGRAMs. Technical diagrams depict mining equipment, trucks, backhoes, and haulage equipment as the center of focus. The more cartoon-like FATALGRAMs depict an oddly comic distortion of the laws of time and space, suspending figures in mid-air during a roof fall

---

[39]Anon. (Canterbury Coal Refresher Training, Interview with coal miner), April 1997.

[40]The abstracts provide a biennial update of all chargeable mine fatalities categorized according to subject area (underground coal mining; surface coal mining; underground metal mining; surface metal mining). Fatalities in each category are grouped by accident type. A statistical analysis categorizes accidents according to the victim's age, accident classification, victim's occupation, and the size of the mine where the fatality occurred. The abstracts are intended primarily as educational tools. A disclaimer warns the reader that they are not necessarily accurate representations of the accident:

The abstracts included in this program are intended to be used for instructional purposes. Efforts are made to base this data on MSHA's final accident investigation reports. However, preliminary investigation information may have been used for these abstracts to meet the deadline for this publication. (U.S. Dept of Labor, MSHA, 1991)

(Fig. 5.1) or depicting victims on their knees as an 80 ton block of roof falls on their heads (Fig. 5.2).

At first, I thought that the extreme distortion and cartoon-like quality of the FATALGRAMs could not possibly provide effective training. After interviewing miners about their interpretations of FATALGRAMs and observing FATALGRAMs in the context of safety training and education, however, I have come to a rather surprising conclusion. Despite their apparent deficiencies from a theoretical perspective, FATALGRAMs work in the context of training to encourage miners to articulate the complex interaction of events and conditions in a workplace accident.[41] Despite the apparent oddity of the representation and even the grim humor of the name FATALGRAM itself (suggesting fatal attraction), miners report that these cartoon-like reminders of recent fatalities do, in fact, improve safety awareness in the mines. As one miner commented, "They make you think." Another commented: "When you get back to the mine and see that belt, you think about the accident and how he died."[42]

It is easy to critique any document that presumes to present a snapshot of a disaster. A real snapshot of the moment of the disaster, of course, is technically unfeasible, particularly in the dim light of an underground mine. Post-accident photographs can depict either the horror of the accident (dead bodies, widows, and orphans) or the bleakly dehumanized slides of an empty and devastated mine, but not the moment of disaster. The artist of the FATALGRAM must thus imagine victims and equipment at the moment of disaster and reconstruct details of the accident from brief verbal abstracts and preliminary reports of the accident (personal interviews). Fig. 5.3 shows how the real observer (photographer) and object (miner) died together in a mine accident.

Despite their limitations, FATALGRAMs stimulate miners to explore a range of perspectives if they are encouraged to speculate about the contributing factors in the accident. Even when miners critiqued FATALGRAMs, their responses demonstrate the curious ability of FATALGRAMs to stimulate discussion of safety issues, to evoke prior knowledges, and thus, to provide a more complete picture of safety than the snapshot metaphor suggests.

When a British mechanical engineer who had worked in American, British, and South African mines viewed the FATALGRAMs, for example, he

---

[41]Design theorists like Bertin (1981) have argued that designs cannot violate physical laws: People do not hang suspended in the air following a roof fall, with a rock striking their back. FATALGRAMs also violate the principles of effective visual-verbal design set forth by theorists like Horton (1991) and Tufte (1983; 1991).

[42]Anon. (Canterbury Coal refresher training, Interview with miner), April, 1997.

ABSTRACT OF INVESTIGATION

Date: May 28, 1991   (Died: May 29, 1991)
Slide Number: 8   (Fatal Case Number: 26)
Accident Classification: Roof Fall
Type of Mine: Coal - Underground
Location: West Virginia

Age of Victim: 52
Job Title: Roof Bolter Helper
Total Mining Experience: 20 years
Experience at this Classification: 1 day
Experience at this Activity: 1 day
Number Employed at Mine: 7
Time of Accident: 5:30 p.m.
Mining Height: 50 inches

A roof bolting crew was installing roof bolts. A roof bolt hole off the left rib was being drilled when the roof bolter helper, for some unknown reason, moved down the left rib and behind the bolter operator into an area of unsupported roof. A piece of roof 6 feet long, 5 feet wide, and 6 to 8 inches thick fell. The roof bolter helper was transported to the hospital where he died the next day.

Means of Prevention
Stay under supported roof and follow the roof support plan.

FIG. 5.1. FATALGRAMs depict an oddly comic distortion of the laws of time and space. Source: U.S. Department of Labor, Mine Safety and Health Administration, Metal and Nonmetal Mine Safety and Health (1991b).

ABSTRACT OF INVESTIGATION
Date: February 13, 1991
Slide Number: 2   (Fatal Case Numbers: 8-11)
Accident Classification: Roof Fall
Type of Mine: Coal - Underground
Location: Virginia

	Case #8	Case #9	Case #10	Case #11
Age of Victim:	37	30	26	50
Job Title:	Section Foreman	Roof Bolter	Roof Bolter	Section Foreman
Total Mining Experience:	19 years	11 1/2 years	2 1/2 years	23 years
Experience at this Classification:	Unknown	Unknown	Unknown	Unknown
Experience at this Activity:	Unknown	Unknown	Unknown	Unknown

Number Employed at Mine: 19
Time of Accident: 4:00 p.m.
Mining Height: Unknown

A massive roof fall occurred resulting in the deaths of two roof bolters and two section foremen. The fall measured approximately 115 feet long, 28 to 35 feet wide, and 3 to 15 feet thick.

Means of Prevention
1. Always comply with the roof control plan.
2. Provide supplemental roof support such as timbering and cribbing where excessive widths exist.
3. Provide proper alignment and directional control.
4. Conduct adequate preshift examinations.
5. Post danger signs where excessive widths exist.

FIG. 5.2.   MSHA's FATALGRAM foregrounds two miners (on their knees) and presents two others as shadow-like objects in the background as a massive block of rock falls on their head. Source: U.S. Department of Labor, Mine Safety and Health Administration, Metal and Nonmetal Mine Safety and Health (1991c).

ABSTRACT OF INVESTIGATION
Date: October 12, 1990
Slide Number: 41   (Fatal Case Number: 48)
Accident Classification: Explosives and Breaking Agents
Type of Mine: Silica Flux - Surface
Location: New Mexico

Age of Victim: 31
Total Mining Experience: Unknown (Contractor)
Total Experience at this Mine: Unknown
Total Experience at this Job: Unknown
Number Employed at Mine: 4
Number Working at Time of Accident: 4
Time of Accident: 5:00 - 7:00 p.m

A part-time helper had drilled and loaded a series of blast holes. His mother and sister were there to observe the blasting. After preparing for the blast, his mother and the other workers left the area. The sister wanted to take pictures of the blast, so she remained. They backed off 150 feet from the blast site to take pictures. When the blast was detonated, the helper was hit on the head by a large piece of flyrock and was killed instantly. His sister suffered massive chest injuries and a punctured lung.

Means of Prevention
All persons shall be cleared and removed from the blasting area unless suitable blasting shelters are provided to protect persons endangered by concussion or flyrock from blasting.

FIG. 5.3. In this FATALGRAM, both observer (photographer) and object (miner) died together in a mine accident. Source: U.S. Department of Labor, Mine Safety and Health Administration, Metal and Nonmetal Mine Safety and Health (1990).

first made some obscene remarks about the potential value of the paper they were written on.[43] "They tell you nothing about what's going on," responded another former National Union of Mineworkers (NUM) official and head of numerous safety committees.[44] Yet both miners proceeded to speculate about what might have happened. They criticized and expanded upon MSHA's recommendations for improving mine safety, and they pointed out details in the diagrams that raised questions about mining practices and safety violations that could have led to the disaster. Each miner told a different story based upon different experience. No two miners told the same narrative. Their narratives revealed an acute knowledge of labor relations, politics, economics, mechanics, and the everyday dangers of a hazardous environment.

In the MSHA refresher training course, miners, trainers, and the trainer's supervisor agreed that the level of interest increased substantially when the instructor used FATALGRAMs to discuss the contributing factors in 10 selected accidents. According to the miners, the slides were easy to visualize and think about. More important, they could remember details when they returned to the mines.

Part of the success may reflect the ways that trainers encouraged miners to recall prior experience, use prior knowledge, reason through the sequence of events in the disaster, and look beyond the particular accident to more general behaviors, attitudes, and decisions that contribute to accidents. In the following excerpt, the trainer repeatedly encouraged miners to "speculate full blast" about the cause or causes of the disaster:[45]

> **Trainer:** I'm not sitting here saying that I have the answers because I have like three or four sentences for each fatality. I don't know. Let's *speculate*. Its' O.K. *Let's speculate full blast*. I'm not sitting here holding the trump card to see if you can find it, but you know I have enough here to give you background data on each fatality, and then it's up to us to *speculate* and take that information and invest it in such a way that when you go back to work—what can you have gained from this? [Italics added]

The trainer also admitted that he had no additional information about the accident. As a result, he too was forced to speculate about the contributing causes of the accident:

> **Trainer:** I'm *speculating* that the reason for the steel mat is support, but I'm *speculating*. I've seen that. Surface coal mining areas. A lot of people complain.

[43]British miner, (Personal interview with the author), August, 1994.
[44]British miner, (Personal interview with the author), August, 1994.
[45]Trainer. (Canterbury Coal refresher training, Transcripts of training sessions), April, 1997.

I have no idea. Steel mats and electricity don't go together. It says . . . It says . . . How do you think he got out. He probably did jump clear and from here on out, the whole story will play out without the mat. (Italics added)[46]

In reasoning through the sequence of events in the disaster, miners drew upon their own practical knowledge of mechanics, physics, and psychology as they theorized about why the miner died when he attempted to jump from the cab, what strategies they could employ in other situations, and how the miner could have prevented the disaster (by turning uphill, for example, or using other parts of the truck to prevent a mechanical failure from turning into a human disaster). They drew upon their own experience as truck mechanics, fire fighters, and military maintenance. They frequently referred to the laws of physics or mechanics to reason about the cause of the miner's death and the probable consequences or outcomes of other possible actions. Trucks do not roll uphill, they reasoned. If you jump from the truck, gravity will pull you downhill into the path of the truck. Only in a "far-fetched fairy tale world," one miner argued, would a miner escape the laws of physics.[47] Such knowledge was particularly surprising in contrast to their lack of enthusiasm and responsiveness when discussing scientific theory in other modules.

Miners discussed maintenance procedures for trucks, the relationship between the drive shaft and brake failure, the nature of the slope of the road and other roads in the vicinity, driver training, seat belts, the time of the accident (related to lunch breaks and tiredness), pressure on miners to work faster, criminal tampering with the brakes, and the construction of the berm (edge) of the opencast pit. They talked about possible training methods for contractors and miners and how this incident might indicate other problems with maintenance, training, construction, and safety at the mine. Miners also showed a surprising depth of knowledge of mathematics when it was related to calculations derived from practical experience. Former truckers spent about 10 minutes talking about the derivation of the slope described and depicted in the drawing.

The most active speculation, however, was most directly applicable to miners' experience in a coal processing plant. They could visualize the belt in their own plant, and the answers clearly derived from their own experience. Here, miners described problems in the plant when workers attempted to shut down the belt for maintenance, the location of the guards, lock-out tagout procedures (isolating or shutting down the electrical supply), the location of platforms in the mine, and problems that occurred because of the lack of staff. They described how 10 years ago, you'd see lots of men in the plant, and

[46]Trainer. (Canterbury Coal refresher training, Transcripts of training sessions), April, 1997
[47]Anon. (Canterbury Coal refresher training, Transcripts of training sessions), April, 1997

you could always ask for help. Now, they said, you sometimes have to do it yourself. They argued that back injuries often result when miners work alone or try to invent equipment. They joked about using a shovel as a tool, about losing shovels rather than fingers, and about why a long shovel might have been better than a short-handed shovel.

Ultimately, each discussion produced lots of backtracking, speculation, and memories of previous training courses, supporting the notion that this was truly a refresher safety training course:

Trainer: It goes back to another question.

. . .

Miner: It reminds me of the Three Finger Joe, shake hands with danger [a common miner joke and also a training film miners had watched previously].[48]

The discussion of FATALGRAMs changed the dynamics of interaction between trainers and miners. FATALGRAMs drew comments from nearly all miners in the room, in sharp contrast to the lecture, which was clearly controlled by the instructor's questions, and in contrast to other modules in which a single miner (an older and more experienced miner) acted as the unofficial spokesperson for the group.

Ultimately, FATALGRAMs demonstrate how documents can function epistemically as inventional tools that encourage the production of new knowledge even when the documents provide a limited perspective of events. These documents evoke miners to recall knowledge *outside* of the document, helping them retransform the agency's sterile conclusions into living body of experience and expertise that can help other miners think about safety when they return to work.

Ironically, FATALGRAMs are not—in any literal sense—illustrations, since they bear no resemblance to any real mining situation. They do not therefore make a claim to correspondence or pictorial credibility. Instead, FATALGRAMs show how even the most cartoon-like documents can function as rhetorical *topoi* when miners are encouraged to draw upon their memory of mathematics, physics, and mine safety in order to understand an accident.[49]

---

[48]Anon. (Canterbury Coal refresher training, Transcripts of training sessions), April, 1997

[49]As classical rhetoricians recognized, there are two kinds of memory, nature and artificial. Natural memory is "embedded in our minds" ([Cicero], p. 207). Artificial memory could be embedded in vivid scenes and images that could prompt the natural memory to recall a sequence of ideas or images. Medieval theorists constructed elaborate memory systems to prompt the artificial memory. The more striking the image, these theorists speculated, the more likely that it would adhere in the memory. Vivid images evoke the memory of the thing (memoria ad res) that is present in the reader's imagination; the structure and arrangement of images helps audiences to see new or forgotten aspects of the remembered situation.

This collective body of expertise and experience remains invisible, however, unless we have the means to retransform this collective knowledge in writing.

THE WELSH MINER'S PERSPECTIVE: FACED WITH COMPLEXITY, INDIVIDUALS CAN MANAGE A SURPRISING NUMBER OF DIVERSE PERSPECTIVES

In positing the difference between expert and ordinary knowledge, Paradis (1990) defines a model of normal communication that defines complexity as technical complexity (mechanical complexity) which is understandable and distinct from common sense. In this model, experts must "take charge" of the task of rationalizing knowledge to prevent disaster. Paradis writes:

> The rhetorical process of preparing the operator's manual obliges the expert to imagine the consequences of operation and to lay these out for the user. It is a crucial step in the socialization of expertise. We can only maintain that individuals are responsible for their actions if we enable the rational individual to take charge of the growing presence of technology. We can only insist that operators retain legal responsibility for their actions if we provide them the means to understand the human consequences of their behavior.[50]

Paradis (1991) posits a hierarchy of "higher-level experts"[51] or "expert advisors"[52] who "set the conditions of instrument assembly, operation, and maintenance" and serialize them in repeatable unit human actions.[53] In this model, the lay user is "increasingly . . . faced with a situation in which he or she no longer understands the potential consequences of human action."[54]

More recently sociologists of science and risk specialists have argued for a more reflexive approach to risk communication which reflects the viewpoints of diverse stakeholders, both lay and scientific. As Nelkin (1987) argues in the quotation at the head of chapter 3, the uncertainty of risk analysis "allows for a range of perceptions" about the nature of hazards in the workplace. Ultimately, Nelkin argues, "Each view has implications for responsibility and control."[55]

---

[50]Paradis, 1991, p. 275.
[51]Paradis, 1991, p. 275.
[52]Paradis, 1991, p. 258.
[53]Paradis, 1991, p. 275.
[54]Paradis, 1991, p. 275.
[55]Nelkin, 1984, p. xv.

In the following discussion, we look at how one Welsh miner described one particularly controversial technology: the newly introduced practice of roof bolting. This miner is an expert roof bolter, selected to be educated in the United States when R.J.B. Mining introduced the practice in British mines. Because he is a strong labor advocate, we would expect him to make a strong argument for or against the new practice. What is surprising is the degree to which this individual is able to manage many different perspectives in his response.

As I have described elsewhere, the problem of roof control is particularly complex and highly controversial in British mines, where British miners have only recently adopted the so-called American system of roof bolts as a primary means of support. Not surprisingly, the practice evokes a wide range of perspectives, each of which has implications for the health and safety of miners.

When British Coal introduced the American system of roof support in British mines, miners accustomed to more traditional methods of roof support were forced to adjust their viewpoint both geographically and institutionally. The old system of arch supports had protected miners in the event of a fall; the new system of roof bolting forced them to turn their attention to signs in the roof of the mine that might indicate a potential failure. Institutionally, the technical concerns that miners expressed about the new method of roof support overlapped with the social and political history of mining in England. (In one miner's view, all events eventually go back to the 1984 strike that split the NUM into two opposing camps and spawned a new union, the Union of Democratic Miners.) When British miners talked about roof-bolting practices, they also reconstructed a history of management-labor relationships in Great Britain.

British Coal introduced roof bolts in British mines to save costs and reduce the number of miners involved in shaft development and construction.[56] In theory, roof bolting offered the same level of support as traditional metal arches, but British miners distrusted the method—and management's motives—in part because they could not see the workings of the roof bolts inside the roof strata. Management used a variety of methods to persuade British miners that roof bolts were safer. They sent miners to the United States, paying their travel expenses, to learn methods of support from experienced U.S. roof bolters. They also introduced bolts slowly, often in combination with other methods of support, until miners were persuaded to trust roof bolting as the sole means of support in a mine.[57] Then a fatal accident at the Bilsthorpe

---

[56]John T. Boyd Company. Pittsburgh, PA, 1993. See also Sauer, 1996.

[57]Anon., Personal interviews with the author, Health and Safety Executive, Bootle, England. July–August, 1992.

colliery seemed to confirm many miners' fears that this new system of roof support would reduce the level of safety in British mines.[58]

The Welsh miner is able to accommodate all of these perspectives in a single narrative. He describes "how the union feels" at both the national and local level and how miners at his mine decided to support management's decision to install roof bolts. In the following passage, the miner explores the full range of social and institutional perspectives that define his own ambivalence toward this potentially risky technology. Italics in the following excerpt indicate shifts in perspective:

> **H:** The thing is, how the *union miners* got involved with the rock bolting obviously it got introduced and then really, we knew it was coming. *Nationally, the union said*, "We don't believe in rock bolting as a primary support." But it was happening, so *at local level*, we decided the best way for to be safe is get straight into it and *go along with the management*, not go along with, but be part of and insist that you're involved in every stage of what's going on, so that's what we did. *Back at [British mine name]*, we insisted at every stage, or every development, that we had lads taken to America and board paid for that, and (pause) *[names]*, he sent people to America, so they can have a look at the different method. (pause) And insisted that that was vital, that *you were fully aware* all the way through.[59] (Italics added)

The miner's narrative also takes into account the technical complexity of the problem. The miner recognizes the tension between risk and control that accompanies any form of roof support. Like many of his colleagues, this Welsh miner distrusts roof bolts "unless they are properly monitored." The term "monitored" has a literal meaning in the context of roof support. Mines use tell-tale monitors to test the stability of the mine roof. But the term also means the kind of care that miners must exert in monitoring the monitors—that is, in checking the monitors at regular intervals to see whether the roof is sagging. In another sense, miners must also monitor the number of roof bolts and pay attention to determine whether the roof bolter is following safe practice.

For the Welsh miner, roof bolting is safe when closely monitored but disconcerting when it fails. But the Welsh miner extends the meaning of monitoring to include all phases of labor-management communication involved in keeping miners informed about roof bolting. The miner thus contrasts monitoring with the hushed conversations between two people (whispering) that causes problems.

---

[58]Langdon, 1994.
[59]Welsh miner, Personal interview with the author, Italics added, January, 1992.

In the following excerpt, the miner expresses two different perspectives simultaneously. He's happy as long as the practice is monitored, but uncomfortable when there's too much whispering:

> I'm *happy with roof boltin' as long as it's closely monitored.* And there's a close allegiance with the union on the monitoring side, and the facts is (pause) well made accessible to the men. It's when it's start being whispering or not enough communication. *I think that's when things start, problems arise.*[60] (Italics added).

As this example suggests, local knowledge is heterogeneous and complex. Individuals can integrate many different perspectives into a single narrative. Individuals talk about their work in many venues—inside of and outside of the locations they describe—in pubs, in training sessions, in academic conferences, at union meetings, in ordinary conversations, at parties, and more formally, in public hearings. When these individuals talk about their work, they store and reconstruct this collective experience in their narratives.[61] They can move outside of their own geographic and institutional viewpoint rhetorically to see events from a new perspective.

In an ideal framework, individuals acquire a repertoire of responses in training sessions that provide them with the rhetorical flexibility to see events from many different viewpoints. But not all trainers encourage this rhetorical flexibility or use actual narratives to demonstrate how particular perspectives might affect safety underground.

REVISITING A FEMINIST PERSPECTIVE: CAN RHETORIC ACCOMMODATE MULTIPLE PERSPECTIVES?

When regulatory agencies talk about involving stakeholders in the regulatory process, they seek to increase the number of viewpoints represented in the regulatory decision-making process. Although no agency actually limits the number of viewpoints that a single stakeholder might hold, there appears to be an unstated assumption that neither agencies nor stakeholders will contribute more than one viewpoint at any given point in time.

When agencies increase the number of viewpoints represented in regulatory decision making, they also increase the potential for disagreement and

---

[60]Welsh miner, Personal interview with the author, Italics added, January, 1992

[61]These complex viewpoints may disappear in other contexts. Many women miners, for example, complain that their most intimate male buddies in the mines resume traditionally gendered social roles—and social distance—above ground.

regulatory indecision. We have seen the disagreement that occurs within agencies about the interpretation and meaning of previous experience. We have also seen how the problem of precision can turn the regulatory process into a regulatory nightmare and how the process of review and re-review can paralyze agency decision making. In chap. 4, we saw how Nagy summarized and extracted information from many sources to construct a single coherent narrative. But we have not examined how individuals might also represent more than one viewpoint simultaneously as they construct and negotiate an understanding of their work. Nor have we considered why we might encourage individuals to conceive more than one viewpoint

In mining communities, individuals frequently have multiple affiliations with labor unions, local miners' groups, political parties, and work teams. Experienced miners have heard colleagues and management express competing viewpoints—in union halls, pubs, and training sessions. When miners talk about hazards in their work, they frequently represent more than one viewpoint. As this chapter demonstrates, however, agencies may inadvertently silence these diverse viewpoints when they reconstruct events in an accident.

Rhetorical analysis, informed by feminist theory that attempts to examine the silences in written texts, can make visible the viewpoints lost at critical moments of transformation. Such an analysis, we argue, can help us resolve the dichotomy between incommensurable post-modern relativism (all viewpoints are equivalent) and the limitations of a perspective that privileges a single distanced perspective (the so-called androcentric viewpoint of the Archimedian observer).[62]

Feminist theorists like Harding (1991), Haraway (1991), and Fox Keller (1985) have argued that theories of knowledge production must take into account the situated viewpoints of individuals within economic and social communities. Their research raises epistemological questions about how we value information when it comes from sources outside traditional disciplinary and intellectual frameworks. For rhetoricians, feminist theory helps us see (a) how culturally constructed notions of masculinity have influenced how we think about risk and safety in the workplace and (b) how the discourse practices of science reflect institutional and cultural assumptions that may inadvertently or deliberately silence the voices of men and women who labor in risky occupations.

Feminist theory thus raises questions about the practices and costs of documentation and report writing that exclude or silence the embodied experience and local knowledge of workers—male and female. As I have argued in

---

[62]Harding, 1991.

"Sense and Sensibility in Technical Documentation and Accident Reports: How Feminist Interpretation Strategies Can Save Lives in the Nation's Mines" (Sauer, 1993), feminist theory demands that rhetoricians acknowledge the silent and salient power structures that govern discourse, not because we are interested in theoretical constructs about language, but because those power structures affect the fabric of technology on which we all depend.[63] The problem, of course, is to find ways to make those silenced voices audible and visible in the documents that influence policy and procedure in the workplace.

The problem is not how to increase complexity, however, but how to manage these viewpoints within documents that persuade individuals to act safely in the workplace. While the complexity of this accumulated body of these multiple viewpoints may improve the adequacy of documentation, the sheer volume of information creates rhetorical problems when individuals must determine a course of action. If documents represent too many potential viewpoints simultaneously, readers can easily become overwhelmed and overcautious. In a crisis, the number of competing demands, and the unpredictability of outcomes, increases exponentially. In hazardous environments, the day-to-day problem of safety places great demands upon the workers. To work safely underground, individuals must maintain their focus on normal hazards to prevent new losses even as they adjust their focus to handle rapidly changing and frequently unpredictable hazards.

The present chapter demonstrates how individuals are able to manage multiple perspectives in speech alone. Chapters 7–9 show how individuals employ both speech and gesture to represent two and sometimes even three separate viewpoints simultaneously. This research has important implications for workplace discourse as well as ethnographic studies based upon speech (or its transcription) alone.

---

[63]Sauer, 1993.

6

Warrants for Judgment: The Textual Representation of Embodied Sensory Experience

No consistent sociology could ever present knowledge as a fantasy unconnected with our experiences of the material world around us.[1]

—Bloor, (1976)

As we have seen in the previous chapter, miners frequently take risks that seem inexplicable to engineers and management.[2] While miners rarely admit that they themselves circumvent or violate safe practice in the mines, they describe how colleagues retrieve tools from under an unsupported roof (a very dangerous practice) or drill bolts through their hands when they hold a bolting device with their bare hands. Miners have also been injured when they rested their arms on the bolting arm of the roof-bolting machine—a massive machine that drills and positions roof supports (roof bolts) as work advances.[3]

Agencies like MSHA have attempted to understand why experienced miners continue to expose themselves to unnecessary hazard.[4] As we have seen in the previous chapter, MSHA has developed training programs to teach workers to fear the consequences of unsafe practices and they have de-

---

[1]Bloor, 1976, p. 29.

[2]U.S. Department of Labor, Mine Safety and Health Administration, National Mine Health and Safety Academy, 1989. See also Peters & Randolph, 1991.

[3]Personal interviews; Peters and Randolph, 1991.

[4]U.S. Department of Labor, Bureau of Mines, (n.d.), p. 4.

veloped videos to help miners learn from the experience of others. But they have not investigated the role of embodied experience in local sites.[5]

In risky environments, events and conditions change rapidly. To manage risk, management and workers must observe, evaluate, and interpret rapidly changing sensory information in the environment. Safety engineers can prepare a general plan, but local conditions ultimately define when and how individuals manage risk in local sites. The problem of risk thus challenges conventional rhetorical notions of instructions and procedures as a generalizable set of practices and procedures that can be formulated prior to an understanding of material conditions in local environments.

This chapter argues that individuals and agencies can potentially deploy three different types of warrants grounded in embodied sensory experience:

1. **Pit sense (embodied sensory knowledge):** Direct physical sensations felt or perceived in highly specific local environments. Miners feel or sense this knowledge as physical signs or sensations in their bodies.
2. **Engineering experience:** Physical signs or indices embodied in objects and materials. Engineers observe and record this information as the material history of particular sites.
3. **Scientific (invisible) knowledge:** Physical forces, particles, materials, and interactions which are sensed or perceived as data in language, physical tracings, written reports, and inscriptions. Scientists read and interpret data in order to formulate knowledge that is literally invisible to the physical senses.

In using familiar terms to categorize these warrants, it is important to distinguish knowledge representations from the individual decision makers who negotiate among competing warrants to assess and manage risk. Knowledge of risk cuts across the distinctions between expertise in science and lay (or local) understandings. Decision makers need scientific knowledge because it helps them understand how invisible forces interact within the strata to pro-

---

[5]Sociological and rhetorical analyses have investigated the disparity between expert and public understanding of science, the problems of communicating expert knowledge to lay audiences, and the difficulty of "extracting" and articulating knowledge presumed to be tacit (Collins, 1990). Historical studies have demonstrated how miscommunication produces disastrous consequences, but these studies have not investigated how the features of expert and/or lay communication affect risk decision-making in local sites. In creating a dichotomy between expert (scientific) and lay (local) knowledges, many researchers have not investigated how expert representations incorporate sensory information in the environment. See, for example, Perrow, 1984; Petroski, 1985; Trento, 1987.

Evidence is gathered from several sources during the investigation.

FIG. 6.1.   This mine safety pamphlet shows one mining inspector as he assumes three different roles in one inspection. Source: U.S. Department of Labor, Mine Safety and Health Administration, National Mine Academy (1990).

duce hazard; they need engineering experience to anticipate the site-specific problems that result from previous engineering practices in specific local conditions; and they must pay attention to changes in the sensory environment to detect the most immediate signs of danger. In Fig. 6.1, one mining inspector assumes three different roles in a single inspection: (a) he uses a flame safety lamp to check levels of methane; (b) he receives data by telephone to check engineering systems; and (c) he draws on scientific principles to predict outcomes and analyze data. (The manual advises him that "in some investigations, a particular physical or chemical law, principle, or property may explain a sequence of events.")[6]

Experts (in the broad sense that I have used the term) frequently deploy more than one type of warrant to make judgments as conditions warrant. But these warrants are not interchangeable. Scientific knowledge and engineering experience are closely related to the extent that they operate most visibly in the domain of the verbal and analytic (e.g., knowledge perceived and interpreted through instrumentation, written reports, and inscriptions). But this apparent similarity may mask differences in the nature of these warrants and the implications of each type of warrant for future policy and procedure.[7]

The present analysis focuses on representations of roof support because roof control represents one of the most dangerous and uncertain aspects of safety in a coal mine. To help readers understand the uncertainties inherent in roof control, I have provided a brief overview of three methods of roof support in U.S. and British mines. This overview will set the stage for a more theoretical discussion of the nature of warrants grounded in experience, the effect of warrants on risk decisions and risk outcomes, the rhetorical incompleteness of written instructions and procedures, and finally, the textual dynamics of disaster. In the conclusion, I look at the implications of this analysis for writers and rhetorical theorists.

As this chapter argues, individuals can potentially deploy all three types of warrants strategically when they assess risk in local sites. But institutional and disciplinary conventions, local habits, and textual practices constrain the ways that individuals and agencies negotiate competing warrants in their writing. As a result, written representations alone provide an inadequate basis for risk judgments unless decision makers understand the critical role of embodied experience as guide to decision making in local sites.[8]

In acknowledging the role of embodied experience as a warrant for judgments about risk, this chapter does not argue for a return to some romantic past when workers learned safety at their fathers' knees. Nor does it argue against funding for research to improve engineering controls and scientific

---

[6]U.S. Department of Labor, Mine Safety and Health Administration, 1990, p. 11.

[7]In using familiar names to categorize knowledge representations, it is important to distinguish the types of knowledge from the individual decision makers who construct knowledge representations. Because written accounts privilege analytic representations of risk (e.g. engineering experience and scientific knowledge), we may not perceive the degree to which engineers and scientists draw upon embodied experience to assess and manage risk. Chapters 7 and 8 show how miners integrate scientific knowledge, engineering experience, and embodied experience simultaneously and sequentially in speech and gesture.

[8]Chapters 7–9 describe the process of rhetorical transformation in detail.

knowledge to improve conditions in risky environments. In examining the interdependence of scientific knowledge, engineering experience, and embodied sensory experience, this chapter demonstrates how scientific and technical rhetorics must reflect what Kaufer has called "a knower's spatial and sensory understanding of the material environment."[9]

THE PROBLEM OF ROOF SUPPORT
IN U.S. AND BRITISH MINES

The problem of roof support in U.S. and British coal mines illustrates the very real material presence of risk and danger in the context of a hazardous environment.[10] Unfortunately, no method of roof support can provide 100% safety. All roof support systems eventually fail. To predict failure, miners must learn to recognize signs of imminent danger in whatever method of roof support they employ—just as they must sense other signs of danger in their environment.

Because roof conditions differ from mine to mine, individuals must make highly local judgments about risk. No single plan of roof support will work for all conditions in a mine. When miners perceive a high level of risk, they must increase the level of roof support—adding truss bolts, straps, or additional methods of roof support. When miners encounter extremely unstable conditions, the process of installing methods of roof support can actually increase the risk of failure if miners encounter loose rock that breaks or falls as they install new support.[11] Despite improvements in technology, the process is extremely dangerous. One U.S. document explains:

> Over the past 70 years, more than 35,000 miners have lost their lives in underground coal mines from roof and rib falls. In recent years, some excellent meth-

---

[9]Kaufer, personal communication with the author, March, 1996.

[10]I am grateful to Chris Mark, U.S. Bureau of Mines, for his contributions to my understanding of roof support technologies and to George Schnakenberg for his assistance with robotics and automated mine safety systems. In acknowledging their specific help with this project, I am aware of the generous assistance of personnel at the U.S. Bureau of Mines, the Mine Safety and Health Administration, Her Majesty's Safety Executive, the Coal Employment Project, and countless miners in the United States and Great Britain. Their anonymous contributions have helped me form a deeper understanding of the technical and human aspects of mine safety.

[11]Kettlebottoms (sandstone channels in the rock strata) provide a good example of roof conditions whose risk becomes apparent only in hindsight. The process of roof bolting can destabilize the channel, causing a roof fall. If the roof is highly unstable, miners may decide to mine quickly past the area of unstable rock. But this decision may have disastrous consequences if—during the mining process—miners encounter a roof fall. See Sauer, 1992.

ods and programs have been developed to prevent roofs from falling. However, it is still a very serious problem and the majority of fatalities in underground mines are still caused by roof and rib falls.[12]

Not surprisingly, roof-bolter helpers, who set temporary supports in advance of permanent methods of roof support, suffer the highest levels of risk underground—nearly 40% of all mining accidents and deaths underground (figures vary from year to year). Overall, roof falls account for nearly half of the fatalities in U.S. underground mines. The Pittsburgh Bureau of Mines notes that "during the 5-year period 1985–89, 92 coal miners were killed by falls of roof and rib, 4,178 miners were injured, and 11,031 ground falls were reported in which no one was injured."[13]

While the numbers may seem small compared to the number of deaths in the Gulf War or the yearly death toll on the nation's highways, analysis of the cost of mine fatalities demonstrates the significance of these figures. The Bureau of Mines' accident cost indicator model (ACIM) estimates the total cost of all lost-time injuries and fatalities during the 1985–89 period to be $556 million.[14] Interviews with personnel at the Bureau of Mines in Pittsburgh suggest, moreover, that mine injuries account for more lost days per person on average than in any other occupation in the country.

Mines could theoretically improve roof support, using the same methods of support as a tunnel that might be expected to remain stable for hundreds of years. But mines are constantly changing environments, and it would be economically unfeasible to invest resources in permanent support structures. Mines frequently add additional supports if there is any evidence of instability. But roof support is as much an art as a science, particularly when mines encounter unexpected faults (large cracks of discontinuities) in the rock strata.

All rock contains faults. Faults can range from inches to hundreds of feet vertically and from inches to miles horizontally. These faults are particularly dangerous because they are, by definition, present in the rock before mining starts, but mines may not know a fault exists until they have committed to a mining plan. At other times, nearby mines know about faults but do not communicate information to other worksites.[15] Mining practices can also disturb these faults, causing a massive roof fall. If engineers know the layout of faults, they can plan a roof support system in advance that will stabilize the strata

---

[12]U.S. Department of Labor, Mine Safety and Health Administration, 1986, p. 1.
[13]Peters & Randolph, 1991, p. 2.
[14]Peters & Randolph, 1991, p. 2.
[15]Molinda, 1992, p. 61.

and prevent roof falls. But roof support engineers must be prepared to make on-the-spot decisions when miners discover faults as they cut rock.

To prevent roof falls and to reduce the number of deaths and accidents in the mines, mines employ a combination of one or more of the following three types of roof support systems:

Timbers are wooden towers constructed like log cabins that support the immediate roof. Timbers cannot prevent a roof fall, nor can they protect a miner completely. When timbers begin to fail under the pressure of a roof fall, however, they yield or crack, giving miners warning that enables them to escape. While timbers cannot protect a miner completely, experienced miners describe how even a 13-inch space can keep a man alive after a roof fall. Timbers are bulky, however, and block entrance ways. As wood supplies decrease, particularly near western mines, researchers at the Bureau of Mines in Pittsburgh are testing other materials that can "give" like wooden cribs as the roof begins to "work" (or move, often with loud accompanying noises) before it falls.

Full metal arch support, as the name implies, uses metal arches at prescribed intervals to support the roof. As in a tunnel or underpass, arches provide permanent support even when the roof falls above them. Unfortunately, arches are labor-intensive and place more workers at risk in the construction phase of mining. The more workers underground, the greater the potential for disaster. Arch support also creates its own risks: Many miners are injured hauling metal arches. Other miners are injured when arches obstruct roadways. Finally, creating an arch support system is a slow and expensive process. As mines in Britain look for cheaper methods of support, operators have increased their use of so-called "American" roof bolting methods.

Roof bolts (British: rock bolts) are long metal rods that hold together the layers of the immediate roof. In theory, they create a stronger roof by laminating layers of the immediate roof like plywood. Most mines combine roof bolts with resin glues that seal cracks in the rock and, it is argued, glue or seal roof bolts to uneven strata. In many cases, mines also use straps or mesh between rock bolts to support a fragile or uncertain roof. U.S. mines have used roof bolts since the 1970s. British mines introduced roof bolting (around 1994) in response to the John T. Boyd report, which encouraged British mines to find ways to cut costs.[16] British labor unions have generally opposed rock bolts because they allow management to reduce the labor force in mines and because they are statistically less safe than traditional British methods of roof control.

Unlike arch supports and timbers, roof bolts act internally within the physical environment of the mine roof in order to control and prevent roof

---

[16]John T. Boyd Company, Department of Trade and Industry, 1993.

falls. To be successful, roof bolts must bolt the invisible strata of the immediate roof (the layers of rock above the working sections of the mine) to the solid overburden (the mountain above the immediate roof). Thus, the workings of a roof bolt are literally invisible and inaccessible to miners except through sensors (sensometers) that record data about strata movement and changes in the immediate roof that might indicate an immediate fall.

Many U.S. and British miners distrust roof bolts because they cannot see whether the bolts have been installed properly. Some miners point to examples of "chandeliers"—rough outcroppings of roof bolts with little roof material in between—as evidence that roof bolts provide little effective support in the event of a large roof fall. They trust timbers as a means of roof support because timbers produce sensory information that can save their lives. Although timbers cannot support a roof in the event of a fall, their yielding and cracking can warn miners to escape. Miners depend upon this constant flow of sensory information to save their lives.

When roof support fails, it is easy to attribute this failure to (a) tacit knowledge unarticulated in reductive rule-based procedures or (b) incomplete knowledge. Thus, one British accident report blames management for applying roof bolts in an inappropriate situation and assumes that a deeper understanding of strata movement will enable engineers to improve roof support in the mines. In calling for a deeper understanding, however, the report does not make explicit the kinds of understandings that can help experts reduce risk in the mines.

There are, of course, any number of potential answers to the question: (a) greater precision of testing; (b) additional categories of measurement; (c) better rules and heuristics for interpreting sensory data; (d) better machines and instruments to gather data; and (e) more finely grained measurements of roof conditions in the mines.[17] If we assume, for the moment, that researchers can potentially develop more accurate and precise measurements and heuristics for interpreting sensory data in the mines, we are still confronted with ethi-

---

[17]"The Engineer as Rational Man" (Sauer, 1992) examines these problems in the context of imminent danger. As my analysis suggested, the dynamic material environment of the mine challenges mathematical calculation and mathematical analysis. Roof bolting adds another dimension to the problem. As MSHA statistics suggest, roof bolters themselves are at greatest risk in the mines as they advance the face. Just as it is impossible to know the proper procedure for operating a machine prior to its operation, so it is impossible to test and observe conditions within the immediate roof prior to any cut. Scientists would need roof bolters to precede them in the mines in order to prevent disasters during testing. As the discussion of robotic miners suggests, moreover, intelligent machines and human analysts would share the same risk at the greatest moment of danger.

cal, social, technical, and, ultimately, rhetorical questions about the feasibility
of uncovering such knowledge and the problems of relying on written texts to
convey the information that decision makers need to assess and manage risk.

THE NATURE OF WARRANTS
GROUNDED IN EXPERIENCE

In risky environments, warrants for risk decisions are literally grounded in
material sites where knowledge is "mined" or "extracted" from local experience and transformed into writing.[18]

Embodied Sensory Experience (Pit Sense)

Embodied sensory experience provides the most direct warrant for judgments
about risk. In the darkness of a mine, miners employ all of their senses to hear,
see, feel, and smell hazards around them. Miners describe the ability to sense
these physical phenomena as pit sense—an embodied sensory knowledge derived from site-specific practice in a particular working environment.[19] Pit
sense is physical and sensory knowledge in the most literal of senses. It exists
in the ability of the human body to feel changes in pressure and to hear differences in sounds. Thus, pops indicate the pressure of methane; bumps indicate
yielding pillars. When timbers fail, miners hear cracks that warn them of a
working roof.[20]

For outside observers, pit sense may seem to belong to the realm of folklore
or old wives' tales that combine a grain of scientific truth with highly selec-
tive storytelling.[21] But engineers and scientists also share an embodied understanding of risk. At a recent visit to one of the most advanced British mines,

---

[18]See chap. 1.

[19]British miners call mines "pits."

[20]The information in this analysis is based upon 31 depositions from miners following an explosion at the Southmountain Coal Company, Incorporated, Number Three Mine, Wise
County, Virginia, December 7, 1993. Depositions were conducted by the MSHA and the Virginia Department of Mines, Minerals, and Energy, January 12–21, 1993, at the Norton District
Office, Norton, Virginia. I am grateful MSHA for their assistance with this project.

[21]Women in mining communities pay attention to the amount of dust on men's clothing,
the number of washloads, and the amount of dirt carried in when miners return home. Because
they do not formulate their observations in the discourse of science (in parts per million, for example), their observations may be discounted or omitted from official investigations (Sauer,
1993).

one highly educated rock-bolting engineer tested a core sample of rock strata by licking the sample to determine the presence of grit.[22]

Even non-human sensors in this domain develop sensory knowledge or pit sense. Canaries, for example, learn to feel changes in mine ventilation systems. Inexperienced canaries fall to the floor the first time they are exposed to high levels of carbon monoxide. As they gain experience in the mines, however, they learn to sense changes in the environment in order to move to the floor to avoid falling when they faint in the presence of carbon monoxide. Miners also speak reverently of the flame safety lamp, which lives and dies in the same range of oxygen as the miner. Outside the range of oxygen necessary to support human life, the flame changes color and eventually goes out. Without oxygen to support life, humans gasp for breath, suffocate, and die. In the presence of too much carbon monoxide, they develop headaches. These physical responses of the body to changing material conditions produce sensory knowledge that enables miners to take action to save their lives.[23]

Embodied experience plays an important role in local risk decisions. As interviews with miners reveal, miners violate written instructions and procedures when those procedures do not provide the sensory information they need to assess when and how to implement rule-based practices in local situations. Miners hold roof bolts with their hands so that they can feel the bolt as it penetrates different layers of strata. Miners place their hands on the arms of the roof-bolting machine to sense the vibration and resistance of the machine. To the extent that miners' habits provide important sensory information, management and engineering approaches that distance miners from sensory knowledge (e.g., designing a machine that requires miners to place two hands on the controls) will meet active resistance from miners.[24] In South African mines, miners actively resist—and, in fact, vandalize—robotic equipment because it takes them away from embodied sensory knowledge.[25]

As the history of both large- and small-scale technological disasters suggests, however, embodied experience comes at a tremendous cost in lost productivity and human tragedy.[26] Ideally, miners learn from near misses. The gruesome hero Three-Fingered Joe ("shake hands with safety") reminds workers that a coal mine's record of experience is literally embodied in blue scars (from the presence of coal dust in the cut), soft tissue injuries, missing fingers

[22]Testimony in the Southmountain disaster demonstrates that investigators value miners' representations of embodied experience as an index of conditions prior to the accident (chapter 8).

[23]Personal interviews, U.K., 1994.

[24]Peters & Randolph, 1991, p. 27.

[25]DeKock & Oberholzer (July, 1997). Personal interviews, Council for Scientific and Industrial Research, Auckland Park, Johannesburg, SA.

[26]Cf. Nelkin 1984; Petroski 1985; Sauer 1993; Sauer 1992; Trento 1987.

and joints, and missing and broken limbs that signify a miner's embodied experience in the mines.

Engineering Experience

Engineering experience provides an historical narrative of local conditions and the outcomes of engineering decisions and practices. Unlike scientific knowledge, engineering experience produces an account that emphasizes the visible structural units rather than the less visible geologic or physical forces and interactions beyond human sensory perception. Engineering experience describes the institutional, structural, and geographic history of a particular site. The site-specific and historical character of engineering experience distinguishes engineering experience from the more particular embodied experience of individuals and from generalized scientific knowledge grounded in experimental analysis and observation. Unlike laboratory testing, engineering experience emphasizes visible structural units rather than the geologic divisions within strata that are invisible to human observation. The focus is on those discernible features of rock strata that create problems. Molinda and Mark (1997) write:

> In the past, geotechnical evaluations of coal mine roof rock have been based largely on laboratory testing of rock properties, such as uniaxial compressive strength and direct sheer. However, laboratory testing fails to consider the field-scale discontinuities, particularly bedding planes (28), that usually control the structural competence of the roof.[27]

Because engineering experience reflects the history of material conditions and practices in specific sites, engineers place clear limits on the utility of engineering indices as a guide to risk decision making. Engineers recognize the limited applicability of experience in highly specific material situations. The term engineering experience is itself a term of art, but its link to local experience is made visible in its contrast with scientific knowledge that attempts to make generalizable claims about the behavior of systems without reference to specific sites.

Scientific Knowledge

Scientific knowledge provides an explanatory account of the invisible physical forces, molecular interactions, geological divisions, and chemical interactions that are perceived as data in and through instrumentation and analysis.

---

[27]Molinda & Mark, 1992, p. 2.

In a coal mine, scientific knowledge can provide a rationale for practices and outcomes that are literally and temporally beyond the range of human observation and perception.

What scientists recognize as invisible danger, of course, differs from what miners see in the mines. Experts recognize stress distributions, horizontal secondary principal stresses, rock-mass behavior, cell response, and stress distribution in data accumulated from test borings, sensometers, and mathematical analysis. Like roof bolts, the workings of science are literally invisible to human perception.

Scientists themselves frequently deploy more than one type of warrant as a guide to risk decision making. Despite its apparent distance from material experience in local sites, scientific knowledge makes sense only if scientists construct physically probable objects that conform to actual conditions underground. If representations fail to meet these expectations, both mathematical calculations and verbal conclusions must be adjusted to conform to the material possibilities of density, specific gravity, and the like.

Outside of the laboratory, sensory data gathered from multiple tests in specific environments may resist scientists' notions of logical sense. In one experiment, scientists disagreed with the results of mathematical analyses of data. When data predicted an overburden pressure that would be "unrealistically dense, having a specific gravity of approximately 3," one group of scientists deployed "several unproven assumptions" grounded in their own experience to make the data fit their notion of sense.[28] Ultimately, these scientists were forced to acknowledge that the rationality of scientific evidence must sometimes yield to the logic of common-sense reasoning grounded in local observation and experience. (In the present case, for example, objects with a

---

[28]Hanna, Conover, and Haramy (1991) conclude:

The stress distribution in figure 16 shows that the average postmining stress is approximately 500 psi and that the stresses are relatively uniform. It is expected that all cells were located within the confined core; thus a disagreement exists between the measured stresses, 560 psi, and the predicted core stress, 400 psi. Possible explanations of this disagreement are that the vertical stress over the measurement points is greater than the overburden pressure, or that the stresses are not transferred as expected into the tail zone, but are distributed uniformly throughout the core. Vertical stresses at the test site may have been redistributed around the slip zone, which extends through the intersection or around the sandstone channel over pillar B. . . . It is unlikely that the overburden pressure is significantly greater than expected. To generate an overburden pressure of 560 psi at a depth of 360 feet, the overburden material would be unrealistically dense, having a specific gravity of approximately 3. (p. 33).

specific gravity of 3 would be unreasonably more dense than they actually feel in the material environment.)[29]

In order to produce conclusions that make sense, scientists must often treat laboratory results with a certain skepticism until results can be tested in field measurements that can help resolve the discrepancy between theoretical models and real-world experience. Hanna et al. (1991) explain:

> Because of the *uncertainties involved* in determining the pillar stresses and speci-fying the yield zone parameters, particularly σ_o *the above comparison should be viewed with some skepticism.* Although the hybrid distribution was made to *fit the data through manipulation of certain parameters,* the reverse procedure of predict-ing pillar loads and stress distributions is considerably more difficult and uncer-tain. Predictions may be possible, if sufficient field measurements are obtained to calibrate theoretical models of stress distribution and that predictions are limited to areas having similar conditions of load, geology, and geometry [Italics added].[30]

Writers can potentially represent all three types of warrants in a single document, but agency conventions may influence how writers negotiate among competing warrants in their writing. As we shall see in the following section, writers can also deploy warrants strategically to call attention to par-ticular features of risky environments or to deflect attention from technically unwarranted decisions. These strategic choices affect risk judgments and risk outcomes throughout the entire Cycle of Technical Documentation within Large Regulatory Industries.

THE EFFECT OF WARRANTS ON RISK DECISIONS AND RISK OUTCOMES

To determine the best method of roof support, engineers and workers must constantly monitor conditions in the mine and make changes as conditions

---

[29]From this perspective, invisible knowledge may be construed as socially constructed repre-sentations—"immutable mobiles" (Latour, 1990)—that can be manipulated and acted upon without reference to the immediate material environment that occasioned them. Invisible knowledge thus reaffirms the social-constructedness of science as rhetoric, supports the notion that scientists use modalities to control the facticity of knowledge claims, and raises epistemological questions about the reality of represented objects outside of the limits of lan-guage. Because invisible knowledge is least accessible to human sensory experience, it is also the most heavily rhetorical. Unlike the direct sensory experience of pit sense, sensory knowledge embodied in the invisible strata of the mine, must be mediated through language, models, as-sumptions, and argument to allow experts to recognize its invisible material presence.

[30]Hanna et al., 1991, p. 33.

warrant. If supervisors encounter extremely difficult roof conditions, they may stop work entirely and re-commence work at a new location. The notion of a warrantability places a high value on written documentation to support argument about risk (cf. chap. 1).

Warrants grounded in scientific knowledge locate the source of uncertainty in physical and chemical interactions, principles, and laws beyond the limits of human perception; they argue for the extension of scientific research, increased instrumentation, and increased attention to data collection and analysis in risky worksites. Warrants grounded in engineering experience argue for more analysis and data collection to understand trends in specific sites. Warrants grounded in embodied experience argue for increased training to help workers learn from experience without endangering their lives.

Warrants can also affect how agencies and institutions deploy particular technologies to manage risk. Roof bolts are warranted by scientific assumptions about the behavior of strata in response to stress. Arches are warranted by a long history of engineering experience in tunnel-development dating back to Roman times. Timbers are warranted by their ability to provide sensory warnings that enable miners to hear the workings of an unstable roof.

Accidents like the fatal roof fall at Bilsthorpe Colliery in England raise questions about the warrantability of risk decisions prior to the accident. Could decision makers foresee the consequences of their decisions? How predictable were the outcomes? What evidence did they use to warrant decisions? Did conditions warrant changes in practice?

The Bilsthorpe disaster was a particularly vivid reminder of the ways that the strategic choice of warrants—in this case scientific knowledge—influences how writers locate liability and responsibility for risk judgments in risky environments. As Fig. 6.2 demonstrates, the roof fall occurred above the level of the bolts, beyond the limits of either technical control or scientific prediction. As a result, the entire roof fell as a massive block weighing more than 80,000 pounds, crushing miners, who had little or no warning of the impending disaster despite the presence of tell-tale monitors (sensometers) embedded in the mine roof.[31]

In highlighting the unpredictability of scientific knowledge, Her Majesty's Safety Executive (HSE) attempted to deflect attention from management's lack of engineering experience with regard to the new system of roof bolting. The HSE's final report describes the cause of the disaster as a sequence of roof movements and changes in the rock strata that led to a massive failure along a fault in the roof of the mine. The report notes that sensory experience may have saved some miners' lives. The report blames management for applying roof bolts—pointing to the mine's lack of engineering experience with regard

---

[31]Sauer (1992) examines the limits of rationality in the context of imminent danger.

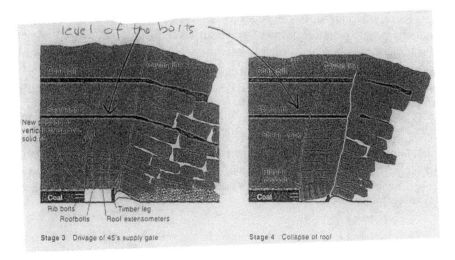

FIG. 6.2.   In the Bilsthorpe disaster, the roof fall occurred above the level of the bolts, beyond the limits of either technical control or scientific prediction. Source: Langdon (1994).

to roof bolting. But the report ultimately concludes that more scientific knowledge is required to understand how strata behavior is influenced by factors "which, while identifiable, cannot be quantified or their effects predicted."[32]

In the conclusion, the report describes the lack of scientific knowledge about strata movement and argues that scientists must undertake additional research to help management understand the workings of mine strata. The report emphasizes that lay readers may fail to appreciate the dynamic uncertainty and unpredictability of a working mine:

> Common features of these falls were *their unpredictability*, the suddenness of collapse giving little or no chance of escape to those in the immediate vicinity. *This point is highlighted because it may not be appreciated by the layman that there is no absolute clear understanding of strata (rock) stability* and the stresses associated with mine workings. Strata behavior is not an exact science and more work needs to be done to obtain a greater understanding and to be able to measure more accurately the effects on mine workings of coal pillars and gateside packs [Italics added].[33]

---

[32]Langdon, 1994, p. 20. In its final report of investigation, the HSE describes the cause of the disaster as a sequence of roof movements, strata deformation, and vertical stresses (p. 14–15, items 44–72) that led to a "combination of remobilisation of goaf blocks about the caving line and failure along the shear crack zone [that] would destabilise a block potentially up to the seatearth" (p. 15, item 53).

[33]Langdon, 1994, p. 16.

The report underlines the unpredictability of roof support and suggests that there were no indications to warn miners to escape. The writers argue: "From *the evidence obtained from those witnesses involved in the fall*, the lack of warning indications on the 'tell-tale' monitors and an examination of the nature of the strata overlying the seam, it is clear that the fall occurred suddenly and as a solid mass"[34] [Italics added].

But other evidence in the report suggests that miners who survived did recognize indications of disaster. As survivors' reports suggest, miners who died failed to recognize signs of danger because they relied upon sensometers to sense invisible strata movement in the immediate roof. Miners who escaped heard the working of the roof and ran for safety. Miners with earplugs were crushed:

> The chargehand, Clifford, and the undermanager, Shelton, appeared to have been the only people who appreciated that the roof was moving as the fall started. *Clifford heard the noise of the wood props creaking loudly and then banging as they broke.* He realized that this was not simply as a result of the heading taking weight but was something far more serious. However, before he could move, the roof lowered (approximately 600 mm) as a block, and it appears that there was then a slight pause in roof movement which gave Clifford the opportunity to run outbye [sic] and he shouted and grabbed at Shelton and James as he went. *Grattage, who was wearing noise reducing ear defenders at the time, appears to have heard nothing of the fall as it started* and it was Clifford shouting and grabbing him that alerted him and enabled him to make his escape [Italics added].[35]

Ironically, sensory information saved another miner, even though he misinterpreted the source of the noise. The HSE report concludes: "At the inbye [sic] end of the heading, *Kocij heard a noise but thought that the compressed air range had been turned on again* and, in fear of being struck by the flailing hose, he ran towards the face. In the event *this action probably saved his life* [Italics added].[36]

The final report is a slick document by U.S. standards. Most U.S. reports have thin paper covers and follow a standard format that pays little attention to design. U.S. reports have a limited audience that rarely extends beyond the agency and interested mining communities. The HSE report, by contrast,

---

[34]Langdon, 1994, p. 14.
[35]Langdon, 1994, pp. 6–7, item 21.
[36]Langdon, 1994, p. 7, item 22.

has a professionally designed cover and layout that suggests that the document was produced for public consumption in the larger sense of the term.[37]

Writers admittedly faced problems of public perception when they constructed the agency's official version of the disaster.[38] The accident was the first fatal accident since British mines had denationalized in 1994. It was also the first disaster since British mines had introduced the American system of rock bolting, a method of roof support opposed by British labor unions who favored the traditional method of arch support. Unions had feared that the new system would reduce safety and increase the potential for disaster. The accident confirmed their fears that the motivation for the practice was strictly economic.

Within the HSE report, miners' voices are woven into the fabric of a narrative that self-consciously describes the heroic efforts of miners in both the accident and the subsequent rescue. The foreword includes a personal letter of thanks to the mineworkers for their support in the investigation, although the miners in question belonged to the new Union of Democratic Mineworkers (UDM), not the NUM, which had lost ground during the 1984 strike when the UDM had agreed to the government's contract.

Ignoring its own evidence that some miners used embodied sensory knowledge to detect the indications of disaster, the HSE uses miners' testimony to argue for increased research to understand rock behavior. In focusing on the unpredictability of scientific knowledge as a warrant for risk decisions, the HSE's conclusion fails to acknowledge how new technologies affect the warrantability of risk decisions in local sites.

---

[37]According to HSE inspectors, the investigative report—which seems particularly polished compared with similar U.S. documents—attempts in part to quiet political and ideological criticism of British Coal's approval of roof bolting at Bilsthorpe (Personal interviews 1995). In addition, the report indirectly addresses discontent resulting from British Coal's relations with the Union of Democratic Mineworkers, a Nottingham-based union which broke from the NUM to end the 1984 coal strike. Thus, the introduction of the document pays tribute to the heroic rescue efforts of "BCC management, colliery personnel, rescue teams, Inspectors of Mines and all others who strove manfully and with great fortitude, often at no small danger of themselves, to effect a rescue of the trapped men and recover the bodies of the deceased in the most extreme conditions" (Langdon, 1994, Foreword).

Because the HSE report is such a carefully constructed public representation of the accident, we must be cautious about any conclusions we draw about the document as a more generalized argument about risk. Nevertheless, the HSE report illustrates how writers can integrate all three forms of knowledge into a single representation.

[38]Statements in this section are based on personal interviews I conducted with HSE personnel at the agency in the summer of 1994.

Within the Cycle of Technical Documentation within Large Regulatory Industries, major accidents like Bilsthorpe become powerful warrants that can persuade decision makers to act to manage risk. Without any prior evidence to the contrary, decision makers may become complacent, producing the conditions that lead to a disaster. Following an accident, they take action to manage risk. These judgments are warranted by the similarity between prior experience and local conditions in the mine. When mines introduce new technologies, decision makers must predict the effects because they cannot make claims grounded in prior experience. Even so, the effects of their decisions influence future policy and action throughout the Cycle of Technical Documentation within Large Regulatory Industries.

The history of risk decision making at Bilsthorpe prior to the disaster shows how estimates of risk were continually revised as conditions warranted. In the early stages of mine development at Bilsthorpe, management used a combination of roof support practices. They had a large body of engineering experience to draw upon that helped them adjust roof support "as conditions warranted."[39] When roof conditions deteriorated, the mine added additional roof bolts "to remedy the situation." When conditions improved, the mine increased production.[40] When the mine began to rely on roof bolts as the sole means of support, management could no longer draw upon the same body of engineering experience to evaluate problems that were occurring within the strata, beyond the reach of human perception. They called in consultants to help them assess local conditions, but the monitoring systems in place at the time created a false sense of security that created the conditions for disaster.

Because the mine did not have adequate experience with roof-bolting technologies, management relied heavily on a simple monitoring system to assess changes in conditions that might indicate problems. This monitoring system indicated that the rock bolts were providing a competent method of support because they did not show signs of change. The monitoring system also "gave confidence in the integrity of the roof bolting system" in the absence of visible evidence to the contrary.[41] On this basis, management decided to use roof bolts as the sole means of support (although consultants rec-

---

[39]Langdon, 1994, p. 5.

[40]Langdon, 1994, p. 5.

[41]The report describes how "further rockbolts were installed in a more random manner in those areas not previously bolted" "as roof conditions warranted" (Langdon, 1994, p. 2). The #45 entry was developed "immediately adjacent" to #43 with no intervening pillar. In this section, the report notes that the support system "was designed to use rockbolts as the principal support" with additional passive supports for a distance of about 300 feet. During this period, the strata were closely monitored and conditions seemed to warrant the use of roof bolts as the sole means of support. Mining consultants recommended that the mine set up additional timbers between the roof straps (Langdon, 1994, pp. 2–5).

ommended additional wooden timbers as a secondary means of support). At 9 a.m. on the day of the collapse, management checked the telltale monitors en route to the site of the accident and observed "nothing untoward" until the roof fell at 10:58 a.m.[42]

The final report does not reveal what HSE personnel described in an interview: The monitoring systems gave a false sense of security because the roof bolts had bolted the strata into a massive block of solid rock that broke above the level of the roof bolts. Both the monitors and the roof bolts were still in place after the fall.[43]

As the Bilsthorpe disaster demonstrates, new technologies interrupt the cycle of texts and action within the Cycle of Technical Documentation within Large Regulatory Industries. When new technologies emerge, decision makers cannot draw on a body of engineering experience to apply the new technology in local sites. Workers' embodied experience must be translated to new situations. Scientific knowledge provides a rationale for using the new technology, but scientific knowledge cannot provide the site-specific experience that decision makers need to detect and manage risk in local sites.

For rhetoricians, the problem of warrantability raises questions about the rhetorical assumptions that influence how writers negotiate and construct competing warrants grounded in experience and—more importantly—how writers construct warrants in the absence of prior experience. As we shall see in the following sections, written communication may inadvertently privilege scientific knowledge at the expense of embodied experience, rendering representations inadequate for risk judgments. If we focus solely on textual practices, however, we may fail to see how written texts work in tandem with embodied experience or how decision makers themselves deploy all three types of warrants outside of written texts in speech and gesture.

THE RHETORICAL INCOMPLETENESS OF WRITTEN INSTRUCTIONS AND PROCEDURES

Within the Cycle of Technical Documentation within Large Regulatory Industries, the processes of rhetorical transformation that shape the construction of instructions and procedures are designed to strip away the site-specific

---

[42]Langdon, 1994, p. 5.

[43]Telltale monitors can reveal separation in the strata, but the roof in the Bilsthorpe mine was bolted together so that the whole block fell as a single piece. The roof bolts were successful to the extent that they held the strata together, but they were not long enough to bolt the strata into the solid (supporting) rock called the overburden. As Bilsthorpe revealed, no humanly feasible bolt can ever be long enough to hold up an entire mountain.

features of local experience that characterize representations of risk in local sites. As a result, trainers must develop their own strategies of transformation to re-embody instructions and procedures with the local knowledge that workers need to work safely in local sites. But risky environments also call into question the underlying assumptions that ground conventional notions of instructions as a set of generalizable and carefully ordered procedures that can produce predictable outcomes.

In questioning the adequacy of conventional models of written communication in the context of risk, we raise questions of liability and control within risky environments. As Paradis argues, if workers are to be held responsible for human error or failures of training, they must be provided with the means to understand and control technology in the workplace.[44] But training texts frequently send mixed messages to miners about their own stance in relation to risk and responsibility in the workplace. On the one hand, instructions by their very nature have an implicit force that persuades workers to follow orderly procedures—why else create instructions? But instructions frequently warn workers to attend to changing conditions that warrant changes in practice, giving workers the freedom to determine when and how to comply with instructions as conditions warrant.

The following examples demonstrate how written texts direct decision makers to sensory knowledge outside of written texts. While it is impossible to determine from textual evidence alone the degree to which the rhetorical inadequacy of these texts result from deliberate manipulation, carelessness, or slavish adherence to institutional practice, these texts show how the standards for assessing familiar and conventional practices like instructions and procedures must be refigured in the context of risk.[45]

The Rhetorical Incompleteness of Written Instructions

Before miners install roof bolts, they must inspect the roof, testing for loose rock and other signs that the roof might be unstable. Chapter 3 of the U.S. Department of Labor's *Roof and Rib Control Manual* provides a carefully ordered list of instructions for examining the roof and ribs:

1. First, make a visual examination of the area. Examine closely for cracks, slips, kettle bottoms, horsebacks, or irregular formation. If you can see the roof is bad, stand clear and don't try to test further.

[44]Paradis, 1991.

[45]U.S. Department of Labor, Mine Safety and Health Administration, National Mine Health and Safety Academy, 1989.

2. Always stand under supported roof when testing.

3. Always use an appropriate testing tool in making these tests.

4. Test only far enough to erect the next supports in unsupported areas.

5. Always start from supported roof and examine toward the face.

6. Always test the ribs in addition to the roof.

7. Never turn your back to the face or ribs while testing.

8. Always use bare fingers against the roof with your thumb pointed toward you. (This will make it difficult for you to walk unintentionally under unsupported roof.)

9. Start tapping the roof lightly at first with your sounding device, increasing your stroke to better hear the sound of the roof and/or feel the vibration.

10. Always use goggles to protect your eyes.

11. Examine the roof frequently while working. (Roof conditions may change suddenly.) Machinery should be turned off while examining and testing the roof.

12. Use a sturdy bench or long testing tool in high places.

13. Always make sure of a safe line of retreat.

14. Take your time and be sure that you have made a thorough examination.[46]

Despite the apparent certainty of the numbered, rational, rule-based procedures, this seemingly ordered list of procedures provides insufficient information for making judgments about risk in an uncertain environment. What will miners see when they make a visual inspection? What do cracks, slips, kettlebottoms, horsebacks look like? What is the sound of the roof? What does the vibration feel like? What signs should observers pay attention to when they examine the roof? What constitutes a thorough examination? When can they stop?

In a conventional model of instruction, user testing would increase the level of detail necessary to perform the tasks listed in the procedures. But risky environments challenge the entire premise on which instructions are based. In a risky environment, safe practice does not follow an ordered sequence; nor can a simple set of instructions specify all of the features that miners must pay attention to in local sites. In a risky environment, these instructions provide a checklist for ongoing and constant activity that instructs miners to pay attention to all 14 tasks simultaneously, even as each individ-

---

[46]U.S. Department of Labor, 1989, pp. 95–96.

ual instruction instructs miners to turn their attention to specific problems identified in the list. Even the instruction to wear goggles is not as simple as it seems, since goggles are difficult to manage and make visual inspections difficult if they become scratched or covered with coal dust, dirt, or water. Given the changing dynamics of the mine environment, the unpredictable nature of risk in the workplace, and the arbitrary standard of safety itself, no single set of rational instructions will be sufficient to direct the worker's actions in an orderly sequence of tasks and outcomes in a risky workplace.

Unlike conventional instructions, these instructions actually command the miner to use common sense and experience to interpret sounds and visual cues in the environment. Paradoxically, the miner must either have experience or other knowledge outside of the document in order to obey the rationally ordered commands within the written text. Thus, while the surface features of the written text seem to silence and exclude the embodied sensory experience and site-specific local understandings of the miners, the instructions actually command miners to draw upon pit sense in order to assess changing conditions in the mine. The imperative does not formulate a specific instruction; instead, the instructions command miners to use their own judgment when they examine roof conditions because "Roof conditions may change suddenly."[47]

The uncertain and unpredictable environment of the mines thus challenges the rhetorical relationship between expert and actor presupposed in the imperative mode of more conventional notions of instruction. Ordinary instructions presume that the audience is less knowledgeable than the writer—else why would the audience need instruction? In a risky environment, workers must assume the role of expert in the broader sense of the term as we have used it in chap. 2—making judgments about the proper order and outcome of the examination. To obey the commands, the worker must become an active, decision-making, empowered agent. In the context of risk, proper procedure is only a partial imperative (similar to the caution to take care), since it relies upon the worker to make the specific decisions and judgments that will produce a legal standard of adequate and sufficient instruction in any particular case.

Because the arrangement of information in the enumerated list of commands is logical, orderly, and precise, the imperative maintains an appearance of institutional authority and control that minimizes the sense of risk. The careful rhetorical arrangement reinforces the notion that nature and humans can be ordered and controlled; proper procedure prevents disasters. The rhetorical strategies of parallelism, numbered lists, and plain speech suggest

---

[47]U.S. Department of Labor, 1989, p. 95.

an orderliness in discourse that reflects an orderly—and presumably natural—arrangement of material environment and human will.

Even as it seems to silence local experience and judgment through the authority of carefully constructed procedural control, the text must also command miners to utilize non-textual knowledge and experiences outside the frame of rational knowledge, institutional control, and generalized expert systems. The manual thus authorizes experience at the same time that it validates proper training and procedure.

A complete representation must help miners incorporate the embodied sensory experience that warrants each step in the procedure. But the manual does not specify how miners might develop this knowledge or how they might develop such experience except in practice.

The Interdependence of Scientific Knowledge and Local Experience

Both mining itself and miner training have distanced miners from direct sensory information in the mines. Robotic devices, for example, can already perform routine, hierarchical procedures at a higher standard of efficiency, cost-effectiveness, and safety than humans. Ultimately, robotic devices will replace most of the routine tasks of mining currently performed by miners underground. At the same time, federally mandated surface training programs have replaced traditional apprenticeships under the careful supervision of experienced miners.

Both of these trends might thus suggest that pit sense—like picks and shovels—belongs to a former era when miners were more directly exposed to conditions of risk in the mines. As researchers at the Bureau of Mines suggest, however, humans will always be needed in situations of profound risk because of the variable conditions and changing flow of sensory information about the environment.

In describing the problem of developing robotic sensors that can imitate human decision making in the mines, for example, developers of a robotic mining machine acknowledge the problems that occur when new technologies distance miners from sensory input in the mines: "Currently, operators use their senses to control equipment and evaluate geological conditions. If the operator is removed from the immediate bolting area, most of the operator's sensory input will be lost."[48]

To develop a robotic roof-bolting machine that imitates the human being's ability to record and analyze sensory data from the immediate material

---

[48]Hill et al., 1993, p. 495.

environment, researchers must provide specific information to control the drill, measure bit sharpness, and sense conditions in the strata. To make decisions involving roof-bolt placement, robotic devices must be able to sense underlying geologic faults such as weak rock and geological anomalies that can result in roof falls. Machines must thus be able to imitate humans. Researchers note: "An experienced roof bolt operator can often tell by the 'sound and feel' of the drill and by observing the rock response to drilling whether layers, fractures, and voids are present, as well as the hardness and type of material being drilled."[49]

To imitate the human miner, robotic researchers must create a smart drill that can recreate and even enhance a human's senses. Ideally, researchers project: "By placing monitoring instruments on the drill of a roof-bolting machine, these 'senses' can be regained and even enhanced to collect more precise information about conditions in the immediate roof strata."[50]

These ideas remain conceptual, however, and researchers note that current systems lose valuable sensory information when expert systems reduce variable conditions to a single number. As a result, the embodied sensory information that might be helpful in detecting underlying faults is lost and intelligent drilling becomes impracticable except in remote and extremely hazardous sites "such as underground repositories for radioactive waste, subsurface regions of the moon and Mars, and ocean bottoms."[51] Given the limitations of robotic systems, the researchers suggest, mines must still count on sublunary miners to use their senses to monitor changing roof conditions.

The Rhetorical Presence of Tacit Knowledge

In defining tacit knowledge, Polanyi distinguishes between formal, scientific rules and procedural rules. Thus, he notes, the formal dynamics of balance on a bicycle do not comprise the rules of riding.[52] To the extent that experts can articulate explicit rules (e.g., lean into curves), knowledge moves from the domain of the tacit into the realm of scientific expertise.

Collins builds upon Polanyi's notion of tacit knowledge to explain why procedural instructions do not contain sufficient information to enable apprentices to replicate their operation in a new lab. For Collins, tacit knowledge is cultural knowledge that is not made explicit in rules and procedures. In the case of the tea laser, for example, he describes how some apprentice

---

[49]Hill et al., 1993, p. 496.
[50]Hill et al., 1993, p. 497.
[51]Hill et al., 1993, p. 498; p. 501.
[52]Polanyi, 1983. In Collins, 1990, pp. 108–9.

builders could reproduce the master's short leads without knowing the exact length. Apprentices whose lasers failed to work would be counted "incompetent," though no one knew why.[53] Ultimately, however, designers could achieve a "decontextualized rule" ("make the top leads less than 8 inches") that was "universally applicable" and "independent of the cultural foundations of knowledge."[54] Scientific expertise (as generalized rules) thus builds upon tacit knowledge but seems to move beyond the craftsman's skill to formulate an increasing articulation of the world. As Collins notes, science itself is far messier than written rules and procedures suggest, and science "talk" about procedures is "far greater in volume and content than program rules."[55] Such talk is also "far more revealing of doubts, qualifications, and uncertainty" than the apparent simplicity of "textbooks, expert systems, and schoolroom science in general,"[56]

Like the concept of tacit knowledge, pit sense describes information that is not articulated in written rules and procedures. There are, however, important differences.

First, the term "tacit" implies silence. The term describes a set of understandings and behaviors that are coordinated and understood in particular social settings prior to their articulation in language. Indeed the term depends upon a notion that certain things may be quietly understood, but not spoken, as a function of group activity. Pit sense, on the other hand, is not understood until it is felt or perceived by individuals in particular material sites. Pit sense may be articulated as a set of commonplace, but the activity of sensation resides in the individual knower who feels or senses changes in material conditions.

Second, Collins's notion of the tacit presumes that apprentices will replicate a previously working artifact. New lasers are said to fail if they fail to work in similar, predictable, and controlled laboratory environments. In the mines, however, apprentices may replicate previous procedures, follow fully articulated rules and procedures, and still fail because new situations and environments change the conditions that enabled previous models to succeed.

Third, Collins's apprentices can trust that expert rules can guide future behavior once tacit knowledge is articulated as scientific expertise. As one MSHA training manual reminds miners, however, no single plan or rule for roof bolting can guide behavior in all situations: "The plan was prepared to meet normal conditions in the mine. The miner may very likely encounter

[53]Collins, 1990, p. 114.

[54]Collins, 1990, p. 114.

[55]Collins, 1990, p. 151.

[56]Collins, 1990, p. 151.

conditions that require more support than the plan called for and in such cases he should tell his supervisor. *Above all, the plan must be followed, and this is everyone's responsibility.*"[57]

Fourth, Collins' apprentices are judged competent to the extent that their tea lasers work in future situations. In the context of profound risk and uncertainty, however, even the most experienced miner may misjudge conditions without being judged incompetent. Miners, for example, must test the soundness of roofs by tapping with a rod while holding a bare hand to the roof, a technique that may seem incongruous in the modern mine. As MSHA's manual warns, however, roof testing can only be approximate—a thud can be mistaken for a ringing sound. "Unfortunately," the text notes, "sounding is not always foolproof. Some roofs may produce sounds which are hard to recognize. Other roofs, for various reasons, may sound safe when they are not, or vice versa. Therefore, it is very important not to take chances when testing roofs."[58]

For writers, tacit knowledge represents what cannot be expressed in words. But writers presume that such knowledge exists within particular cultures—at least for good (competent) apprentices. Ultimately, competent writers may articulate this knowledge in good instructions as generalizable rules. This line of reasoning takes for granted the assumption that instructions—or at least knowledge—exist prior to action.[59]

The notion of sensory information, on the other hand, does not presume that knowledge exists prior to action, sense, or response. Within the mines, for example, miners know that someone must be the first to try out equipment in new circumstances. Miners may apply previous knowledge and experience in knowing how to react, but their reactions must respond to unpredictable and unknown variables. Instructions, then, follow action. The next time miners will know to hold tighter, watch their heads, bear harder, and feel the vibrations. Observers can observe what has passed, but not what will come. Thus, instructions function as a kind of hindsight for the next time, when conditions may or may not be the same. Unfortunately, such knowledge comes at a cost: accident rates are highest for miners with the least experience on particular jobs, regardless of their total experience in the mines.

In distinguishing between tacit knowledge and scientific expertise, Collins' model ultimately privileges science knowledge—most notably in his

---

[57]U.S. Department of Labor, 1989, p. 63, their emphasis.

[58]U.S. Department of Labor, 1989, p. 72.

[59]In one particularly vivid interview, one miner described the ways that a new miner can be thrown by new piece of equipment if he does not hold tightly to the controls. As he spoke, his gestures illustrated the experience. For nearly a minute, he held tightly to an imaginary wheel, white knuckled—even after he had finished his spoken description of the event.

oppositions between craft and science and in the visual models he constructs to represent the experiential foundations of scientific expertise. Collins' model does not examine the ways that textual practices may constrain how individuals represent tacit knowledge in writing. Nor does it investigate how individuals might communicate tacit knowledge outside of texts in speech and gesture.

THE TEXTUAL DYNAMICS OF DISASTER

As Petroski concludes, trial and error is fundamental to a notion of engineering experience that builds upon past events in order to design structures that will solve problems in the future.[60] In order for engineering experience to provide a reliable index of future action, engineers must assume that responses in the future will be predictable; that actions in the past can predict future consequences; and thus, that particular features of the material environment can act as predictable indices of future actions. MSHA investigates accidents to identify the technical cause (or causes) of accidents so that agencies can prevent similar occurrences in the future. But engineering experience must be adapted to new sites and situations in order to provide a warrant for action.

In risky environments, even the best plans meet unseen—and unforeseen—hazards. As Mark and DeMarco (1993) argue, sound engineering must be based upon past experience and site-specific knowledge to develop optimal solutions.[61] As conditions change, engineers must cope with and provide "unique design solutions" in dangerous and often increasingly deteriorating conditions.[62] The following passage describes the changing geologic conditions in one section of a typical U.S. mine:

> Gate conditions steadily worsened as panel extraction advanced toward the third panel. Excessive abutment loading on the gate, due to poor overburden caving and a relatively low pillar support capability, resulted in severe floor heave, a deteriorating roof mass, and pillar dilation to the extent that cribs

---

[60]Petroski, 1985.
[61]Mark and DeMarco (1993) write:

Severe and unpredictable natural conditions provide some of the most difficult challenges that longwall planners face. . . . Solutions which work at one mine can seldom be applied, cookbook style, to other operations. None of the specific designs [described in the paper] has an inherent superiority. Sound engineering, based on past-experience and site-specific knowledge, is needed to develop optimal solutions. (p. 37)

[62]Mark & DeMarco, 1993, p. 33.

were being rolled into the tailgate escapeway. Remedial efforts to salvage the tailgate entry included the use of trusses, super bolts, steel I beams bolted to the roof and trussed together, wire mesh, butted cribs down the entire length of the entry, "Big John" timbers, and steel "doorframes" between cribs to keep them from rolling and closing off the escapeway/aircourse.[63]

Despite this dynamic uncertainty, engineers must construct plans to control risk and predict hazard in the everyday working operations of the mines. Roof-support engineers, for example, must map underground faults, determine stress and load levels, and construct plans that outline the minimum support methods and requirements for each section of the mine. These support requirements include a broad plan for overall roof support as well as site-specific rules for pillar width and additional support. But engineers must also be prepared to respond to change.

Engineering experience depends upon highly descriptive accounts like the description of worsening conditions in the mines previously mentioned. Molinda and Mark (1994) write: "Historically, accounts of mine roof geology have been highly descriptive and required interpretation for use in engineering design and support selection and . . . these reports depend upon specific local observation and thus have highly local application."[64] But technical reports may not provide sufficient information about an accident to warrant arguments for change. In the following MSHA internal report on deep-cut or extended-cut mining, Davis (1991) complains that the agency could not obtain an "accurate determination" of conditions prior to the accident from the information in the record:

> Thirty-four other non-fatal, injury type roof fall accidents were identified in which rock fell from between roof bolts striking persons standing under permanently supported roof during mining. . . . Remote-control continuous miner operators, miner helpers, section foremen, and others observing the mining procedure were involved in these accidents. This type of accident, however, is not necessarily related to the depth of cut. *From the information in the record, and discussions with roof control supervisors in the districts involved, an accurate determination could not be made as to whether extended cuts were used at the time of the accident* [Italics added].[65]

When accident reports fail to provide adequate data, inspectors must "read into a report" to get the data they need.[66] MSHA's publications echo this sen-

---

[63]Mark & DeMarco, 1993, p. 36.
[64]Molinda & Mark, 1994, p. 2.
[65]Davis, 1991, pp. 2–4.
[66]Anon. (Interview with agency inspector, U.S.), April, 1992.

timent: "We take what's given about the accident, read into it what we need to make sense out of it, trust our instincts and when we're finished, our margin of error will be well within the acceptable limits."[67] As in "all fast screening mechanisms," the authors of MSHA's Coal Accident Analysis and Problem Identification Course confess, "the guidelines are willing to trade off a degree of accuracy for expediency. If we are 10% off in our placement of accidents into groups, this is certainly understandable and acceptable considering the lack of information on many of the accidents. . . . The result may have made Group A contain 75 accidents instead of 76. *So what?*" [Italics added][68].

To obtain what one document calls "premium data," engineers must frequently draw conclusions and read between the lines to determine the cause of the disaster.[69] One analyst admitted that few reports gave him the exact data he needed. To determine the number of accidents related to longwall mining, he was forced to draw conclusions from details in the report—the depth of the cut, the number of miners on the section, or the type of injury—since few reports he consulted actually categorized the type of mining conducted on the section.[70]

To transform these individual experiences into the collective history of a mine, engineers must develop observational techniques and methods of data collection that will enable local operators to assess conditions in local environments. To measure change, engineers develop indices that enable both machine and human sensors to recognize moments of change in physical conditions. These moments of change indicate potential hazards like underground aquifers, mineral seams, or changing strata conditions. These indices can provide remarkably simple visible evidence of underlying problems. Sames and Moebs (1992) show how engineers can employ a simple test (like striking a rock with a ball peen hammer) to estimate complex scientific concepts like compressive strength:

> Several index tests have been proposed to estimate compressive strength. The CMRR employs an indentation test proposed by Williamson (42). The exposed rock face is struck with the round end of a ball peen hammer, and the resulting characteristic impact reaction is compared to the chart shown in figure 1 . . . It is the nature of the indentation, not its magnitude, that is important.[71]

Unfortunately, the process of rhetorical transformation that produces engineering knowledge is slow and often results in inadequate and incomplete

---

[67]U.S. Department of Labor, Mine Safety and Health Administration, 1988, p. 44.
[68]U.S. Department of Labor, 1988, p. 43.
[69]U.S. Department of Labor, 1988.
[70]Personal Interviews, 1992.
[71]Sames & Moebs, 1992, p. 7.

accounts of conditions.[72] From an ethical standpoint, engineers cannot wait six years to build a body of engineering experience that is large enough to help them make everyday decisions about safety. Within the mines, they need quick and reliable indices to help them evaluate changing conditions. Molinda and Mark (1994) describe the tension between the inefficiency of engineering experience and the daily demands of safety:

> Historically, the mining industry has depended on quick observations of strata deformation and failure . . . bolt failure, and lithologic changes . . . to evaluate stability. These preliminary evaluations have been used to make significant changes in mine layout and support systems, which may affect the economics of the mine. Because of the subjectivity involved in these visual observations, the mining process is slowed, creating inefficiencies and safety concerns during the time a decision is being reached.[73]

Because they cannot willfully experiment with human lives in local sites, engineers have turned to scientific methods of analysis and experimentation to provide generalizable warrants that can be applied across a variety of sites and situations.

THE LIMITS OF SCIENCE
IN THE CONTEXT OF RISK

To understand (and thus trust) the invisible workings of roof-bolt technologies as the primary means of roof support, miners must be able to visualize the structure of the rock strata above the working sections of the mine. MSHA's manual on roof and rib control (1989) advises miners: "In order to understand why roofs fall it is necessary to understand some things about the nature of the earth above the mine."[74] The manual provides a basic primer describing rock strata, faults, and unusual structures, roofs, and ribs. A more recent revision of the manual includes a cross-sectional view of geological formations and a brief discussion of strata movement, geologic pressures, and roof-support methods. In the 1992 revision, the authors list the following goals for the text:

---

[72]MSHA's accident investigation following the Wilberg mine fire (Huntley et al., 1992) reveals the limitations of engineering experience as a guide to risk judgments. Investigations are slow and often hazardous; writing the report can take many years. The Wilberg investigation was particularly slow because conditions were hazardous and MSHA called off the initial investigation when two miners died in the recovery effort. As a result, the official report of the disaster was not published until six years after the disaster.

[73]Molinda & Mark, 1994, p. 117.

[74]U.S. Department of Labor, MSHA, 1989, p. 5.

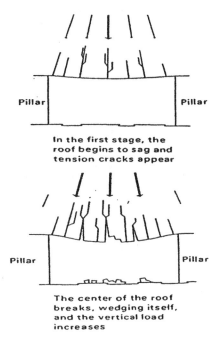

FIG. 6.3. This American diagram presents an idealized picture of the strata or rock layers above the coal seam in the mine. Source: U.S. Department of Labor, Mine Safety and Health Administration, National Mine Health and Safety Academy (1989).

The student will *recognize* the basic geological formations in roof strata and will describe how they were formed.

The student will *be knowledgeable of stresses* in the strata and the effect they have on mining operations [Italics added].[75]

To help miners visualize the invisible strata above the immediate roof, instruction manuals and training texts also include diagrams. The diagram in Fig. 6.3, for example, shows a cross-sectional view of a mine entry and an idealized picture of the strata or rock layers above the coal seam in the mine. Pillars of coal on either side of the entry provide walls or columns of support that help prevent roof falls. According to this diagram, the immediate roof of the mine shifts or resettles in response to the tremendous forces or pressures from the weight of the overburden (the mountain above the immediate roof) when a layer of rock or coal is removed. The British diagram in Fig. 6.4, on

---

[75]U.S. Department of Labor, Mine Safety and Health Administration, National Mine Health and Safety Academy, 1992, p. T-i.

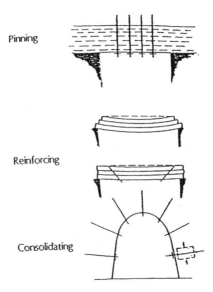

Pinning

Reinforcing

Consolidating

FIG. 6.4.   Unlike the American diagram, this British diagram depicts both vertical and horizontal pressures in the mine roof and ribs (pillars of coal left standing to support the mine roof). Source: British Coal, Nottinghamshire Area, Lound Hall Training Centre (n.d.).

the other hand, depicts both vertical and horizontal pressures in the mine roof and ribs (pillars of coal left standing to support the mine roof).

As I have argued elsewhere, differences in culture and politics can explain in part the differences in the ways that British and U.S. diagrams help miners understand roof-bolt technology. Thus, British diagrams more closely resemble the more traditional arches they were designed to replace. American diagrams, on the other hand, use the analogy of a wall anchor to help miners visualize the invisible workings of a roof bolt. The American representation thus more closely resembles a room with a ceiling and walls in which roof bolts can act like simple anchors.[76]

Despite their geometric differences, however, both British and U.S. diagrams depict an abstract, clean-boundaried, cross-sectional perspective that bears little resemblance to the visible, dark, ragged-edged, coal-dust-filled, noisy three-dimensional haulageways of a working cut. The diagrams show no men, machines, overhanging wires, trolleys, belt lines, or overhanging

---

[76]Sauer (1996) discusses in detail the flaws with this analogy. The three most obvious flaws are: (a) differences between the condition of a ceiling and the scaling or crumbling roof a mine; (b) the different effects of gravity in wall anchors and roof bolts; and (c) the weight and mass of supported materials.

ledges. Instead, these diagrams depict the working sections of the mine as a vacant and blank void. They thus construct a stance toward the material in which humans, machines, and coal are invisible. The knowledge that miners need to achieve safety, according to these diagrams, lies beyond the sensory realm where it is invisible to the human eye.[77]

More important, neither British nor U.S. manuals explain how the miner might actually recognize the invisible forces within the immediate roof. In the context of a training course, for example, with programmed text and multiple-choice answers, the ability to recognize geological formations may simply involve pattern matching between sample representations in a classroom setting. In dimly lit context of a mine, miners who focus on more immediate problems (like a wet or muddy floor) may fail to recognize more serious hazards in the overhanging roof.

IMPLICATIONS FOR RHETORICAL THEORY

Writers live in a world of texts and believe in the power of texts to reconstruct or construct the material world as representations. Writers also live outside of the material sites I have described in this text, but they are called upon to (a) theorize about the ways that this world works; (b) represent its successes and failures in accident reports and investigations; (c) analyze and comprehend failures of language and failures of technology; and (d) use their analyses to construct instructions, policies, and procedures to prevent disasters in the future.

Ideally, we could argue, a complete training course should teach miners the rules and procedures they need to apply in their daily work. If we view knowledge as unembodied—unconnected and abstracted from the material environment, immutable and mobile, miners should be able to read and understand instructions as easily in a coal mine as in a classroom (ignoring the obvious differences in visibility caused by poor lighting and dust in a mine).

The notion of embodied sensory information, however, does not presume that knowledge can exist prior to sensation. More importantly, this notion recognizes that much of the information miners will need cannot exist within texts prior to action, sense, or response. Thus, while texts can provide guidelines, indices, and tips for recognizing predictable signs of risk and hazard, miners must observe, sense, and recognize hazards within highly specific local

---

[77]Cf. Lynch, 1990.

environments—outside of texts—in order to make decisions to prevent disaster. Miners may apply previous knowledge and experience, but they must react to unknown and unpredictable variables. In this context, instructions follow action. Next time, miners will know to hold tighter, watch their heads, bear harder, or feel vibrations. As they react, they may draw upon previous training and experience, but they do not call upon texts at the moment of action to help them react. Nor do they record their reactions in written communication, although they may be called upon to testify to their experiences in legal depositions and training sessions.

When miners use gesture to talk about these experiences, their bodies still bear the imprint of the physical presence of embodied sensory information. In one interview, an expert miner described the first time he held a bolting machine—and was nearly thrown by the unexpected force of the machine. During the taped interview, this miner continued to grip an imaginary wheel with white knuckles as he physically reenacted the experience. Novice miners, on the other hand, use few gestures in comparison to experts, particularly when they describe risk and hazard in the mines.[78] These findings thus suggest that much of the information miners need is tacit only in relation to a paradigm of knowledge that privileges information articulated as unembodied textual or verbal communication.

The dynamic and uncertain material environment of a coal mine thus raises ethical questions about textbook notions of instructions as systematic procedures designed to translate expert knowledge to lay audiences or to prescribe safe behaviors for highly local and unpredictable environments. In questioning traditional notions of instructions, however, we must face the uncomfortable question of practice and an even more problematic question of liability. How can we convince miners to follow safe practice if we cannot prescribe safe practice? And more important, how can we safeguard miners' rights to a safe and responsible workplace if we reject the notion that experts can predict and prescribe safe practices in the workplace?

The problem with valorizing pit sense, of course, is that it inscribes both the power and responsibility for safety in individual miners. Miners with pit sense escape hazards. Those without pit sense—the unlucky, the accident prone, the disbelieving—encounter more than their just measure of injury and loss. Many miners believe this. One apprentice miner attributed his own recent string of accidents—a torn ligament in the knee, a broken finger, and lacerations on his arm—to a problem of luck.

---

[78]Sauer & Franz, 1997.

From a legal perspective, moreover, a belief in pit sense may shift responsibility from management to miners, who may not have the equipment, authority, or responsibility to protect themselves. Both British and U.S. laws hold management ultimately responsible for safe practice in the mines. Safe practice includes both engineering practice—decisions about roof support, for example—and behaviors. (Miners, for example, are prohibited from working beyond the last row of supported roof.) If management were to allow miners to make individual decisions about safety based upon unwritten rules of pit sense, this argument suggests, miners would lose the statutory rights they had won in a long labor struggle to hold management accountable for safety in the mines.

In examining how risk communication incorporates the three types of warrants we have described in this chapter, the present argument suggests that engineers, scientists, and workers can work together to construct equipment, policy, and procedures that meet the needs of users in complex and difficult material environments. As DeKock, Kononov, and Oberholzer (n.d.) conclude, engineering systems must assist miners, not control them. In developing a robotic control system, they note, one of the most important requirements that was identified was "that the system must have two way communication between the operator [human miner] and the continuous miner [mining machine]."[79] In arguing for improvements in engineering technologies and a better understanding of the indices of safety, we might argue, we must also draw upon the expertise and experience of miners. Only then will miners' actions become explicable as rational responses within a continuing flow of embodied sensory information in the mines.

The present account thus expands upon previous accounts of scientific rhetoric in the laboratory and technical discourses within large research and development organizations. As this account suggests, generalizations about science-in-the-laboratory or scientists-in-laboratories may have limited application in local and dynamic material environments. For each local site, knowing and recognizing involve different sensory and visual perceptions, differences in stance in relation to the material environment, and differences in the consequences that result from knowing. Within a classroom, knowing means getting it right. In a coal mine, knowing may involve the body's response to sensory knowledge in order to avoid catastrophe. Because scientific knowledge is both distanced from the material conditions of particular sites and generalized (or idealized) from multiple data points, scientific knowledge can provide a rationale for particular decisions, but it cannot supply site-specific rules and procedures that will apply generally to all conditions in the material environment.

---

[79]DeKock, Kononov, & Oberholzer, n.d., pp. 20–21.

If embodied information is not present in written communication, we might ask, where is it to be located? In the following chapters, we show how miners deploy both speech and gesture to construct representations of risk that integrate theory and practice. These interviews allow us to examine the rhetorical transformations that occur outside of written communication at critical moments within the Cycle of Technical Documentation within Large Regulatory Industries.

III

DOCUMENTING EXPERIENCE

7

Embodied Experience: Representing Risk in Speech and Gesture

The "imagistic" component of language coexists with the linear segmented speech stream and suggests that the coordination of speech and gesture may provide fresh insight into the processes of speech and thought.[1]
—McNeill, (1992)

I am not unaware how great a task I have undertaken in trying to express physical movements and words and portray vocal intonations in writing. True, I was not confident that it was possible to treat these matters adequately in writing. Yet neither did I suppose that, if such a treatment were impossible, it would follow that what I have done here would be useless, for it has been my present purpose to suggest what ought to be done. The rest I shall leave to practice. This nevertheless, one must remember: good delivery ensures that what the orator is saying seems to come from his heart.[2]
—[Cicero], (1989)

As the previous chapter suggests, miners learn to understand danger as instincts, hunches or feelings. Miners describe the ability to sense these physical phenomena as pit sense, an embodied sensory knowledge derived from site-specific practice in a particular working environment. When miners describe their experiences, their gestures convey the physical presence of this embodied sensory experience.

---

[1]McNeill, 1992, p. 125.
[2][Cicero], 1942, p. 205.

Expert miners gesture more frequently on average than novice miners and use more gestures in which they use their entire bodies to portray a character in their narratives.[3] Gesture thus provides additional information about the spatial and temporal dimensions of risk—information not available in speech alone. The knowledge embodied in gesture may become invisible, however, if writers transcribe only the speech portion of oral testimony as evidence in scientific and technical accounts of the disaster.

These observations raised questions about the kinds of information individuals might represent in speech and gesture:

1. How might knowledge represented in speech and gesture reflect individuals' situated experience in risky environments?
2. How might individual miners use speech and gesture to represent the uncertain knowledge that characterizes risky situations?
3. What might a study of gesture reveal about the ways that knowers construct and negotiate strategic representations of risk outside of written scientific and technical texts?

To answer these questions, this chapter analyzes videotaped interviews with expert and novice coal miners over a six-month period from July 1995, to January 1996, in the United States and Great Britain.[4] As the following discussion suggests, miners' gestures are not merely pictorial or emotional embellishments to speech. Like speech, gesture can express both semantic content and a speaker's viewpoint.[5]

Speakers can use gesture to represent features of manner and motion not possible in English verbs.[6] They can use gesture to model the size and shape of

---

[3]Sauer & Franz, 1997.

[4]This research was funded by a National Science Foundation Grant for research in Ethics and Values Studies in Science and Technology (Grant # sbr-9321470). I would like to acknowledge the generous support of the Foundation and program director Rachelle Hollander for her continued support of the project. I would also like to thank David Kaufer and Linda Flower of Carnegie Mellon University for their careful readings of early drafts of this project; Laura Franz who worked with me to code and analyze gestures in the preliminary phases of the project; Susan Lawrence, who accompanied me to South Africa and made valuable suggestions in regard to character and observer gesture; Amy Cyphert, who has worked with me for four years to code and interpret gesture; Laura Martin, whose precise coding has helped me see new meanings in the data; Terri Palmer, who has worked with me to understand the translation and transformation of science in gesture; and, most of all, Martha Alibali, of the University of Wisconsin, Madison, whose work with gesture in children inspired this project.

[5]See Alibali & Goldin-Meadow, 1993; Alibali, Bassok, Olseth, Syc, & Goldin-Meadow, 1995; McNeill, 1992; McNeill & Duncan, 2000.

[6]McNeill and Duncan, 2000.

three-dimensional objects and environments. They can show how conditions and processes change over time. They can demonstrate procedures like installing a roof bolt or shoveling coal, streamlining and highlighting salient features of the process; or they can use their bodies to show how roof bolts work invisibly within the strata. They can use their hands to demonstrate the rolling force of explosions, or they can show the magnitude of a roof fall with their entire body. Speech and gesture together thus provide a richer representation of risk than either speech or gesture alone.

Recent research in psychology shows that speakers can express two different viewpoints in gesture—reenacting events as characters in their narrative (character viewpoint) or representing events as an observer (observer viewpoint).[7] This chapter extends previous rhetorical and psychological studies of gesture in order to understand how speakers use gesture rhetorically when they represent their experiences of risk.

This chapter argues that speakers can assume two different viewpoints in gesture and speech: mimetic and analytic. When individuals assume a mimetic viewpoint, they reenact events and conditions as if they were characters inside the spaces they describe. When individuals assume an analytic viewpoint, they assume a position outside of, above, or at a distance from the events, actions, and situations they observe. This distance is frequently figured in the gesturer's body. Mimetic gestures frequently take place in the plane of the body. Analytic gestures take place literally at arm's length from the observer. These viewpoints are figured in three-dimensional gestural space in front of the speaker's body, in contrast to the flat plane that previous analyses of gestures have used to figure gestural space.

Speakers can shift viewpoints to describe events and conditions from a new perspective. When speakers described their experiences of risk, they used mimetic gesture to reenact their own experiences and to imitate the actions and gestures of colleagues, management, and others—both human and nonhuman (including the mountain itself). In some cases, they used linguistic markers ("others may think"; "my boss said") to identify other speakers' voices and perspectives in their narratives. Speakers employed analytic viewpoint gestures to represent abstract processes, objects, and conditions that would be literally beyond the reach of human perception. As they talked, they shifted between mimetic and analytic gestures, directing the audience's attention to new aspects of the events, conditions, and processes they described.

One surprising finding of this analysis is that speakers can describe two distinct viewpoints simultaneously—one in speech and one in gesture. (One ex-

---

[7]McNeill, 1992.

ceptionally articulate miner expressed three viewpoints simultaneously—one in speech and one with each hand separately.)[8] But not all individuals gesture all the time. As this chapter suggests, both the presence and the absence of gesture can reveal important clues to a speaker's understanding of risk.

Analysis of gesture suggests that the viewpoints that miners assume in speech and gesture play an active role in the rhetorical transformation of experience. Like narrative viewpoint in speech, the viewpoints that speakers assume in speech and gesture determine the kinds of information that speakers can represent within that perspective. Writers accept the idea that narrative viewpoint places constraints on the content of fictional and non-fictional narration. But they may not be aware of how viewpoint in gesture also affects the semantic content of warrants embodied in gesture. When speakers shift between analytic and mimetic gestures, they change the spatial and temporal distance of audience and speaker in relation to experience. In chap. 7, we examine how speakers use viewpoints strategically, foregrounding and rendering visible aspects of experience not possible if we view the scene from another viewpoint. In the process, both speaker and audiences can develop a more complete understanding of risk as they work to capture and interpret knowledge embodied in gesture.

The examples in this chapter suggest that the coordination of speech and gesture can affect the rhetorical force of risk representations in local sites. When speech and gesture are concordant, gestures can reinforce speech, increasing the rhetorical force of individual representations of risk magnitude as a warrant for future action. When speech and gesture are discordant, audiences may force on only one channel of information, creating uncertainty in training and in written representations of risk. In a noisy environment like a coal mine, individuals may focus on representations embodied in gesture, losing the viewpoints and semantic content of speech.[9] Obviously, individuals cannot observe or experience events from more than one embodied location, but they can assume a new viewpoint rhetorically to imagine events from a new perspective. When individuals assume a new viewpoint, they turn their attention rhetorically to new features of the environment. They see events and conditions differently and make different judgments about risk.

Ideally, agencies might use new digital technologies to capture gesture in video. But the full potential of what speakers can express may be lost if writers

---

[8]This miner was interviewed by Amy Cyphert, a student at Carnegie Mellon University. I am grateful to Amy for her help in this project.

[9]McNeill, 1992, p. 125.

within agencies do not have the tools to capture knowledge embodied in gesture in writing; if writers cannot interpret the semantic meanings conveyed in gesture; or if writers do not understand how viewpoint affects the semantic content and rhetorical force of individual representations of risk. Risk communication may fail if speaker and audience do not have the cultural and linguistic resources to develop a common spatial and temporal referent.[10] The problem of finding commensurate referents is exacerbated when speakers must translate risk across difficult linguistic and political boundaries, when audiences do not share a common language, or when audiences do not have a vocabulary for scientific terms and concepts. In South Africa, where miners speak as many as 11 different languages, trainers must pay particular attention to the viewpoints they convey in speech and gesture.[11]

For rhetorical theorists, analysis of gesture extends our notions of what counts as literacy and suggests that we must examine what it means to define literacy in terms of multiple modalities—written, spoken, visual, or gestural. U.S. miners are trained to think about the world beyond the limits of their own experience; they can describe events and conditions from more than one viewpoint. But their scientific knowledge may not be apparent if researchers do not pay attention to gesture. South African miners, on the other hand, received little formal training, especially under apartheid. They know little about the structure of their environment and the effects of local changes with the system. Yet these miners have an acute understanding of the physical spaces in which they work; they know how gases operate and how fires spread. This knowledge becomes visible when we analyze both speech and gesture.

Graf suggests that "gestures do not perform a function totally different from language, they work together in a speaker's argument to underline and amplify the message of language by stressing the emotional, non-rational elements."[12] The following discussion expands upon previous studies of gesture as an emotive feature of discourse by providing a general taxonomy that identifies the mimetic and analytic viewpoints in both speech and gesture that characterize an experienced miner's understanding of risk. In learning to classify and interpret these viewpoints as a rhetorical feature of scientific and technical discourse, researchers in both rhetoric and risk can begin to de-

---

[10]Levinson, 1996. See also Jackendoff, 1996; Levelt, 1996.

[11]Sauer & Palmer, 1999. Efron's (1941) early research in gesture suggested that culture influenced the production of gesture. My own research suggests that miners who share a common material culture may employ similar gestures to represent similar objects and situations.

[12]Graf, 1991, p. 4.

velop more complete understanding of the knowledge that individuals express in speech and gesture.

HAND AND MIND: WHAT GESTURES REVEAL ABOUT THOUGHT

Recent psychological studies of gesture demonstrate that gesture communicates not only a continuing stream of semantic information but also the speaker's viewpoint in relation to the events, objects, and processes embodied in gesture. These studies suggest that the "imagistic component of language" coexists with speech and suggests that the coordination of speech and gesture—and its mismatch—can provide new insight into the nature of knowledge in dynamic and uncertain environments (McNeill, 1992).[13]

Psychological studies of gesture focus on the idiosyncratic gestures that speakers produce in ordinary speech, as opposed to the more formalized gestures (like handshakes or peace signs) that we associate with social convention. In ordinary speech, gestures (also called gesticulations to distinguish them from the more deliberate signs and conventionalized social gestures) occur in streams or strings. They have a beginning, a middle (or stroke phase), and an end.[14] Speakers may repeat gestures for emphasis or connect sequences of gestures. Ordinary gestures serve a variety of purposes in speech. Some gestures imitate motions (a rolling ball, for example) or model objects (as when a technician's hands seem to form an object in space). McNeill theorizes that beats (like a tapping finger) provide coherence and indicate changes in subject focus. Unlike handshakes and other conventional social gestures, gestures in this sense reflect idiosyncratic, spontaneous, and imagistic meanings that pre-exist speech.[15] Chawla and Krauss (1994) demonstrate, moreover, that ordinary viewers can distinguish the gestures of actors from the natural and spontaneous gestures of individual speakers.

To study these ordinary gestures, researchers in psychology have developed elaborate coding schemes that categorize the shape and movement of

---

[13]McNeill, 1992, p. 125.

[14]McNeill, 1992.

[15]McNeill, 1992, p. 12. Roodenberg (1991), for example, describes how conventionalized gestures like handshakes emerged in the Dutch republic. Historical studies have focused their attention on highly conventional gestures in portraits and illustrations because we have no archive to recover the idiosyncratic and individual gestures that occur in ordinary conversation. See also Bremmer, 1991; Dreissen, 1991; Schmitt, 1991.

the hand, its position in relation to an imaginary quadrant drawn in relation-ship to the body, and its duration in relationship to speech.[16]

In *Hand And Mind: What Gestures Reveal About Thought*, McNeill (1992) categorizes five types of gesticulation: iconic, metaphoric, deictic, cohesive, and beats. Iconic gestures are pictorial and "bear a close formal relationship to the semantic content of speech."[17] According to McNeill, iconic gestures reveal not only the speakers' memory images, but also their points of view as characters or observers in their narrative. Metaphoric gestures are also picto-rial, but the pictorial content refers to an abstract idea rather than a concrete object or event. Deictic gestures are pointing gestures. Cohesive gestures con-nect ideas and narrative events. Finally, beats are not pictorial gestures, but rather short, quick motions of the hand that move in a rhythmic beat with the speech. Beats have the same form regardless of the content. They are used to emphasize words, not for semantic reasons, but for discourse-related rea-sons. Beats mark speech that departs from the narrated chain of events, and provide a structure within the plot as the story is told.

McNeill (1992) argues that gestures reveal not only the speaker's "memory image" but also "the particular point of view that he had taken toward it."[18] According to McNeill, speakers can assume two distinct viewpoints in ges-ture. In the observer viewpoint (OV), the speaker keeps some distance from the narrative. In the character viewpoint (CV), the narrator is inside the story. The difference is apparent when a speaker describes someone running. In the character viewpoint, a speaker might move arms and legs like a runner. In the observer viewpoint, a speaker could represent another person running by wiggling the index and middle finger in rapid succession. These view-points accord with studies of memory, which suggest that speakers remember events as either characters or observers in their narratives.[19]

---

[16]More recent studies have used computer-aided modeling techniques that measure the vec-tors of motion or the force of the hand movement. Some studies employ gloves to mark move-ment, but these studies cannot get at the shape and meaning of the hand. There are several problems with computer-aided analysis: Gloves impede normal gesticulation and may make speakers self-conscious; computers cannot recognize the difference between the left and right hand. This problem of recognition produces the so-called crossover problem—when speakers cross hands, computers interpret the gestures as moving to the center and out again. Computers can track movement, but they cannot interpret the difference between (for example) a pointing gesture and an identical gesture that is meant to represent the flight of a bird.

[17]McNeill, 1992, p. 125.

[18]McNeill, 1992, p. 13.

[19]Schachter, 1996.

McNeill demonstrates that speakers may also present more than one viewpoint in their gestures. He characterizes dual-viewpoint gestures that simultaneously present a scene from two viewpoints (as opposed to shifts between character and observer roles). These dual-viewpoint gestures enable the narrator to take on a character's role and to observe it simultaneously. According to McNeill, adults use dual-viewpoint gestures to diagram the abstract dimensions of interpersonal communication, to indicate irony, or to indicate when the speech of one character is quoted. These dual-viewpoint gestures occur in the speakers' hand movements within the frame defined by the speaker's body. McNeill does not describe the ways that speakers might employ speech to represent a second viewpoint, distinct from the viewpoint presented in gesture.[20]

McNeill hypothesizes that children use dual-viewpoint gestures because they may not differentiate the viewpoint of the observer from the events of the story itself. In adults, he suggests, dual-viewpoint gestures may fill a logical gap in meaning, create new meanings, or allow a speaker to express two meanings simultaneously.[21]

Goldin-Meadow and Alibali (1995) have extended McNeill's analysis to study speech-gesture mismatch in children (where what a child says is different from what he or she gestures when explaining a problem). These studies show that speakers can represent two different semantic meanings or viewpoints simultaneously in speech and gesture and that these two distinct meanings may represent different or transitional approaches to a problem.

Goldin-Meadow and Alibali (1995) have used gesture to study how children use gesture when they talk about their solutions to simple mathematical problems. This analysis shows that speech-gesture mismatch can provide insight into the learning process as students move from one level of understanding to another. Goldin-Meadow and Alibali theorize that "the discordant state [where speech and gesture frequently mismatch] appears to be transitional in that it both predicts openness to instruction and is sandwiched between two relatively stable states."[22] They thus conclude that the "within-problem variability" (when speakers employ two or more distinct strategies for problem-solving) in speech-gesture mismatch "is indeed associated with periods of transition."[23] In these studies, discordance refers to the semantic meaning rather than to the speaker's viewpoint on the event.

Goldin-Meadow and Alibali (1995) note that their study "leaves as an open (and testable) question whether gesture-speech mismatch identifies

---

[20]McNeill, 1992, pp. 122–3.
[21]McNeill, 1992, p. 318.
[22]Goldin-Meadow & Alibali, 1995, p. 125.
[23]Goldin-Meadow & Alibali, 1995, p. 125.

learners in transition throughout the life span."[24] In fact, one of the assumptions of this research is that children follow a developmental path as they learn and embody more and more abstract gestures in written and oral problem solving. They thus raise questions about the ways that adult learners might represent problems as they encounter new and unfamiliar social and technical contexts. These studies have also only recently focused on the ways that rhetorical situation can affect a speaker's use of gesture.[25]

THE ROLE OF GESTURE IN SPEAKERS' REPRESENTATIONS OF RISK

Psychological studies of gesture provide an analytic framework for understanding how gestures function rhetorically in individuals' representations of risk. These studies raise questions about how adults learn gestures as they enter new discourse communities and suggest that gesture may provide a method for understanding how individuals represent risk outside of written texts.

In situations of risk, knowledge is uncertain. To warrant judgments about risk, individuals must negotiate among many competing representations. As we have seen in previous chapters, this notion of risk goes beyond a simple relativism (each person has his or her own viewpoint or interpretation) in order to argue that risk knowledge requires knowers to understand that each viewpoint provides a situated but incomplete view of the whole.[26] In these environments, having access to more than one viewpoint—the viewpoints of both unions and inspectors, for example, or the systems approach of an engineer—may provide decision makers with a greater range of problem-solving strategies than any single representation from a single viewpoint.

When miners talked about risk, the viewpoints that they assumed in speech and gesture differed from laboratory subjects' viewpoints in several important respects. These differences reflect the uncertainty of risky environments and the nature of miners' embodied experience in local sites. Because laboratory studies of gesture ask subjects to recall events they have observed in cartoons, laboratory subjects cannot describe themselves within the space of the narratives they recount in speech and gesture. When speakers remembered events they themselves had experienced and spaces they themselves had occupied, they could reenact and analyze their own experiences within the spaces they describe in both speech and gesture.

---

[24]Goldin-Meadow & Alibali, 1995, 117.
[25]Littleton & Alibali, 1997.
[26]See Code, 1991; Harding, 1991; Haraway, 1991.

The complexity of risk experience that speakers described also affected our analysis. Researchers in laboratory studies have access to the narratives that speakers reconstruct in speech and gesture. But miners described a range of experiences—some familiar and some quite complex. Because we did not have access to the original events that speakers describe, context was critical in understanding a speaker's semantic content in gesture—to distinguish when a miner was representing the twisting motion of installing a roof bolt rather than a light bulb, for example.

Analysis of miners' speech and gesture suggests, first, that individuals can represent themselves and others as characters in their narrative (mimetic viewpoint) and second, that they can move outside of this embodied experience to observe and analyze events from a distance (analytic viewpoint). When speakers assume a mimetic viewpoint, they enact events directly with no rhetorical distance between themselves and the action. When speakers assume an analytic viewpoint, they place temporal and spatial distance between their current position (as observers or narrators) and the events they describe.

Mimetic and Analytic Viewpoints

Mimetic and analytic viewpoints provide the building blocks that enable miners to integrate multiple viewpoints simultaneously to represent risk. Speakers can employ both speech and gesture to deploy two viewpoints simultaneously or sequentially. Speakers sometimes gesture during pauses in speech; few speakers gesture all the time. As a result, there are moments when there are missing viewpoints in their representation. These missing viewpoints call attention to larger problems of representation or absences in a speaker's understanding of risk.

The terms *mimetic* and *analytic* allow us to categorize viewpoints so that we can understand how speakers negotiate among competing viewpoints as they work to construct representations that capture the geographic, institutional, and temporal complexity of risky environments.

I describe these viewpoints briefly in order to lay out the conceptual distinctions between the mimetic and the analytic. The hypothetical examples in this discussion are merely illustrative and do not reflect actual examples of miners' discourse. In the sections that follow, I describe these categories in detail and provide real examples from my interviews with miners.

Mimetic Viewpoint. Speakers can represent themselves as characters in their own narratives or they may mimic (imitate) the viewpoint of an other—including both human and non-human others. When narrators assume a mimetic viewpoint in gesture, they reenact events as if they were present at the scene they describe. Thus a speaker might wave her hands frantically to depict herself "flaggin' and screamin'" as she remembers herself trapped against the mine wall. A speaker who imitates an arch would raise her arms over her head to represent the arch with her arms.

When speakers assume a mimetic viewpoint in speech, they imitate the voice, tone, and language of the character at the time of the event. They frequently speak in the present tense and recreate dialogue like an actor. When narrators employ a mimetic viewpoint to represent colleagues and buddies, they may actually change the tone of their voices. One woman miner, for example, imitated the deep bass tone and perfunctory manner of her boss. In some cases, narrators may mark the shift in their speech. When narrators assume a mimetic viewpoint, audience and gesturer stand in a different temporal and spatial positions in relation to events in the narration. Audiences observing the gesturer see the gesturer reenacting events. Gesturers reenact events as if they were again present in the scene they describe.

Analytic Viewpoint.   Although they cannot change their literal position in relation to the events they describe, narrators can move outside of their own embodied experience in their representations in order to analyze events from a distance. When narrators assume an analytic viewpoint, they observe and evaluate action from a distance. Analytic viewpoints help narrators make sense of actions, events, and situations they describe. They can comment on their own action from the distanced perspective of another observer, or they can look back upon and comment upon past actions, events, and conditions.

When narrators employ an analytic viewpoint in gesture, they frequently gesture "at arm's length" in front of their bodies so that they can observe the action from a position outside of and above the gestures they depict. Thus, instead of representing her own actions, a miner would use her hands to depict what she saw from a distance as she stood pinned against the wall—the operator of the machine, the machine coming at her, or her buddies running to get help. To represent an arch from an analytic viewpoint, the miner would draw a small arch with her fingers.

When narrators assume an analytic viewpoint in speech, they represent events in the past tense. They explain events and situations, add asides to orient themselves and the audience, and frequently use third person in their narratives: "The machine had nearly pinned me against the wall. There was very little space between me and the continuous mining machine. He [the operator of the machine] was coming at me full force."

TABLE 7.1

Eight Possible Combinations of Speech and Gesture

	MM	MA	AA	AM	MØ	AØ	ØM	ØA
Speech	Mimetic	Mimetic	Analytic	Analytic	Mimetic	Analytic	—	—
Gesture	Mimetic	Analytic	Analytic	Mimetic	—	—	Mimetic	Analytic

Audiences observing analytic gestures stand in the same relation to the narration as the speaker. In this case, both speaker and audience observe and analyze events from a distance.

Missing Viewpoints.   If we focus on those moments when individuals use both speech and gesture, we may fail to account for those instances when individuals use only one modality—either gesture without speech or speech without gesture. Narrators may use no gestures when they describe past experiences or rehearse information they have memorized but not experienced. Narrators may also gesture without accompanying speech. When one viewpoint is missing, audiences receive only one channel of information instead of the two channels provided in speech and gesture together. Miners sometimes gesture during a pause in their speech. In this case, miners may lack a vocabulary to express their visual and embodied experiences of risk.

Multiple-Viewpoint Representations

Speakers can use mimetic and analytic viewpoints as building blocks to construct representations that integrate more than one viewpoint simultaneously and sequentially. Speakers can potentially represent eight different combinations of viewpoint in speech and gesture (Table 7.1).

Speakers can use these combinations to represent their own experience or they can assume the viewpoint of others. Like the Welsh miner in chap. 4, speakers can integrate the viewpoints of many different individuals and organizations into a single narrative. Analysis of viewpoint in speech and gesture adds a new dimension to the analysis of viewpoint that we presented in chap. 4.

When miners represent their boss, they can imitate his actions, manner, and hand motions. When they represent a roof bolt mimetically, they hold their arms at an angle—like a wing nut. When they represent a roof fall mimetically, their arms embody the rock. They can represent the boss analytically as a small figure at arms length; they can depict a roof bolt with two fingers; and they can show how the rock falls with the flick of a finger. The taxonomy in this chapter helps us categorize each viewpoint in turn so that we can see how individuals shift between mimetic and analytic viewpoints in their narratives.

Speakers can deploy both speech and gesture to represent more than one viewpoint simultaneously and sequentially in their narratives:

Simultaneous Viewpoints. Speech and gesture provide two channels of communication that enable speakers to present more than one viewpoint simultaneously. Miners can employ mimetic and analytic viewpoints concordantly, or they can employ different viewpoints simultaneously in speech and gesture. The miners we observed employ mimetic viewpoints simultaneously in speech and gesture, analytic viewpoints simultaneously in speech and gesture, or two different viewpoints simultaneously in speech and gesture.

Thus, a miner can employ a mimetic viewpoint in speech ("Here I am flaggin' and screamin' ") while she represents the oncoming mining machine with her hands in an analytic viewpoint gesture. Or, she might wave her hands wildly in a mimetic gesture (reenacting herself flagging and screaming) as she calmly described the scene from the analytic viewpoint of a distanced observer: "He was coming at me with the continuous miner [mining machine]."

Sequential Viewpoints. Individuals frequently link analytic and simultaneous viewpoints in a sequence of contrasting or thematically linked viewpoints. Or, they may describe a single event from a variety of viewpoints. These complex narratives may help narrators integrate the complex and frequently contradictory viewpoints that characterize risk. When speakers deploy more than one viewpoint in sequence, audiences must continually redefine their interpretation of the narrator's position in time and space. The following hypothetical narrative combines several viewpoints in sequence:

Speech: Here I am flaggin' and screamin' (mimetic viewpoint in speech). He was coming at me with the continuous miner (analytic viewpoint in speech). There was very little space between me and the mining machine (analytic viewpoint in speech). My boss screamed (analytic viewpoint), "Get out" (mimetic viewpoint).

Gesture: Holds hands in front to measure space (analytic gesture). Waves hands as if flagging the operator (mimetic gesture—representing self in narrative). Imitates posture of boss (mimetic gesture—representing another character in the narrative). Points to audience (mimetic gesture imitating boss).

Psychological studies of gesture suggest that most gesticulations are spontaneous and idiosyncratic. But when speakers are conscious of gesture, they can deliberately shift viewpoints, like actors who coordinate specific forms of gesture with speech.

The following examples show how miners employ mimetic and analytic viewpoints when they observe and remember risk. (Appendix A provides a brief overview of the subjects we interviewed and our research method.) Chapter 8 will discuss the ways that miners can use these viewpoints strategically and deliberately to produce a new understanding of risk.

MIMETIC VIEWPOINTS IN MINERS' REPRESENTATIONS OF RISK

Unlike two-dimensional or flat representations in a written text or illustration, mine space is dynamic, uncertain, and multi-dimensional. Miners must orient themselves and their equipment in relationship to other miners working in the same section. As miners advance the face, they must move machines and equipment forward. As they cut coal, they change the dimensions of their environment. As they remove coal from the mine haulageways, rock strata shift in response to the changing forces, producing roof falls and eruptions. To prevent roof falls, miners must install temporary supports and, later, permanent roof supports.

When miners work underground, they lose their ordinary sense of space and time. They must understand their position in relation to the coal face and in relationship to the three-dimensional patterns of rooms, haulageways, cross-cuts, and exit ways beneath the surface of the mine. They must orient themselves inby (inside of) and outby (outside of) the coal face so that they can describe their positions to others and know the means of escape in an emergency.[27]

When miners describe their experiences, they depict themselves as characters within these spaces. As they speak, their concentrated focus suggests that speech and gesture work together to produce a highly emotional and intense memory of their work. Sometimes, experts also assume the viewpoint of another character in their narrative. These mimetic gestures suggest that they can put themselves in the place of an Other because they can imagine themselves as Other in the spaces in which they work. Even when miners represent themselves, however, this Self represents another past self, located in another historical space, but vividly present in the miners' narratives.

---

[27]Chap. 4 defines these terms in detail. As we have seen in chap. 4, miners inby a fire may have no means of escape in a single-entry mine.

When speakers employ mimetic viewpoints, they reenact their embodied memories of risk. They also reenact their sensory and physical memory of the spaces, social relationships, and institutional relationships they encountered as they worked. As the following discussion suggests, their gestures depict more than the mere physical dimensions of objects and spaces in their environment.

Imitating Self.   Miners must learn from experience to prevent risk in the future. The role of embodied experience is particularly evident when experienced miners learn to use new equipment and machinery. For each miner, generalized instructions and procedures can provide general guidelines for using equipment, but each miner must test his or her individual strength against the physical force of the machine. These embodied experiences build a body of knowledge miners can use to prevent disaster in the future. Next time they will hold tighter, press harder, or react more quickly. In many cases, however, a miner's experience is literally embodied as scars, bruises, and missing fingers.[28]

Expert miners use mimetic gestures to depict these harrowing experiences. In the following example, E3 uses mimetic gestures to describe the difficulties he experienced the first time he operated a roof-bolting machine. Because he is small (less than five feet three inches tall), E3 must struggle to control the machine.

E3 frames this experience within the physical and institutional uncertainty of the mine. Each location in the mine literally produces a new experience. As a result, E3 must work hard to control the machine if it hits "a load of hard stuff," "if it hits something hard," or "if you hit a lump or rock, or a lump of hard stuff."

E3's gestures place him as a character in the narrative, reenacting past memories even as he focuses intently on the imaginary wheel. His gestures move his narrative back to the time of the remembered action as he describes the problem to his audience. During each segment of the narrative, E3 holds tightly to an imaginary wheel (Figs. 7.1, 7.2).

Imitating Spaces.   Miners also use mimetic viewpoints to depict the size of the physical spaces they must work in. Unlike conventional fish stories in which speakers use their hands to depict the size of objects from a view outside of the object, miners use their entire bodies to depict the magnitude of hazards they have observed.

---

[28]Sauer, 1998.

FIG. 7.1.   E3 holds tightly to an imaginary wheel in this mimetic gesture.

FIG. 7.2.   E3 holds the wheel intensely. In this mimetic viewpoint gesture,
he reenacts his physical struggle to control the machine with his entire body.

In Fig. 7.3, E5 uses her entire body to describe the extent and magnitude of
a roof fall underground. Her gestures fill the entire frame and suggest the
emotional as well as physical magnitude of the event. This miner describes
the magnitude of the event from a position located within the space of the
hazard she describes. In her gesture, her entire body falls within the plane of
the space she describes.

Imitating Others.   Many miners imitated the voice and posture of others, including fellow miners and management. In these mimetic gestures, expert miners' gestures reflect highly local social relationships.

In Fig. 7.4, E5 imitates the voice and actions of her boss, who encouraged
her to become a roof-bolter operator. In this case, gesture and speech work together to reinforce the positive—but also emotionally charged—interaction
between the miner and her former boss. Although E5 insisted that she faced

FIG. 7.3.   E5 uses her entire body to describe the extent and magnitude of a roof fall.

FIG. 7.4.   E5 imitates her boss's gestures. In the narrative, she imitates his voice. This speaker uses her entire body (note shoulder position) and head to represent another speaker talking.

no sexism in the mines, she frames the following narrative in terms of race (her boss was Black, the only Black boss at the mine), gender (she is a girl), age (he was young), and class (she has worked as a helper and wants to move into the more respected position of operator). The scene must also be framed by an understanding of institutional hierarchy and procedure. As both E5 and her boss know, letting an apprentice operate the machine will violate mine procedure:

E5:   When I first went to evening shift I'd been there ninety days when I went to evening shift and I was on a pillar section. My boss was Black . . . and young

... the only Black boss at the mine, and was young and he just—you know he knew all my older sisters, went to school with 'em and stuff, and he would make me go up and bolt during dinner time ... now mind, it was illegal for me to be on the equipment.

.....

He'd say, <u>when you goin' to learn</u> [voice of the boss]?[29]

Ironic Imitation. Imitating others can produce irony when speakers do not share the same institutional position and ideological understanding of the persons they imitate. Miners have a long history of adversarial relations with management. They are suspicious of management's motives, particularly in regard to the economics of safety. They have heard the reasons why safety is expensive, and they know that profit and production drive managers' decisions in regard to risk. But miners also know that profit and production keep mines operating, especially in the face of mine closures and economic uncertainty in the mining industry. Miners' individual institutional encounters with particular managers are thus fraught with the kind of emotional intensity that might be expected to produce mimetic gestures.

The following example supports this hypothesis. In this case, E5's tone is critical to understanding both the semantic meaning and the viewpoint of the gesture. In this example, E5 uses direct quotation to mimic her manager, who claims that providing adequate roof support is "too expensive" and "not profitable." In this case, hearing the speaker's tone is critical to understanding the irony implicit in her claim that adequate safety measures are "not profitable."

The miner's speech is also critical to understanding the discordance between the viewpoint she assumes rhetorically and her own viewpoint as a miner. If E5 assumed a miner's viewpoint for the phrase "stick some bolts in," we might expect her to use her arm to depict the torquing motions of a miner inserting a roof bolt in the overhead strata. When E5 assumes the viewpoint of her manager, she does not depict the torquing motions and embodied knowledge of an experienced roof bolter. (Miners usually raise one arm and cup their hands as if they were screwing in a light bulb to indicate in a characteristic motions of installing a roof bolt.)[30] Instead, her gestures reflect her manager's lack of embodied experience (Fig. 7.5):

E5:
It's extremely expensive,

---

[29]Underlining in the text indicates the stroke phase and duration of the gesture. See McNeill, 1992.
[30]Sauer & Franz, 1997.

FIG. 7.5. E5 imitates the gestures of her manager. Although she talks about installing roof bolts, she does not employ the characteristic torquing motions (like screwing in a light bulb) that experienced roof bolters employ when they describe how to install a roof bolt. Her gestures demonstrate that manager and miner do not share similar viewpoints in relation to risk and safety.

It's not profitable; we can't afford to support the top adequately. <u>Stick some bolts in it and we'll all be happy.</u>

Imitating Non-Human Others. U.S. and British miners are frequently animated storytellers who live in communities where stories keep alive the memory of those lost in mine accidents, the unwritten history of labor politics, and the near-misses and accidents that characterize daily life in the mines. In the following example, E3 uses mimetic gestures to imitate the poor parrot in the mines who discovers the presence of gas and wants to get out quickly.

I have included the following example because it illustrates two points: First, miners use mimetic viewpoint gestures to inject humor into their narratives, and second, E3's humor may be a more general form of gallows humor that enables miners to make light of the seriousness of the risks they faces. The humor thus masks the seriousness of E3's sympathetic understanding of the parrot's situation.

In this sequence, the parrot doesn't fall off the perch—as most might people assume he would—and simply passively die in the presence of gas. Instead, the parrot appeals to the gas man (literally, the miner in charge of methane checks) to get him out of the cage (where presumably he would die from the gas). The implicit moral of the story points to the shared experience of parrots and miners, who are figuratively caged together underground:

> E3:   92. The parrot doesn't fall off the perch. It goes, "<u>Get me out of this bloody gas will you . . . gas man . . . gas</u>."
>   *Mimetic Viewpoint Speech:* Imitates voice and tone of parrot.
>   *Mimetic Viewpoint Gesture:* Miner raises hands, grasps bars, shakes them.
>   Body becomes part of gesture as he imitates a parrot jumping in the cage.

Imitating Geography.   Expert miners have observed and experienced near misses and—in some cases—disastrous accidents.

In Fig. 7.6, E5 describes a massive roof fall. In this gesture, E5 raises both hands above her shoulders, drops her hands, palms forward, fingers up, to imitate the crashing rock in a roof fall. This gesture is a typical roof fall gesture that is common among in miners in the United States, Great Britain, and South Africa. In the complete sequence of this gesture, E5 uses her entire body to represent the falling rock.

ANALYTIC VIEWPOINTS IN MINERS' REPRESENTATIONS OF RISK

Expert miners frequently describe objects and processes they have observed from a distance—the curve of an arch in a mine roadway, the subtle torque miners apply when they install a roof bolt, or the ways miners might use their

FIG. 7.6.   E5 depicts classic "roof fall" gesture.

hands to measure the spacing of roof bolts. Movements of the hand illustrate features of dimension, magnitude, and proportion missing in conventional English language descriptions in speech. When miners assume an analytic viewpoint, they can evaluate and describe hazards without directly experiencing risk. In this sense of the term, the analytic viewpoint is passive and detached, more abstract than mimetic gestures, distanced in space and time, and thus outside of the speaker's direct sensory and emotional experience.

To assess risk, workers in risky occupations must actively move outside of their own embodied experience. When workers assume an analytic viewpoint, they can speculate about the causes and outcomes of events they have not directly experienced and assume a viewpoint rhetorically outside of and above their own institutional and geographic location. When miners describe risk, they can employ this viewpoint to reflect upon and make sense of the cause of events and procedures they have experienced. They can move outside their geographic location in order to observe and analyze events and processes from the viewpoint of colleagues, apprentices, and management. The analytic viewpoint in these narratives may thus reveal a miner's perception of the (frequently watchful) social and institutional relationships among management, apprentices, helpers, and team members underground.

In the following examples, narrators represent an active accounting of risks they have processed, analyzed, and, in some cases, actively come to terms with.

Analytic Detachment. Miners often assume an analytic detachment when they describe highly emotional risks they have experienced and observed in the past. In the following example, E5 remembers one of two incidents in response to the question: What is the most dangerous thing that's ever happened to you? In this sequence, E5 describes herself "flagging and screaming" as she is caught between two equally dangerous spaces—the unsupported "top" (the roof of the mine) and an oncoming mining machine that will crush her if the operator doesn't hear her. If E5 runs out from under 40 feet of unsupported top, she risks death in massive roof fall. If she stays, she'll be crushed.

The emotional intensity of this narrative might suggest that E5 would produce mimetic gestures to depict this scene (particularly when she describes herself "flagging and screaming"). Instead, E5 employs an analytic viewpoint that suggests an emotional detachment as she views the oncoming machine as it pins her to the rib. Her hands depict just how close the machine came to her own ribs.

FIG. 7.7.   E5 looks into the space created between her hands. In this case, E5's hands depict the two-foot pads of the t-bar which pinned her to the rib and nearly crushed her.

Throughout this scene, E5 calmly assesses the problem in spatial and institutional terms. In this narrative, E5 is in the wrong space in relationship to the driver, who is behind the machine. To escape danger, she must communicate her position to the driver above the noise of the cutting machine. Her calm tone and analytic detachment throughout this sequence suggest that she focuses on the size and position of the pads that have pinned her to the rib. In telling the story, she carefully analyzes the operator's failures (in cutting off center and in failing to maintain communication with his partner) and her own compromised location in relationship to the machine. Her body position and focus recall E3's gestures when he represented himself staring into a dangerous situation (Fig. 7.7):

E4:   About a year ago, a guy I was boltin', he trams the pinner from place to place, he's on the other side of it, and it was cut off centers and was kind of curved around, and when he came up into the place he couldn't see me and I was standing at the last row of bolts, and one end of the t-bar—both ends have big pads, two foot square, . . . and where I was standing against the rib, the t-bar itself pinned me to the rib, and it was trammin' [moving forward] . . . and my only choice was to run out from under forty foot of unsupported top or be crushed <u>and here I am, flaggin' and screaming</u> and he shut the pinner off and said, are you flagging me?[31]

[31]Libby. (Interview with female miner, U.S.), NIOSH Conference, Morgantown, West Virginia, March, 1996.

FIG. 7.8.   E5 traces shape of an arch with flat palm of hand, fingers splayed. In a mine, these steel arches are overhead—like the metal supports in a tunnel. As she describes the arches, E5 assumes a viewpoint that is above the arches, looking down upon them, as she traces their shape with her hands. To assume a mimetic viewpoint of the same space, miners hold both hands overhead in the shape of an arch.

Architectural Landscape.   Expert miners frequently use gesture to create dynamic, three-dimensional maps that help them locate events and decisions in relation to particular equipment and structures in the mine.[32]

In the following sequence, E5 describes arch support, a method of roof support commonly employed in British mines. In this sequence, E5 assumes an analytic viewpoint that does not place her within the space of the arch (as she might have depicted herself in a mimetic gesture, holding her arms above her head in an arch) or inside of a narrative of risk. Instead, E5 is safely above and outside of the environment she describes as she traces a tiny arch with her finger. The following passage is purely descriptive. E5 is an observer—an American visitor who is the guest of British miners in a British mine. As a result, E5 does not depict herself as a character in the narrative (Fig. 7.8):

E5:   <u>Big steel arches</u>, you know, like I-beams and they're arched and this big metal looks like valley tin but it's thick, you just bolt to it.[33]

---

[32]McNeill notes that observer-viewpoint gestures are more likely to accompany static and intransitive verbs. This feature of observer-viewpoint gestures may reflect the type of verbs speakers generally employ to describe physical dimensions of objects they observe (the roof is high; the arches are like I-beams). See McNeill, 1992.

[33]Libby. (Interview with female miner, U.S.), NIOSH Conference, Morgantown, West Virginia, March, 1996.

Analytic Judgment.   In risky workplaces, experts must analyze human performance to assess safe work practices, to determine the sequence of tasks for particular jobs and positions, or to set general performance standards. To construct a body of risk knowledge, experts must imagine past events they have neither seen nor experienced. To determine the cause of an accident, for example, investigators must reconstruct victims' actions from physical evidence after the accident and from their own knowledge of mine practice, particularly if there are no survivors or witnesses to the actual event.

In the following example, E1 explains to apprentice miners why their proposed safety switches would not work on manriding conveyor belts. He describes a previous accident in which a worker lost an arm when he was thrown ("chucked up") on the blind side of the belt. In this example, E1 has not seen the accident, although he has ridden belts himself. As a result, E1 lacks the embodied experience that might enable him to reenact events as a character in the narrative.

From an institutional perspective, E1's narrative serves a rhetorical function as both example and warning of the consequences of the victim's carelessness getting on the belt. E1 frames this narrative with an institutional warning that even the best lockout system will not prevent an accident if miners fail to follow safe practice. In these analytic viewpoint gestures (indicated within the narrative), he depicts the miner with his hand and fingers only—not with his entire body. The action appears in front of E1—at arm's length:

E1: Even with the best lockout system in the world, would that have helped him. It was the ways that he was getting on the belt that were

> (no gesture)

the problem. He jumped <u>on, rolled over like a ball,</u>
>   (1)   (2)

> > (1) Quick presentation of both hand forefingers pointing towards each other (analytic viewpoint gesture)
> >
> > (2) Both hands, fingers pointing toward each other, imitate rolling motion, repeat till end of phrase (analytic viewpoint gesture)

and <u>was chucked up in blind side,</u> and

> > Thumb out, fingers curled in, arm swings from center to upper left, holds for remainder of thought (analytic viewpoint gesture)

<u>there's no way while he's rolling back</u>

> > Same as (2) but less pronounced (rolling gesture)

he's going to think of that side

>Hand in a fist, bends elbow and quickly brings fist to shoulder (analytic viewpoint gesture)

so he can get to the lockout. It was too late. It was the getting on process that were the problem.[34]

SIMULTANEOUS VIEWPOINTS IN MINERS' REPRESENTATIONS OF RISK

Miners frequently express two different viewpoints simultaneously in speech and gesture. When miners employ two distinct viewpoints in speech and gesture, they can frame their discourse in one viewpoint and reenact their embodied sensory experience in another. They can introduce irony or view themselves and others from a distanced analytic viewpoint. They can analyze and comment upon another's action in speech while they simultaneously reenact events and experiences as characters in their gesture.

In the following example, E5 uses the same gesture that E3 used to depict himself holding the wheel (which we characterized as a mimetic gesture). In this case, however, E5 reenacts another miner's experience—her sister's—while she describes the same events from her own viewpoint from a safe distance on the other side of a ventilation curtain (Fig. 7.9):

>E5: My sister was on the crew, and she was putting supplies on the bolter, and when I got up she was still standing there holding on to the plate.
>
>*Analytic Viewpoint* (Self) in speech: Miner stands a distance from the event as it occurred. As she tells the story, she does not reenact her own experience in speech, but instead, calmly describes what she saw happening in the space in front of her.
>
>*Mimetic Viewpoint* in gesture (Mimetic Other): Miner depicts her sister holding the roof bolt. The audience observes what the miner saw—her sister's embodied fear.[35]

As these examples suggest, miners' narratives of risk are complex, multi-layered, and richer than simple textual accounts may suggest. If we look at gesture alone, we may misinterpret the viewpoint she embodies in her gesture. If we look at transcripts of speech alone, however, we may miss the ways

---

[34]Anon. (Trainer, U.K.), near York, England, July, 1995.

[35]Libby. (Interview with female miner, U.S.), NIOSH Conference, Morgantown, West Virginia, March, 1996.

FIG. 7.9.   In this sequence, E5 represents two viewpoints simultaneously in speech and gesture. In her gesture, E5 holds an imaginary object in her hands. She assumes the character of her sister, who was holding onto the plate during a roof cave-in. In her speech, she describes the event from the distanced viewpoint of an observer.

that experts can reenact the experience of others—both human and non-human—visible in E5's embodied reenactment of her sister's experience.

MISSING VIEWPOINTS IN MINERS' REPRESENTATIONS OF RISK

The absence of gesture in speakers who usually employ a high number and frequency of gestures as a general feature of their discourse suggests that gesture is more than a natural or cultural feature of expert discourse. If mimetic gestures enable experts to reenact events with a high level of sensory input or a high emotional content, these missing viewpoints may signal those topics where experts lack embodied knowledge or emotional attachment. When speeches become rehearsed, through repetition or memorization, speakers may also not repeat the spontaneous gestures that provided the imagistic foundation for their narratives. But missing viewpoints may also indicate that speakers have suppressed their imagistic memory of powerful and threatening events.

While the following examples are by no means exhaustive, they suggest that the absence of gesture might provide clues to the role of gesture as a significant feature of expert discourse.

Missing Experience.   Novice miners use both character viewpoint gestures and observer viewpoint gestures to describe machines and processes underground.[36] But sometimes these miners lack the embodied experience of risk. As a result, they can describe events and procedures in words, but they use no gestures as they speak.

N1 and N2 belong to a cohort of novice miners in a six-month training program in north of England. When N1 and N2 describe the process of roof bolting, they both use analytic and mimetic gestures to depict the arms of the roof-bolting machine. The gaps in these miners' embodied experience become evident as they describe every day processes and situations of risk. These gaps are particularly evident because both miners have shared the same training program for six months. The differences in these miners' gestures suggest that they do not share the same embodied understanding of their environment—particularly when they describe risk.

When N1 describes how he can tell when a situation is risky, he shifts between mimetic and analytic viewpoint gestures. As an observer, he describes how he evaluates the signs of risk: "There's things all over the floor" and "bits of rock" hanging around. In the first gesture, his hand motions sweep back and forth, depicting the flat surface of the mine floor. He shifts to a mimetic viewpoint inside of the mine space as he raises his chin to describe what the listener will see "if you tend to look above." In the final gesture, he returns to an analytic viewpoint as he tells the interviewer "there's bits of rock hanging around." In this final passage, his gesture does not match his speech. Instead, N1 repeats the analytic viewpoint gesture he used to depict the messy mine floor. N1's gestures thus suggest that he understands that "things all over the floor" and "bits of rock hanging" are signs of potential danger:

N1:   It depends . . . If, if it's a messy place, you know for a fact. It's going to be risky. There's things <u>all over the floor</u> . . .

    Analytic gesture: Fingers splayed; hands move away to sides while staying parallel to floor (indicating flat surface), then return to center.

It's . . . you know you can't see as well . . . If it's low . . . <u>if you tend to look above</u>

    Mimetic gesture: Looks up toward ceiling, raising eyes and chin. N1 gestures with hands as if presenting object forward.

<u>there's bits of roof</u> rock hanging around, you know. Look at the signs.

    Analytic gesture: Both hands palm down, fingers splayed. N1 gestures back and forth across an indicated floor area.[37]

---

[36]Sauer & Franz, 1997.

[37]Anon. (Interview with British novice, N1), Doncaster, England, January, 1996.

When N2 was asked to describe how he knew a situation was risky, he used no gestures at all. N2 tells the interviewer that he's supposed to survey the area, but when pressed, he cannot articulate the signs of risk. Unlike N1, he fails to connect "things hanging down" to a bad roof. Instead, N2 attributes "things hanging down" to poor maintenance. His answer explains why the mess is still present, but he does not know where it came from. When he tells the interviewer "it's generally looking with the eyes," he does not raise his eyes or depict his own embodied experience looking upward within the space of a mine.

Many miners are shy and inarticulate (though many also are loquacious). It is possible that N2 was nervous in the presence of the American interviewer. His missing viewpoint suggests, however, that N2 does not have an "imagistic" or embodied understanding (cf. McNeill, 1992) of the spaces he describes:

BS: How do you know when a dangerous situation has arisen?

N2:   Before you enter an area, you're supposed to, um . . . survey the area of work, and most of the time you can see if there's a problem there and if there's going to be a hazard. I mean, there's indicators and things that will go off if there's gas. . . . It's generally looking with the eyes.

BS: What do you see?

N2:   Well, it could be . . . it could be things . . . it depends on what kind of thing it is. There could be things handing down or things . . . It all depends on how well the area's maintained, see.

BS: You can just tell, right?

N2:   I can just . . . You learn . . . You just, you look and [see what] can go wrong.
(no gestures)[38]

Both miners were interviewed in the same rhetorical situation. They shared the same training program and worked together underground. Both miners use a high number and frequency of gestures when they answer other questions in the interview. The difference in their understanding of risk become apparent only in the difference in their gestures.

Missing Illustrations.   When individuals describe mathematical shapes and concepts, they frequently use their hands to form triangles or they may trace a triangle in the air with their index fingers. Goldin-Meadow and Alibali (1995) have shown how children use specific gestures to depict math-

---

[38]Anon. (Interview with British novice, N2), Doncaster, England, January, 1996.

ematical properties in addition, subtraction, multiplication, and division. Mathematics teachers use their hands to construct dynamic graphical representations. Their hands rise as they show the curve of a parabola or a rising slope on a graph. These gestures reinforce learning and show the speaker's internal understanding of mathematical concepts.[39]

In the following examples, E1 and E5 used no gestures when they discussed risk in mathematical terms. Both of these cases are very interesting because of the high number and frequency of gestures these miners employed as a general feature of their discourse—suggesting that they might be otherwise natural gesturers.

In the following two excerpts, E5 describes (a) the percentage of injuries in her mine that result from tripping and falling, and (b) the amount of methane produced as a by-product of mining. In both cases, E5 frames her statements to suggest that she is uncertain about the numbers she cites and the statistical accuracy of her data. E5 may have remembered these numbers without an underlying mental image of the mathematical meaning of the numbers she cites, especially since both figures seem relatively high in comparison to those from other mines. The absence of gestures suggests that she lacks a meaningful schema to frame her data:

E5:   One area of accidents was back injuries, twisted knees, *almost something like* 66–68 percent of the accidents at that operation in a year were from slipping and tripping . . . and one thing they thought might help was to go to boots with a better grip, better tread, because boots were expensive.

(no gestures during sequence; italics added)

E5:   We doubled our methane output in a year. We're putting out *last I knew* three quarters of a million every twenty-four hours out of the fan.

(no gestures; italics added for emphasis)[40]

In the following example, E1 also cites statistics. In this case, E1 is explaining to miners why the company has readjusted accident figures to present a more favorable picture of safety in the mine. In this case, E1's lack of gestures may also reflect the absence of a consistent statistical framework for linking data to meaningful experiences.

E1:   The overall accident rate for the deep mine sector—'cause that's the only one that I'm involved with—is very encouraging . . . uh we . . . we set off

---

[39]Goldin-Meadow & Alibali, 1995.
[40]Libby. (Interview with female miner, U.S.), NIOSH Conference, Morgantown, West Virginia, March, 1996.

from British Coal in January and we used British Coal's full year of accidents
... uh ... for '93–'94 because we couldn't put any credibility ... uh ... on the
accidents we saw for the nine months up to December, so we used the full year
... and the rate was 13.33—so that was the yardstick we were using ... uh ...
we're currently standing at 9.74.

> (no gestures)[41]

Missing Memory.   Although expert miners use a high number and frequency of gestures on average, not all expert miners gesture all the time.
When E5 describes the dust in her mine or mathematical statistics, for example, she uses no gesture.

E4, however, used no iconic gestures at all. During the entire interview,
she kept her hands folded in her lap. Throughout her interview, E4 used only
one type of gesture—moving her thumb in a gentle circle. There are many
explanations for her absent gestures. This lack of gesture might suggest that
E4 is not a natural gesturer. Or, it may suggest that she has been taught not to
gesture as an effect of her gender or social class. The absence of gestures is particularly striking, however, since E4 has experienced what might seem to be a
highly emotional and memorable experience: She was nearly burned to death
when a cable exploded.

If we imagine the kinds of mimetic gestures that E4 might employ in this
situation, her lack of gestures suggests that E4 does not want to remember an
embodied memory of the experience. If she were to reenact herself as she was
burning, she might lose control—and the confidence that enables her to return to the mines. If gestures represent the internal imagistic content of
speech, this miner's gestures reinforce her claim that she "does not think
about it"—though she did have powerful dreams at one time.

The analytical discourse that characterizes her spoken narrative suggests
that E4 has analyzed and processed the experience, considered her options rationally, and—as she describes it—she does not think about it any more.
There are many explanations for why particular miners gesture. This miner's
desire to suppress the emotional details of a highly emotional experience provides one explanation for her lack of gesture:

BS: What's the scariest thing that ever happened to you. ... ?

F5:   Probably the scariest thing I had a cable blew up and I burnt and I was on
       fire.
             No gesture: Hands move together in her lap and remain stationary
             throughout

---

[41]Anon. (British manager, transcript of training session), near York, England, July, 1995.

You know how everybody has a fear of drowning, and that's one of their
biggest fears, and . . .
when it's happening it's a reality.
It's one of the scariest things to happen to anybody. . . I guess that would
be the most scary . . .
I considered not going back. I had dreams afterward. I don't think about
it . . . the situation . . .
That was about six, seven years ago, no I don't think about it.
(No gestures)[42]

Talking Heads. MSHA training films frequently present speakers who
use no gestures. These films may deliberately avoid gesture, we might conclude, when the camera focuses away from miners who are gesturing. Miners,
management, unions, and trainers have expressed concern that these films
are not as effective as they might be. If mimetic and analytic viewpoints provide miners with the kinds of representations they need to assess and manage
risk, then the absence of gesture in these films (and the use of scripted actors
who lack embodied expertise) may contribute to the ineffectiveness perceived in these films.

SEQUENTIAL VIEWPOINTS IN MINERS'
REPRESENTATIONS OF RISK

To understand risk, analysts must make sense of the series of events, situations, and decisions that produce a disaster. When miners describe these experiences, they frequently shift between mimetic and analytic viewpoints—
framing a problem first as a distanced observer, then acting out the event as a
character in the narrative. In these sequential viewpoints, miners may try out
a range of problem-solving strategies or options to manage difficult and rapidly changing situations. Miners can create two different but parallel sequences in speech and gesture. As miners shift viewpoints in either modality,
they frequently produce mismatch in the viewpoints they express in speech
and gesture.

In the following example, E5 describes the process of roof bolting. As an
expert roof bolter, she understands both the mechanics of installing roof bolts
and the underlying theory of the process. In this example, E5 assumes an ana-

---

[42]Anon. (Interview with female miner, U.S.), NIOSH Conference, Morgantown, West Virginia, March, 1997.

lytic distance when she describes the length of the resin pack (used to secure
the roof bolts) and the size of the bearing plates. Within her narrative, she
shifts to mimetic gestures to represent, first, strata and then compression. Her
hands depict the layers of the rock itself. As she moves her hands together,
the movement depicts compression.

What is particularly striking in the following example is the way that E5
integrates the scientific knowledge she has learned in training sessions into
an embodied understanding of her environment. U.S. miners learn the underlying theory of roof control in training sessions that depict the strata as
line drawings and simple textual descriptions.[43] In this sequence, E5 uses her
hands to pull imaginary layers of rock together, as if she embodies the roof
bolts themselves. When she describes how to spin the bolts, she turns her
hand gracefully, imitating the torque in a characteristic roof-bolting gesture:

We use <u>full resin bolts . . .</u>

Analytic viewpoint in speech and gesture. (Analytic perspective). In this gesture, E5 measures length of full resin bolts in front of body. Note that E5
does not gesture for the word "use" in this segment.

which means <u>we put glue in the holes</u>

*Simultaneous viewpoint: Mimetic Viewpoint in gesture (Imitating geography);
Mimetic viewpoint in speech (Imitating self).* E5 holds her palms together in
front of body, hands pull together. In this gesture, E5's hands imitate the
strata as they pull together under the "force" of the roof bolt. E5 speaks in
the present as if she were in the midst of the process of putting glue in the
holes. This is an extremely interesting gesture that anticipates the outcome of the process of roof bolting. McNeill calls this a "resultative" gesture because it predicts the results of actions.[44]

and <u>have bearing plates of different sizes depending on the roof . . .</u>

Analytic viewpoint in speech and gesture. Hands show width of bearing
plates, palms together in front of body, hands flat, thumbs up; repeats gesture three times.

<u>insert the bolts,</u>

*Mimetic viewpoint in gesture (Imitating geography). Mimetic viewpoint in
speech (Imitating self).* E5 repeats gesture pulling strata together. Her hands
imitate the strata.

---

[43]U.S. Department of Labor, Mine Safety and Health Administration, National Mine
Health and Safety Academy, 1986; U.S. Department Of Labor, Mine Safety and Health Administration, National Mine Health and Safety Academy, 1989.
[44]Personal interview, March, 1998.

spin the glue till it hardens

> *Mimetic viewpoint in speech (imitating self). Mimetic viewpoint in gesture (Imitating Others—non-human).* Index finger up, hand in fist imitates the roof bolt as it spins inside the bolt hole. This is a characteristic "roof bolt" gestures.

make the bearing against the roof

> *Mimetic gesture in speech (Imitating geography). Mimetic viewpoint in speech (Imitating self).* Elbows close together, hands slightly apart, palms up, as if supporting the roof of the mine with her hands (like Atlas holding up the world), but in front of her torso.

to essentially (pause)

> *Missing viewpoints: Analytic Viewpoint in Gesture. No speech (analytic perspective).* Fingers form open rings, one hand above the other, about 4" apart; hands gradually come together so that the rings are joined. This is a very complex gesture and suggests that E5 can envision the action of the bolts inside of the (invisible) strata. Her viewpoint in this gesture is outside of the mine environment, above the strata (which would be naturally above her head in a mimetic viewpoint gesture). The absence of speech in this segment suggests that E5's understanding may exceed her ability to describe the process in speech.

pull the layers of top together

> *Mimetic viewpoint in gesture (imitating geography). Analytic viewpoint in speech (analytic perspective).* Hands come together again, palms flat. E5's hands imitate (a second time) the strata as it pulls together under the force of the roof bolt. Her speech explains the action that she depicts with her hands.

to make it a solid glue

> *Missing viewpoint: No gestures.*[45]

These sequential viewpoints suggest that E5 has both a scientific and an experiential understanding of the process—particularly in contrast to South African roof bolters, who could not describe the underlying theory of roof bolting although they could explain how they themselves installed roof bolts.[46] These differences suggest that expert miners' gestures reflect cultural knowledge as well as individual experience within the social and institutional hierarchy of the mine. E5's gestures suggest that expert miners can transform abstract diagrams and visual representations into an embodied representation of risk. Chapter 7 reexamines this sequence in order to investigate how her colleagues interpret this same narrative.

---

[45]Libby. (Interview with female miner, U.S.), NIOSH Conference, Morgantown, W.Va. March, 1997.

[46]Personal interviews, 1997.

INTERPRETING MINERS' REPRESENTATIONS
OF RISK: THE PROBLEM OF AUDIENCE

Analysis of miners' gestures reveals the degree to which individual miners speak with their hands. While the content in this analysis is specific to mining, the distinction between analytic and mimetic gestures allows me to examine in general terms how individuals might represent their embodied understanding of risk. This analysis has important implications for rhetorical theory as well as practical implications for writing within large regulatory agencies.

In risky environments, an individual's understanding of risk is literally situated in his or her environment and figuratively situated in institutional and social relations with others in that specific environment. While many individuals share similar social roles within institutions, no two individuals share the same location within the physical environment, just as no two bodies can share a single space in a single time. Within this environment, individuals must be able to draw upon local knowledge to build a body of risk information that can help them in the future. But each individual's location within the environment limits what they observe and experience.

Each individual's account of risk consists of a wide variety of associations, events, and episodes that reflect both individual cognition and the role of the rhetorical situation in shaping an individual's position in the original remembering or construction of the event (cf. Schachter, 1996). The gestures we have defined as characteristic within the culture of mining reflect similar cultural practices—such as roof bolting—which are observed, analyzed, and embodied in individuals. Individual differences, on the other hand, reflect individual experience that can be mapped and compared when speakers discuss similar topics and practices.

Just as viewpoint constrains the semantic content of risk representations, the semantic context also affects how audiences categorize a speaker's viewpoint. In risky environments, audiences need both rhetorical and visual literacies to enable them to interpret the idiosyncratic and institutional meaning of gestures in technical contexts.[47] As with irony in written texts,

---

[47]Given the complexity of mimetic and analytic viewpoints that speakers present in gesture, it is not surprising that classical Roman and elocutionary theorists might not perceive the value of a speaker's frequent shifts in gesture, multiple and complex perspectives, or simultaneous viewpoints. Without the knowledge to perceive and interpret embodied sensory knowledge, classical rhetoricians constructed a theory of gesture divorced from the technical and semantic content of speech. Without modern video technology to record and analyze gesture, classical rhetoricians could not develop a theory of gesture that encompassed the details of everyday decision making. Iconographic portraits of gesture or the crude representations of elocutionary

audiences must learn to read visual and auditory clues in order to distinguish a speaker's embodied experience from imitation, irony, and mimicry. Audiences must be trained to read gesture as a rhetorical feature of discourse—like voice and tone—as speakers produce gesture simultaneously with speech. Finally, audiences need culturally situated knowledge of scientific and technical contexts, both technical and rhetorical, in order to understand the range of analytic and mimetic viewpoints that are logically and technically possible in a given context. Even when audiences have the interpretative skills, they may misread gestures in real-time conversations because, unlike researchers, they cannot replay videotaped interviews to confirm their initial reading of ambiguous or idiosyncratic gestures.

The problem of interpreting viewpoint in speech and gesture has particular consequences in workplace discourse. What role should viewpoint play in the evaluation of the expertise of workers' embodied and analytic knowledge? Is viewpoint a necessary feature of instructional discourse? Of information transfer? Of authority and expertise? What happens to semantic content that is not conveyed in a given viewpoint when agencies interview observers following an accident? Most importantly, if gesture is idiosyncratic, spontaneous, and cannot be preplanned, as Chawla and Krauss (1994) suggest,[48] how can theorists construct a productive theory of gesture that will enable speakers to represent the types of viewpoints that individuals need to assess and manage risk? Does it matter if speakers perceive that a viewpoint is rehearsed or unnatural if it provides critical information about risk?

While it may seem moot to discuss those situations without either speech or gesture, we may reflect upon the number of scientific and technical training films and representations that lack any visible human perspective and, thus, the embodied knowledge of risk that might come from human speech and gesture. Thus, while speakers may not be able to preplan the timing or spontaneity that accompany naturalistic gestures, technical communicators and trainers can choose those speakers who are more likely to have access to the embodied and analytic knowledges that produce the range of gestural rep-

---

theorists like Bulwer (1644/1974) and Austin (1806/1966) can show speakers performing gestural acts, but these crude portraits cannot represent the details that enable audiences to distinguish speakers reenacting their own experience from mimicry, irony, and imitation. Such portraits render invisible the rhetorical reframing of experience that makes argument, persuasion, and—ultimately—communication with an audience possible.

[48]Chawla and Krauss (1994) studied subjects' ability to discriminate spontaneous and rehearsed gestures. Three sets of naive subjects viewed videos without sound, heard the sound track only, or saw both sound and picture. Audiences were best able to discriminate spontaneous from rehearsed gestures when presented with both sound and picture. The results raise interesting questions about the ways that experts might interpret gestures as well.

resentations we have observed in miners.[49] Video producers, for example, might design scenes that include movements of the hands and body; reassess their use of actors who simply read from scripts; or frame the visible and visual proofs they use in technical training films and presentations to lay and expert audiences. As Alibali suggests, moreover, the fact that audiences can perceive the difference between rehearsed and spontaneous gestures does not negate the communicative value of gesture as a feature of technical discourse.[50]

Analysis of viewpoint in speech and gesture ultimately confirms that individuals build knowledge representations that combine many different types of warrants—though these warrants may not be visible if we focus on speech alone. This analysis confirms that knowledge of risk cuts across traditional distinctions between lay (local) and expert (scientific) knowledge and suggests that miners can deploy multiple viewpoints to construct complex and heterogeneous representations of risk that integrate theory and practice. This research has particular implications for historians who work with oral testimony and raises questions about the strategies that investigators must employ to capture information that is embodied, dynamic, and uncertain.

In the remaining chapters we describe how speakers use these viewpoints rhetorically at two critical moments in the Cycle of Technical Documentation in Large Regulatory Industries.

APPENDIX A: CHARACTERIZING VIEWPOINTS IN MINERS' REPRESENTATIONS OF RISK

The present study draws upon analysis of videotaped interviews and training sessions in order to investigate the ways that miners represent risk in speech and gesture.

Data Set I: Novice and Expert Miners in Training Sessions

In our preliminary study, we videotaped four apprentice miners (N1–4) in their initial twenty-day training session at a large coal mine near York, England, as they described problems they had observed in their underground training. At the end of the session, novices gave a brief formal presentation describing problems they had encountered in their training. Two expert miners (E1 and E2) observed and evaluated the presentations. Our third expert

---

[49]Chawla & Krauss, 1994; Morrel-Samuels & Krauss, 1992.
[50]Personal communication, March, 1997.

(E3) was a miner who was blacklisted because of his political activity in labor unions during the 1984 strike and subsequent mine closures. His commentary provides an outsider's comment on the training. These preliminary studies suggested that experts used more character-viewpoint gestures than novices and gestured more frequently on average than novices. One novice used no character-viewpoint gestures at all (Sauer & Franz, 1997).

These preliminary observations raised questions about the role of gesture in miners' representations of risk. Were character-viewpoint gestures linked to a miners' embodied understanding of risk? Did experts thus use character-viewpoint gestures because they had more direct experience in the mines? Did novices use more observational viewpoint gestures because they had merely observed mining practices in training sessions and in brief visits underground? How did gesture reflect a miners' representation of specific kinds of problems underground? How might we account for the particularly lively gestures of one expert miner or the absence of user-viewpoint gestures in three novices?

Data Set II: Expert and Novice Miners Describe Risk, Roof Bolts, and Safety

In the second phase of this study, we conducted a second set of semi-structured interviews with the same cohort of novice and expert miners (now six-month apprentices) at a small technical college in Doncaster, England, in January, 1996. During these interviews, apprentices described safety practices in general at their home colliery; the conveyor belt system in their home colliery; the dust situation in their home colliery; the method of roof support in their home colliery; the tool they were working on in the machine shop; and roof bolting as a general means of roof support in a mine. To compare their responses with expert U.S. miners, we interviewed two expert female miners (E4 and E5) at a NIOSH Conference in Morgantown, West Virginia, in March, 1997. These questions allowed us to discriminate miners' responses for subjects they had experienced directly (such as riding the conveyor belts or working with tools in the machine shop); subjects they had observed (methods of roof support) or learned in course work (roof bolting as a general means of roof support); and subjects that might reflect cultural practices or training (such as arch support).

Our goal in this study was to document the variety of viewpoints speakers employed in speech and gesture as they described their experiences of risk, not to provide an exhaustive account of miners' gestures or accounts of risk. Analysis of these interviews suggested that speech and gesture together provide a more complete understanding of a speaker's viewpoint than either speech or gesture alone.

8

Manual Communication: The Negotiation of Meaning Embodied in Gesture

Perhaps the clearest example of the robustness of language comes from the fact that language is not tied to the mouth and ear but can also be processed by the hand and eye.[1]
   —Goldin-Meadow, McNeill, and Singleton, (1996)

C'ain't talk about it, honey. Gotta show it.[2]
   —Libby, (1999, Interview with female miner, U.S.)

When miners describe how to insert a roof bolt, they articulate an ordered sequence of steps: Drill a hole, insert the steel, insert the glue, insert the bolt, and spin the bolt to set the glue. But their gestures depict different aspects of the process: They can depict abstract scientific forces like compression or they can demonstrate how to insert a 15- or 20-foot cable. They can spread their arms like a wing nut to show how roof bolts open inside the strata, or they can imitate the rock itself as it falls. As we have seen in the previous chapter, these gestures demonstrate both theoretical knowledge and local practice that are literally at hand when miners describe risk. Speech and gesture together provide a richer representation of risk than either speech or gesture alone.

This chapter argues that gestures are not merely the vehicles of semantic content. Instead, individuals can deploy gesture strategically to depict new

---

[1]Goldin-Meadow, McNeill, & Singleton, (1996), p. 34.
[2]Libby. (Interview with female miner, U.S.). Osage, West Virginia, February 27, 1999.

meanings, to focus attention on features of the object or situation not apparent in other viewpoints, to construct representations that integrate theory and practice, and to create new frameworks for understanding embodied experience. As the examples in this chapter suggest, both production and interpretation are active rhetorical processes in which individuals use gesture to make sense of experience.[3] The viewpoints they assume in speech and gesture assist in this process.

When speakers combine analytic and mimetic gestures, they stand in two different relations to the representations they create. They can observe themselves as others see them and reflect upon the actions they recreate in gesture. The patterns they create in gesture help them organize, dramatize, reflect upon, and understand the nature of their work. When speakers change viewpoints, they also change the patterns they express with their hands. These new patterns help them make sense of the meanings they create in their gestures. As they reflect upon the meaning of their gestures, they also make sense of the world these gestures represent.[4]

This analysis suggests that gesture is both a noun and a verb. A gesture is both an iconic image and an act of rhetorical meaning-making that assists and constructs an individual's knowledge of risk. Coding gesture enables agencies to capture gestures in writing so that gesture, like speech, can become part of the official transcript of an accident. When agencies capture gesture in writing, knowledge embodied in gesture becomes part of the accumulated body of experience that decision makers can draw on to assess and manage risk. Analysis of viewpoint can help agencies discover gaps in the record that might be filled by cueing individuals to assume new viewpoints—as agencies question and encourage miners to elaborate their written and verbal testimony. But we must also understand how gesture also plays a more active role in the rhetorical transformation of experience.

To study how viewpoints affect a speaker's representation of risk, I have focused my analysis on an experienced roof bolter named Libby, who had participated in previous interviews and whose description of the process of roof bolting appears in chap. 6. In a follow-up interview in Osage, West Virginia, in February, 1999, I asked Libby to demonstrate how to insert a roof bolt in three different combinations of speech and gesture. I speculated that different

---

[3]Kelly & Church, 1997.

[4]Linguistic anthropologists like Drew & Heritage (1992) use conversational analysis to examine how particular institutions are enacted and lived through as accountable patterns of meaning, inference, and action (p. 5). They argue that "talk-in-interaction is the principle means through which the daily work activities of many individuals are conducted" (p. 3). My own work examines both speech and gesture as clues to talk-in-interaction in institutional settings. See Duranti & Goodwin (Eds.), 1992; Levinson, 1992; Schegloff, 1992.

genres (demonstration and process description) might produce different combinations of viewpoint in both speech and gesture. I asked Libby to use various combinations of speech and gesture so that she would be free to focus on either speech or gesture in her demonstration.[5] Altogether, Libby produced five different combinations of speech and gesture. Her response suggests that speakers can produce different viewpoints strategically in response to different rhetorical cues.

To study how audiences interpret meanings embodied in gesture, I also videotaped four miners as they watched silent videos of Libby talking about roof-bolting practices. Two of these miners were experienced roof bolters; two were experienced miners with little or no recent experience in roof bolting. I used silent videos so that these miners would also be free to focus on meanings conveyed in gesture when speech did not provide clues to a gesture's meaning—as might occur when speakers presented two or more discordant meanings simultaneously or when speaker and audience did not share a common language or a common cultural background.

Not surprisingly, inexperienced miners could not guess the meaning of Libby's silent gestures. One miner responded, "Cain't read lips." Another (who had not roof bolted for 10 years) recognized that Libby was talking about roof bolting, but he did not elaborate. Experienced roof bolters, on the other hand, used gesture to explore meanings conveyed in gesture, to negotiate meanings with others, and to construct and co-construct new knowledge of risk in a process known as uptake.[6]

While we might easily attribute an experienced miner's embodied knowledge of roof bolting to his or her first-hand experience underground, these examples suggest that individuals integrate knowledge seamlessly from many

---

[5]In an article entitled "Silence is Liberating," Goldin-Meadow et al. (1996) show that gestures without speech perform different linguistic functions than gestures that accompany speech. This research suggests that the viewpoints that speakers construct in each condition would perform different rhetorical functions—changing both the form and content of representation. According to these authors,

> The gestures that hearing individuals produce along with speech form an integrated system with that speech; it is only when analyzed without speech (i.e., when coded with the sound turned off) that these gestures appear to be unsystematic. Thus, we suggest that because the gestures produced by the deaf children's hearing mothers formed an integrated system with their speech and were constrained by that speech, those gestures were not "free" to assume the languagelike qualities of their deaf children's gestures. One might suspect that if the mothers merely refrained from speaking as they gestured, their gestures might have become more languagelike in structure, assuming the segmented and combinatorial form also found in the children's gestures. (p. 40)

[6]Kelly & Church, 1997.

sources.[7] The examples in this chapter take place within a material context where decision makers must ground judgments in embodied experience, engineering experience, and scientific knowledge. The examples show, first, that American coal miners understand the scientific principles that govern roof bolting, and second, that they represent these principles in both speech and gesture. Whether these constructions are strategic forms of communication or merely the visible form of "habitual thought" (McNeill and Duncan, 2000), the following analysis suggests that researchers must pay greater attention to the ways that gesture influences knowledge production at critical moments of transformation within the Cycle.[8]

THE EFFECT OF VIEWPOINT IN MINERS' REPRESENTATIONS OF RISK

Gesture plays an important role in workplace communication. According to Goodwin, gestures organize the "visible, public, and interactive phenomena . . . that constitute the work of a profession."[9] Deictic gestures (pointing gestures) focus an audience's attention on an absent object through pointing, streamlining action, and highlighting critical features of the process. Deictic gestures can also help speakers recognize shapes and sizes that would be unwieldy or impractical to describe in speech. In this simplified model, talk-in-action provides the semantic meaning—the narrative component—for actions demonstrated in gesture.[10]

Risky environments complicate the relationship between speech and action. In risky environments, decision makers must continually monitor con-

---

[7]As a rhetorician interested in the ways that individuals represent their experiences of risk, I do not want to argue that gesture makes visible a kind of internal cognition or mental imagery, although this research is suggestive. Nor do I want to argue for the rhetorical superiority of miners' embodied or experiential understanding.

[8]More recently, McNeill and Duncan (2000) suggest that gestures may represent thinking for speaking. They distinguish the thinking for speaking hypothesis from the Whorfian version of the relativity hypothesis on the grounds that thinking for speaking has a diachronic focus on "thinking" rather than a synchronic focus on "habitual thought" (p. 141).

[9]Goodwin, 1999, p. 29.

[10]Goodwin (1999) shows how pointing gestures function linguistically in workplace discourse. Drew and Heritage (1992) argue that "talk-in-interaction is the principal means through which the daily work activities of many individuals are conducted" (p. 3). These authors use conversational analysis to examine how particular institutions are enacted and lived through as accountable patterns of meaning, inference, and action (p. 5). Conversational analysis has been a fruitful method of analysis for studying talk-in-interaction in institutional settings. See Duranti and Goodwin (Eds.), 1992; Levinson, 1992; and Schegloff, 1992.

ditions that warrant changes in practice. In these environments, speakers can use gesture to demonstrate two different aspects of expert behavior. When miners assume a mimetic viewpoint, they demonstrate how to perform tasks like an expert. In the analytic viewpoint, they demonstrate how experts move outside of experience rhetorically to assess, judge, and reflect upon events and conditions in their environment.

When Libby talked about roofbolting, she produced five different versions of her narrative in five different combinations of speech and gesture: (a) the scientific process narrative we analyzed in chap. 6; (b) a demonstration from the viewpoint of a user; (c) her revision of that demonstration focusing on the movement of her hands; (d) a demonstration without speech; and (e) a description without gesture. In each condition, gesture helps her discover features of experience that might not become apparent in other modalities.[11]

These representations are neither simple representations of objects and actions nor illustrations of her meaning. Gesture is linked temporally to speech, but Libby's gestures help her construct an embodied understanding of risk not apparent—or perhaps possible—in speech alone. Libby uses mimetic viewpoint gestures to demonstrate how an experienced miner follows proper procedures when she installs a roof bolt. But she also uses analytic viewpoint gestures to monitor the events she represents in the rhetorical space between audience and speaker. As she shifts from one viewpoint to another, the viewpoints that Libby assumes in speech and gesture redefine the relationship between speaker and audience. These viewpoints shape the semantic content of her representation and convey meanings not possible in speech alone. They help her integrate theory and practice, and—most important—they transform her understanding of her work.

Viewpoint Redefines the Relationship Between Speaker and Audience

When speakers assume mimetic viewpoint gestures, audiences observe the speaker as an actor who re-presents events and situations that are temporally and spatially distant. When speakers assume analytic viewpoints, speakers and audience share the same rhetorical relation to the objects and situations

---

[11]When she demonstrated the process as a user, her hands assumed a natural rhythm. She repeated the process before she paused to reflect on her action. Once she established the outline of the drill pod, she set the steel naturally in the center of the pod. In the final sequence, she grasped an imaginary methanometer without stopping to shape her hand or reflect on the motion. When asked to clarify the meaning of her compression gesture, Libby increased the number and complexity of gestures in her response.

in the rhetorical space between speaker and actor—observing events literally at arm's length. Speaker and audience share a common temporal relationship, though they see events slightly differently. When speakers assume an analytic viewpoint, audiences see the speaker as an actor engaged in the act of analysis, reflection, and contemplation. The speaker's viewpoint thus works rhetorically to shape the audience's rhetorical attention toward the speaker as a rhetorical agent who demonstrates how good speakers (or good miners) reenact, reframe, analyze, and reflect upon their own practice.

The viewpoints that Libby assumes in each condition demonstrate how viewpoint redirects an audience's attention to features of an object or situation that are not apparent in any other viewpoint. In the scientific process description, Libby uses analytic viewpoint gestures to depict the compression of the strata. McNeill (1998) observed that Libby's gestures anticipate the outcome of her actions and seem to transform the knowledge she has learned in training sessions into an embodied understanding of her environment. McNeill calls these gestures resultative because they anticipate the results or outcomes of actions.[12] These gestures turn the viewer's attention toward the future, anticipating the results of the process she describes in speech and gesture.

In the demonstration, Libby watches the back of her hand from an analytic distance as her arm rises upward like the bolt itself. When Libby assumes this viewpoint, the audience can watch how Libby herself turns her attention to the bolt. In this demonstration, audience and speaker share the same viewpoint in relation to the bolt presented in the rhetorical space between speaker and audience. When Libby pauses to reflect on the motion of her hands in her revision, her audience must once again shift attention, focusing now on the stylized patterns she creates with both hands.

In this sequence, Libby creates a problem space in which individual elements (roof bolts, bearing plates) and actions (insert the bolt, insert the steel, spin) provide the material solutions to the problem. In Aristotle's terms, they are the givens with which she must work (see Table 8.1).[13] Libby frames her actions within the physical space created by her image of the drill pod from a temporal position outside of the representation ("Have I got my hole drilled?"). Her revision moves from this literal space to a new figurative and greatly stylized representation of the problem. Her gestures become more stylized in her revision, not more particularized or more articulated as we might imagine if she were focused on the action rather than on the relation between actions in her gestures. When Libby pauses to reflect, she recognizes the recurrent pattern in her gesture and steps back to identify this pattern for her audience: "takes both hands."

---

[12]McNeill, personal communication, March, 1998.
[13]Aristotle, 1992, p. 37.

TABLE 8.1
Libby Constructs an Imaginary Workspace
Before She Begins Her Demonstration

Speech	Gesture
You've got your drill pod	Both hands outline the shape of the pod: L-shaped, fingers splayed, thumbs to center, form a square in front of the body; palms slightly angled, fingers splayed down. Both hands: analytic viewpoint
You set your steel in	Right hand rises quickly after previous phrase to chin level. Right hand makes a fist, drops suddenly so that the movement is actually straight down → center of "pod" outlined above. Left hand remains in L-shaped position at outside boundary of the pod. Right hand: mimetic viewpoint - user Left hand: analytic viewpoint - size of object
Hold on to it till you get it started up	Right hand: raises fist slightly mimetic viewpoint - user
spin	Right hand: spins fist with index finger raised; index finger makes a small "vortex." Multiple viewpoints simultaneously: RH imitates spin of the bolt, but not the action of a miner spinning the bolt with a spanner. Libby watches her hand at a distance—from an analytic viewpoint of the observer
Insert the bolt	Right hand rises with index finger straight up. Back of hand to audience. Libby is looking at her fingers and pointed index finger as it rises upward to the Upper Center. Small spin at top of gesture. Multiple viewpoints: RH imitates the bolt (mimetic gesture) + Libby observes this action from the viewpoint of a miner/observer (analytic viewpoint gesture).

It is not possible within the framework of this analysis to determine which features of the representation are self-conscious and which reflect deeper "patterns of thought" and habitual action (Cf. McNeill and Duncan, 2000). These representations do not reveal Libby's thoughts at the moment of risk decision making. They are not demonstrations in the conventional sense, though they demonstrate what an experienced miner's representation might look like. These gestures demonstrate how an experienced miner can assume alternative viewpoints strategically to help her think about her work.

Viewpoint Shapes Semantic Content

When speakers assume a viewpoint, their viewpoint limits the range of topics they can logically and consistently represent within the framework of a single narrative. Goldin-Meadow, McNeill, and Singleton (1996) argue that speech and gesture do not always present the same information about the event because the method for conveying meaning is "fundamentally different" from that of speech. Speech is linear and segmented. As a result, speech breaks meaning into segments and reconstructs these meanings by combining segments along an axis in time. Gestures are not limited to the single dimension of time. Goldin-Meadow, McNeill, and Singleton argue that gestures are "free to vary on dimensions of space, time, form, trajectory, and so forth and can present meaning complexes without undergoing segmentation or linearization."[14] Gestures present a whole event that can be combined with other such events to provide more than one "angle" on experience.[15]

The effects of viewpoint on semantic content become apparent when we compare the viewpoints that Libby assumes in each of the five conditions. In the demonstration, she sets up the physical context for the demonstration ("Have I got my hole drilled?"). In the scientific process condition, she explains how the mechanics of roof support (the layers, the bearing plates) produce the intended consequences ("to make it a solid beam"). In the gesture-only condition, she assumes a mimetic viewpoint as though she were following a checklist of instructions: "Check gas . . . Sound the top . . . Drill the hole to the desired depth." Each condition limits the viewpoints that Libby can assume and thus the range of topics she can represent in her narrative.

When we compare conditions by topic, we also discover important absences in semantic content in each condition (Table 8.2). When Libby assumes an analytic viewpoint in her scientific process description, she describes the strata from a position outside of and above the spaces she describes. As a result, she does not talk about the nature of the steel bolt, the depth of the drill hole, or the size of the drill pod—features that would only be visible from a viewpoint inside the spaces she describes. When Libby assumes an analytic viewpoint, she omits the safety checks that she describes in other viewpoints (sounding the top or checking for gas), since these are presumably unrelated to the workings of the resin bolt within the strata.

In her demonstration, Libby does not talk about the depth of the drill hole, the glue, the bearing plates, the working of the bolts or the strata—since these are invisible from the viewpoint of a working miner. When she assumes a mimetic viewpoint in her demonstration, her arm movements convey the depth of the hole while her speech focuses on how to insert the bolt as she was instructed. In her revision of the demonstration, Libby focuses on the patterns of movement in her arms rather than the steps in the process. In this condition, she speaks rhythmically and reduces the level of detail in her dem-

---

[14]Goldin-Meadow et al., 1996, p. 37.
[15]Goldin-Meadow et al., 1996, p. 36.

TABLE 8.2
Distribution of Topics in Libby's Speech

	Scientific Process	Demo	Revision	Gesture Only	Speech Only
Check gas				✓	✓
Sound top				✓	✓
Drill pod		✓			
Steel	✓		✓	✓	✓
Drill hole				✓	✓
Insert glue	✓		✓	✓	✓
Have bearing plates	✓				
Make the bearings	✓				
Insert Bolts	✓	✓	✓	✓	✓
Spin	✓	✓			✓
"to essentially"	✓				
Pull layers together	✓				
Make a solid glue	✓				
Collapse the boom				✓	
Size of pay line				✓	

onstration: "Put the steel in, put the glue in, put the bolt in." Then she shifts to an analytic viewpoint to think about the patterns she has created with her hands.

We might expect that Libby would provide more detail in the condition of speech without gesture to make up for the absence of gesture. Instead, Libby chants in a strange, high-pitched sing-song voice, placing ironic emphasis on the single added detail ("drill the hole to the desired depth"). The following excerpt reveals the formulaic quality of her speech, particularly evident in her repetition of the verb "stick":

L: Sound the top
Check for methane

(pause)
(another miner comments, inaudibly)

BS: Don't help.

L: Drill the hole to the desired depth
Stick the steel in
Stick the glue in
Stick the bolt in
Put the wrench on
That's it
(laughter) . . .

This spare checklist contrasts the rich detail that these ordinarily loqua-cious and experienced miners provide—informally—when they talk about risk. We can suggest a number of explanations. Libby has memorized a rou-tine; she follows patterns established in her revision; she's performing for the audience; she is tired of performing and imitates a safety trainer; or she is locked into a pattern of grammar that controls her speech. Without gesture, Libby stutters and has a hard time beginning her sequence. (This shyness contrasts with her usual gregarious nature. As a female miner in a difficult oc-cupation, she is not a shy woman; she tells us at another point in the inter-view, "I have a serious lack of modesty.")

Following De Saussure (1916/1959), Goldin-Meadow, McNeill, and Sin-gleton (1996) argue that "segmentation," "linearization," and "hierarchy"—such as Libby produces in her sing-song narrative—are "essential characteris-tics of all linguistic systems."[16] Gesture, on the other hand, can represent dy-namic events that change along the dimensions of time, space, and trajec-tory. Each gesture is a "complete expression of meaning unto itself."[17] When speakers employ only a speech without gesture, speakers must find language to express the global, imagistic ideas that are more easily expressed in gesture. It is thus possible that Libby may not have the vocabulary or conceptual lan-guage to express her embodied experience in words. We see this effect again when Libby explains the meaning of her strata gesture.

If Libby lacks a language to express her embodied experience, it is also pos-sible that trainers did not provide a language that could provide an adequate representation of the process. Written instructions without gesture are rhe-torically incomplete—like Libby's sing-song chant—unless writers can cap-ture the dynamic, imagistic constructions of experience that are expressed more naturally in gesture.

Table 8.2 shows the distribution of topics in speech for each of Libby's five narratives. This table reveals the full range of topics that Libby can draw on strategically in any one representation. We can use this table to identify the gaps in any single representation. If we assume that a speaker can shift viewpoints strategically, we can prompt speakers to assume new viewpoints as we might question speakers to elaborate on a missing or un-derdeveloped detail in their narrative. Speech and gesture can thus work to-gether to help speakers overcome the rhetorical incompleteness of a single modality because the two modalities provide rhetorically different but com-plementary types of information.

Absences in semantic content take on greater importance in the context of risk if information necessary for risk judgments cannot be represented

---

[16]de Saussure, 1959. In Goldin-Meadow et al., 1996, p. 36.
[17]Goldin-Meadow et al., 1996, p. 36; cf. McNeill, 1992.

within a single viewpoint. Whatever the reason for the absences in each condition, the effects of viewpoint on the range of topics Libby presents seems to call into question the primacy of spoken language as the sole index of an individual's understanding of risk.[18]

Viewpoint Helps Speakers Elaborate the Meanings Conveyed in Speech

Gestures are not merely a redundant expression of a speaker's meaning in speech. Instead, gestures provide information in combination with speech. McNeill and Duncan (2000) suggest that speakers think in terms of a "combination of imagery and linguistic categorical content."[19] In these combinations, the contents of speech and gesture are complementary but not identical. Gesture provides information not present in speech.[20]

Table 8.3 analyzes the gestural viewpoints that Libby assumes when she represents three components of the process: inserting the steel, inserting the glue, and inserting the bolt. In her four cued conditions (excluding the gesture-only condition), gesture reveals differences in meaning not apparent in speech. In her demonstration, Libby "sets" the steel in the pod outlined by her left hand. In the revision, her right hand drops suddenly up to "put the steel in" the pod. In the demonstration, Libby rotates and raises her fist to "put the steel in." (In the scientific process description, she omits this phase of the process.)

---

[18]Rhetoricians have focused their analyses on text (and written transcripts of conversations) because they are presumably stable objects. Myers (1990) writes: "Written texts have great advantages as research material, advantages that have long been taken for granted by literary critics, but are perhaps not sufficiently appreciated either by them or by social scientists . . . Only written texts allow for such close reading, because they hold still while one goes over them, and hold still until one can come back to them" (p. 6). Latour (1990) argues that this very stability created "immutable mobiles" that transcend time and space. Science was possible because multiple knowers could stand in the same place in relation to texts achieve consensus about what counts as knowledge.

Studies of gesture question the primacy of text and language as the "provide of the oral modality" (Goldin-Meadow et al., 1996). The authors ask: "Why did language become the province of the oral modality? Why is speech the most common form of linguistic behavior in human cultures when it could just have easily been sign?" (p.52). They speculate that the segmented structure of oral communication frees speakers to "capture the mimetic aspects of communication along with speech" (p. 53).

A complete discussion of the transformation from orality to literacy is beyond the scope of this chapter. One of the ironies of my present project, of course, is the problem of capturing meaning in gesture in the linear form of the academic treatise.

[19]McNeill & Duncan, 2000, p. 2.

[20]McNeill & Duncan, 2000.

TABLE 8.3
Viewpoints in Speech and Gesture for Key Actions in Libby's Narrative

Scientific Process	Demonstration	Revision	Gesture Only
	You set your steel in + demonstrates action as user sets steel pod *Mimetic user* Right hand rises quickly after previous phrase (up to chin level). Right hand makes a fist, drops suddenly so that the movement of this gesture is actually straight down to center of "pod" outlined above. Left hand remains in L-position at outside boundary of "pod" (user putting steel in drill pod)	Put the steel in + demonstrates action as user puts steel pod *Mimetic user* Right hand makes a fist. Drops hand from Upper Center (Right hand fist). Lower center. Low position in relation to body. (user inserting steel in pod)	Put the steel in + demonstrates twist as drill steel rises roof *Mimetic user* Right hand fist in front. Rotates fist and raises fist. (drill steel rises)
Insert the bolts + strata/ compression *Mimetic viewpoint in gesture (Imitating geography) and Mimetic viewpoint in speech: (Imitating self). Libby* repeats gesture pulling strata together. Her hands imitate the strata.	Insert the bolt + hand imitates rising bolt & spin *Complex viewpoint:* Right hand: Raises hand with index finger straight up. Back of hand to audience. She's looking at fingers and pointed index finger as it rises upward to UC. Small spin at top of gesture	Put the bolt in. + user inserts bolt *Mimetic viewpoint:* Switches hands. Left Hand rises to Upper Center. Right hand down (positioned at angles at end)	Put the bolt in + user inserts bolt *Mimetic viewpoint:* Switches hands. Left Hand rises high up (higher than previous gesture)

Her statements are semantically similar in each condition, but they convey very different acts when we factor in the meanings conveyed in gesture. In the demonstration and revision, Libby uses a mimetic viewpoint to demonstrate how she sets the steel in the pod. In the demonstration without speech, Libby assumes an analytic viewpoint. In this condition, Libby's arm depicts the steel as it rises. The gesture elaborates on the speech, showing how the drill steel spins as it rises in front of the watching miner. Without speech, Libby is free to focus on gestures and the meanings she conveys in her hands.

Gestures may be particularly important when speakers use English verbs like "do," "put" and "get" in instructional contexts. As McNeill and Duncan (2000) demonstrate, these verbs lack the details of manner and motion that can be supplied by adverbial phrases. In the context of Libby's instruction, verbs like "put" and "set" provide very little information about the manner and motion of her actions. They do not tell, for example, how the steel rises, how the bolt spins, how long the process lasts, or how quickly Libby can finish the task. Yet these details help us understand the process as a dynamic set of actions that take place at specific times and in specific places.

In these instances, Libby's gestures enhance meanings conveyed by similar verbal commands, particularly when miners describe dynamic and spatially located events.

Viewpoints Help Miners Integrate Theory and Practice

When speakers assume more than one viewpoint simultaneously in speech and gesture, they can represent two different sets of events that unfold in time in different locations.

When Libby describes the practice of roof bolting, she represents two different and simultaneous viewpoints in speech and gesture, each of which is equally correct as a representation of the process. Her gestures show the action of the roof bolt and the manner in which it draws together the strata from a position outside of and above the mine. In speech, she describes how resin bolts work to construct a solid beam. Her gestures suggest that Libby's understanding of the process comprises both theoretical knowledge (analysis) and action (mimesis). But theory is not linked to words or gesture to embodied practice.

While we might expect her to employ mimetic gestures to demonstrate how to insert the glue, Libby uses gesture to depict the compression of the strata as the bolt tightens and forms a solid glue. Her hands imitate the strata, the action of the bolt, and the force that holds the bearing in place against the roof. In the middle of her account, Libby pauses in speech, but she employs a complex gesture that suggests that she can visualize the workings of the bolt inside the mine strata. This pause shifts the agency of the action from miners (who insert the glue) to roof bolts (which break seals on the glue to make a solid beam). In the end, Libby draws these narratives together in a single gesture, integrating theory and practice.

At first, these gestures seem to be an example of a mismatch between the viewpoints expressed in speech and gesture. But the notion of a semantic mismatch does not account for the temporal and theoretical coherence of the rhetorical viewpoints she integrates in speech and gesture.

With one exception (the size of the bearing plate), Libby's gestures embody the theoretical workings of the roof bolt invisible to miners underground: the spin of the bolt, the force of the bearing plate holding up the roof, the compression of the strata, and the workings of the bolt inside the strata. These factors explain how resin-tensioned roof bolts work to support the roof. They provide the scientific warrant for using roof bolts as a method of roof support.

Libby's gestures also reflect good practice. These gestures demonstrate that Libby pays attention (visually and dynamically) to those aspects of roof-bolting that the U.S. Department of Labor's training manual calls the "most common mistakes made while installing resin type roof bolts."[21] These include the size of the bearing plate, the length of the bolt, the spacing of the bolts, the amount of torque applied to the bolt, insufficient spin, and "using too much upward thrust while drilling hole."[22] All new miners receive this training, and it is reinforced (though somewhat selectively) in yearly refresher training courses at each mine.[23]

---

[21]U.S. Department of Labor, Mine Safety and Health Administration, National Mine Health and Safety Academy, 1992, p. V-61. It is interesting to compare the training manual's description of the process with Libby's narrative. Like Libby's narrative, the manual first outlines the roof bolter's tasks, then explains how roof bolts work to compress the strata:

> Early resin-tensioned bolt systems used a fast-setting resin to provide the tension anchor. The bolter would insert the resin and bolt, spin to mix the resin, wait for the resin to cure, and then apply torque to the bolt. The resistance provided by the resin causes the dome on the nut to break allowing the nut to tighten, compressing the strata between the plate and the bottom of the resin grout. (T-45)

[22]U.S. Department of Labor, Mine Safety and Health Administration, National Mine Health and Safety Academy, 1992, p. V-61.

[23]The manual depicts three methods of roof support: keying, the beam method, and suspension (pp. V-22 to V-24). In the Keying method, "the roof bolts intersect slip planes of randomly joined rock, forming them into a stronger unit" (p. V-22). In the Suspension Method, the roof bolts tie the lower layer of the roof to a stronger layer located above the main roof (p. V-23). In the Beam method, the roof bolts bond layers of rock together like plywood—making a strong and supple "beam" (p. V-24). A resin or glue fills in cracks in the strata, creating a stronger bond between the roof bolt and the rock.

TABLE 8.4
Libby Explains the Strata/Compression Gestures

Speech	Gesture
L: Bending the layers of top	Left hand flat, palm down, thumb bent slightly up and to center; Right hand palm up. Right hand is Under Left hand but not parallel (not completely parallel or lined up).
Where it. . . .	Fuzzy gesture. Makes a circle with the right hand under left hand.
Roof layers	Hands switch position. Right hand on top. Flat. Fingers splayed. Left hand palm up under Right hand. Right hand makes circle, then moves up slightly in 2 segmented movements → wide position, 15″ apart, right hand on top.
to make a roof – L: different layers of top	
bend (-ing?) it together	Right hand fist rises, comes down in "drill steel" motion. Then back to parallel position → lap
to make it a solid beam	Hands do not compress. Bottom hand moves, then switches hand, then top hand moves.

Table 8.4 shows a transcript of Libby's explanation of the meaning she conveyed in her strata/compression gestures. If we focus only on Libby's speech, we might conclude that she has an inadequate understanding of the theoretical workings she describes. When Libby explains the meaning of her compression gesture, her speech is fragmented and ungrammatical: "Bending the layers of top . . . where it . . . roof layers . . . to make a roof—" In this sequence, Libby's gestures complete the unfinished phrase "layers of top, where it . . ." as her hands move slowly in circles, imitating the unstable layers of rock. Like her original scientific narrative, Libby's understanding of risk is comprehensible because gesture and speech work together to provide a dynamic three-dimensional representation of the underlying instability of the rock strata.[24]

Because she can represent two different warrants in speech and gesture, Libby can integrate theory and practice simultaneously in a single narrative. But her scientific knowledge becomes apparent only when we examine the viewpoints represented in gesture.

In a follow-up interview at Osage, West Virginia, in February, 1999, Libby (1999) confirmed that she had depicted her hands showing how the roof bolt was "bending the layers of top . . . to make it a solid beam."

[24]As we have seen earlier in this chapter, speech and gesture are complementary but different forms of expression. The global, dynamic, and imagistic nature of gesture is not easy to represent in speech—particularly if speakers do not have a vocabulary to express the concepts expressed in gesture.

Viewpoint Transforms a Speaker's Understanding of Experience

Ultimately, the viewpoints that Libby assumes in speech and gesture transform her understanding of experience, creating new knowledge in the process.

In her first demonstration, Libby reenacts the role of a careful miner as she focuses on one critical moment in the process. When Libby demonstrates how to insert the bolt, her hand imitates the action of the bolt in the gestural space between audience and speaker. In this complex gesture, Libby does not dramatize how a miner would insert a bolt. Instead, Libby assumes an analytic viewpoint from which she can analyze, interpret, and reflect upon her own actions—creating a new understanding for herself as well as for her audience.

Libby's self-conscious analysis of her own actions is particularly evident when she revises her first demonstration. In her speech, Libby repeats the sequence of instructions: put the steel in, put the glue in, put the bolt in. Her fisted hands switch place rapidly up and down in harmony with the rhythm in her speech. At the end of this sequence, she pauses and comments, "Takes both hands."

Gesture thus helps Libby organize the recurrent pattern in her gestures within a simple rhetorical frame: "like milking a cow."[25] When she pauses to reflect, she isolates the pattern of hand movements required to insert the bolt, establishes a rhythm for her action, and creates a vivid and comic image that helps her audience remember the pattern in the future.[26] Her embodied actions demonstrate how an expert miner can step outside of her embodied experience rhetorically to observe and analyze her work.

Qualitative studies based upon a single interview cannot determine, of course, whether these gestures are strategic and deliberate. Other features suggest that Libby has acquired patterns of automatic and habitual action. When she demonstrates the process as a user, for example, her hands assume a natural rhythm. She repeats the process before she pauses to reflect on her action. Once she has established the outline of the drill pod, she sets the steel naturally in the center of the pod. In the final sequence, she grasps an imaginary methanometer (an instrument to measure methane) without stopping to reflect on the motion. When asked to clarify the meaning of her compression gesture, Libby increases the number and complexity of gestures in her response.

---

[25]At the end of this sequence, Totten (1999) makes fun of Libby's gestures, which bear little resemblance to actual milking gestures. Libby confesses, "Not that I ever milked a cow."

[26]Early rhetorical texts talk about the function of vivid detail in heightening the memorableness of a speakers' argument. See, for example, [Cicero], 1989.

Libby herself denied that she thinks about much of anything when she roof bolts: "I think it's something that you just do subconsciously 'cause you know when that steel jumps you're getting cracks." Yet she admitted that these things were "in the back of her mind" even when the work seems "automatic." One miner explained, "She's been doing this so long, she doesn't have to think about it. She just does it."[27] But another miner worried: "I don't want to be under a roof with some miner who's on automatic."

THE ROLE OF GESTURE IN THE RHETORICAL
CONSTRUCTION OF MEANING

As activity theorists point out, social systems are composed of many minds that can act in parallel to understand the whole. In complex technological systems like mining or navigation, knowledge may be distributed among many individuals who do not share the same perspective or viewpoint.[28] To manage complex technical and social processes like roof bolting, individuals must communicate this knowledge to other individuals (Hutchins, 1996).[29] If social processes are to be internalized, then individuals must transform this distributed knowledge into the kinds of information that make sense within their own cognitive frameworks.

The representation of workplace practices in gesture, of course, is not the same as expertise embodied in practice. As Geertz (1973) suggests, individuals can wink, imitate winks, or burlesque a wink. These actions are indistinguishable from rhetorical acts of deliberate meaning-making and representation, though they share similar forms. Geertz thus warns about making judgments about knowing on the basis of the external features of representation:

> Culture is public because meaning is. You can't wink (or burlesque one) without knowing what counts as winking or how, physically, to contract your eyelids . . . But to draw from such truths the conclusion that knowing how to wink is winking . . . is to betray a deep confusion as, taking thin descriptions for thick, to identify winking with eyelid contractions.[30]

The following examples show how audiences use gesture to make sense of gesture in a process called "uptake" (Kelly & Church, 1997; 1998). These ex-

---

[27]Totten, (Interview with female miner, U.S.), Osage, West Virginia, February, 1999.
[28]Minick, 1997.
[29]Hutchins, 1996. Hutchins (1996) writes: "The kinds of transformations that internalization must make will be in part determined by the differences between the information-processing properties of individual minds and those of distributed cognition" (p. 60).
[30]Geertz, 1973, p. 12.

amples show how gestures function rhetorically when individuals are free to focus on the meanings expressed in a single modality.[31]

Miners Use Gesture to Interpret Gesture

Uptake refers to the process by which audiences imitate and reproduce meanings conveyed in gesture.[32] When Libby's colleagues worked to figure out the meanings embodied in her gestures, they did not attach semantic meanings to gestures as signs, as one might say, "When Libby holds her hands like this, she means X or Y." Instead, the Osage miners created their own variants of Libby's gestures that isolated meanings embodied in Libby's gestures. They used these gestures rhetorically, as speakers would use questions and assertions in speech. In these examples, speakers elaborate meaning, isolate component meanings, and create revisions that help them reflect upon the meanings they create with their hands.

These gestures do not simply transfer information from one miner to another, nor are they simply copied and recopied in sequence. Both the number of variants and the number of interpretations suggest that new knowledge is the product of "communicative uptake" (Alibali, 1998) as miners work to articulate meaning in and through gesture.[33]

Libby's colleagues isolate three component gestures as they work to figure out the meaning of Libby's gestures: compression, strata, and spacing. These three gestures are related conceptually. Roof bolts compress the strata, making a tighter or more compact beam. The spacing of the roof bolts is critical in preventing the parallel plates of rock (strata) from shifting or moving.

Speakers can represent strata as isolated components or they can add motion to show how the strata are compressed by the action of the roof bolt. These components can be combined in different combinations to produce

---

[31]Without the advantage of digital and video technologies, speakers in live interactions cannot play and replay gesture to see nuances of meaning. They have no method of recording gesture—as they might take notes in speech—to check the accuracy of their visual memory.

[32]Kelly & Church, 1997; 1998. Church, personal communication, Chicago, 1998. I am grateful in every respect to Martha Alibali for her suggestions in this matter. Alibali's work with mathematics learning in children has informed this entire discussion.

[33]Alibali used the example of a triangle to explain the notion of uptake. In a math class, for example, a teacher may depict a triangle with her two index fingers and thumbs. When students respond, they may trace a small triangle with their index fingers. In this transformation, students reproduce the teacher's original idea in a new form. The new variant is smaller and incorporates motion. These changes in size and motion provide an image of the students' conception of a triangle. The gestures help them understand a new concept in their own terms—as speakers might elaborate the meaning of an obscure or difficult sentence in their own words (Personal communication with the author).

new meanings. Speakers can also vary the duration, size, shape, position, and force of each component. Each new combination and each new variant helps them negotiate competing interpretations of the meanings expressed in gesture.

Speakers Use Gesture to Amplify Component Features and Concepts

Libby represented strata with the flat surface of both hands. This gesture is highly ambiguous unless we understand that Libby is talking about some aspect of coal mining. Within the context of coal mining, this gesture could also have many meanings: the height of a tunnel after a roof fall, the width of a supporting beam.

Libby's colleagues produced three variants of her strata gestures: with both hands facing, with hands parallel and palms up, and with hands skewed at an angle. The details that Libby's colleagues added to this gesture suggest that they understood that Libby's gesture represented strata.

Libby's original gesture combined strata and compression in a single hand movement (Fig. 8.1). The first variant (Fig. 8.2) seems to isolate the strata component from the movement associated with compression. In this gesture, Libby's colleague does not compress the strata tightly, as Libby does. Instead, she holds her hands parallel with little movement. The gesture creates a snapshot of Libby's gestures during one segment of her narrative; it lacks the action associated with a verb in speech.

In the second variant (Fig. 8.3), Libby's colleague reproduces the notion of strata in her own terms—with hands parallel. In this gesture, Libby's col-

FIG. 8.1.   Libby's original strata/compression gesture.

FIG. 8.2.   Strata variant 1: This gesture retains the parallelism that characterizes strata, but the gesture does not highlight the compression component of this gesture.

FIG. 8.3.   Strata variant 2: In this variant, Libby's colleague depicts the movement of the strata as parallel plates that shift, but her gesture does not show the compression that Libby depicts in her gesture.

league adds movement not present in Libby's original gesture. This movement seems to show how the strata shift as parallel plates within the mine. (Libby also depicts the same movement when she explains the meaning of her gesture in the follow-up interview). The added detail seems to confirm that Libby's colleague has an active image of strata in her mind.

The third variant (Fig. 8.4) shows how gestures lose some of the visual character of the original as speakers amplify the manner and motion of the gesture. Libby's colleague also isolates the compression component of Libby's original gesture, in a single 13-second gesture (Fig. 8.5). In this gesture, she explores the notion of compression as an abstract concept isolated from the iconic visual representation of the rock strata.

FIG. 8.4.   Strata variant 3: This gesture loses some of the visual character of the strata component as Libby's colleague amplifies the movement characteristic of compression.

Gestures Influence the Production of New Meanings in Gesture

Libby's colleagues isolate components of her gestures in new gestures that influence how they interpret other aspects of her gestural narrative. In Fig. 8.6, Libby's original iconic gesture shows the size and approximate shape of the square, flat bearing plates that are used as supplementary support in roof bolting. As they work to interpret Libby's gesture, Libby's colleagues take up the notion of size in two new variant gestures. But neither of these variants incor-

FIG. 8.5.   Libby's colleague explores the meaning of compression in a new 13-second gesture that demonstrates how speakers take up meanings embodied in gesture. In the background, a second colleague co-constructs a smaller version of the same gesture.

FIG. 8.6.   Libby's original spacing gesture.

FIG. 8.7.   Spacing variant 1: This gesture takes up the notion of size but produces a mismatch in semantic content.

porates the more obscure (verbal) reference to the bearing plates. These new
gestures thus take up the notion of size, but produce a mismatch in semantic
meaning. In the first variant (Fig. 8.7), Libby's colleague copies the spacing of
the original gesture, but she uses he index finger rather than her entire hand.
In the second variant (Fig. 8.8), Libby's colleague rotates her own wide strata
gesture 90 degrees to produce a new spacing gesture. The first variant suggests
that Libby is talking about the length of the stick of glue. The second variant,
however, is far to big to be a stick of glue; it has the expansive quality of the
mimetic gesture that Libby herself uses to talk about a roof fall.

In these two variants, we see how speaker and audience use their own gestural interpretations to reflect upon the patterns, discontinuities, and relationships in Libby's gestures. What is perhaps most interesting is the way that

FIG. 8.8.   Spacing variant 2: This gesture is noticeably wider and more expansive than Libby's original gesture. The gesture was produced by flipping the wide strata gesture 90 degrees.

the miner in the background seems to reject the first variant—by refusing to take it up in her own gestures. But this miner does take up the expanded version of the gesture in the second variant (Fig. 8.8). Both miners then use this expanded gesture as they work together to resolve differences in their interpretation of Libby's gestures. This gesture ends the sequence and provides the ground for their interpretation that Libby is talking about the spacing of the bolts.

While it may seem artificial to ask miners to guess the meanings conveyed in gesture in videos without speech, rhetoricians have not previously found it artificial to rely on speech without gesture. Researchers rely upon written transcripts because they are stable and permanent records of rhetorical interaction and because they can be analyzed and re-analyzed as conditions warrant.[34] But we may also rely on written transcripts because we have not had adequate technologies or methods of analysis to capture and interpret how speakers communicate meanings in gesture.

The role of gesture in miners' representations of risk is particularly striking when we look at a transcript of the interactions I have described without the advantage of gesture. What might a researcher or investigator conclude if they were forced to make an interpretation based only on the speech portion of Libby's narrative transcribed below? What would they conclude about Libby's understanding of the process? What would they conclude about the quality of Libby's demonstration? What would they know about the viewpoints she assumes or the ways she integrates theory and practice?

---

[34]Myers (1990) makes a strong argument for using texts to investigate social interactions.

BS: Tell me how, show me how, show me how to insert a roof bolt
Libby: Drill the hole, spin the glue, insert the bolt, huh, insert the glue, spin
         the glue.

As Libby herself concludes at the end of this sequence, "Cain't talk about it, honey, gotta show it . . ."

MINERS USE GESTURES RHETORICALLY IN COLLABORATIVE INTERACTIONS

When we examine gestures in isolation as snapshots, we do not see how Libby's colleagues interact with each other in gesture, how individuals seem to influence the shape of new variants, or how miners use gesture rhetorically to negotiate the meaning of Libby's gestures in the same way that speakers might vary tone and sentence structure to explore, question, or assert meanings in speech.

These particular miners share an affinity that may be idiosyncratic, but they demonstrate that individuals can use gesture to communicate, interpret, and create meaning. Analysis of gesture thus allows us to see how speakers use gesture to explore, query, and co-construct meaning—collaboratively and individually—as they work to interpret meanings embodied in gesture:

Speakers Use Gesture to Explore Meanings Embodied in Gesture

As Libby's colleague studies the video, she slowly tightens her right hand into a C-shaped gripping gesture—the 13-second compression gesture we described in the previous section (Fig. 8.5). During this entire gesture, she said nothing.

When I asked her what she was doing, she explicitly acknowledged that she was "trying to figure out what Libby is saying." In this gesture, Libby's colleague isolated one element of Libby's strata/compression gesture—focusing on the more abstract notion of compression.

Speakers Use Gesture as Queries to Invite Response

The notion of a query suggests that gestures can be used rhetorically to invite spoken or gestural responses in others—like an embodied question. When

FIG. 8.9.   Libby's colleague (front, foreground) uses gesture as a query to invite response.

speakers query, they produce gestures with emphasis such as bouncing or enlargement, then pause or stop their action as if to invite a response.

When Libby's colleague (Fig. 8.9) shifted her hands from a wide strata gesture (hands parallel and horizontal) to a wide spacing gesture (hands parallel on the vertical), she looked to her colleague for confirmation or response. Her colleague copied this gesture before she responded in speech: "Probably the spacin' of the bolts."

Like Libby's gestures, the query invites her audience to reflect upon the patterns, relationships, and discontinuities she expresses in this new variant. Her audience signals agreement by copying the gesture, then articulating a new interpretation in speech. This pattern is consistent with research in gesture that suggests that gesture precedes speech by an interval that is proportionate to a speaker's verbal fluency (Chawla and Krauss, 1994). In this case, however, speakers work collaboratively to interpret the meanings they have taken up in gesture.

Speakers Co-Construct Knowledge in Gesture

Co-construction occurs in speech and gesture when individuals working collaboratively coordinate their speech and action so closely that one speaker's meaning seems to influence the production of meaning in the other. When speakers co-construct meaning, they seem to coordinate speech and gesture without overt scripting or direction.

FIG. 8.10.   Miners co-construct similar gestures within a fraction of a second. These imitations are so closely coordinated that they seem deliberately synchronized. These gestures show how miners are attuned to the meanings expressed in both speech and gesture. When speakers co-construct gestures, the process of uptake is nearly instantaneous.

The Osage miners paraphrase each other's speech and copy each other's gestures so closely—sometimes within a fraction of a second—that it is sometimes difficult to determine who initiated the gesture and who copied. These miners work together so closely that it seems as if they are deliberately coordinating their speech and action. In some cases, the two miners create very similar gestures. In other cases, these miners create smaller variants that echo the other's larger gesture.

When Libby's colleague (foreground, Fig. 8.5) holds her compression gesture for 13 seconds, the miner in the background creates a small imitation of the gesture for a fraction of a second. In Fig. 8.10, both miners create the same large spacing gesture nearly simultaneously. In this case, the closely constructed gestures seem to bring closure to the differences that arose as they worked to interpret the meaning of Libby's gestures.

IMPLICATIONS FOR WRITERS

When we examine the meanings that these miners articulate in speech, we discover that their speech is distracted and disjointed. If we look only at the grammar and logic of their speech, we might conclude that Libby's colleagues are—at best—inarticulate. Yet these women are otherwise forceful and convincing speakers, devoted union representatives, and safety spokes-persons at their mine. Their speech is disjointed, we might suggest, because gesture provides a more adequate image of the temporal and spatial com-plexity of a notion like compression. They understand experience in and through gesture.

In these examples, spoken language does not easily accommodate the tem-poral and spatial complexity or the features of manner and motion that can be accommodated in gesture. Gesture and speech are complementary but they are not identical. Speakers can use gesture to represent features of space and movement not possible in speech. No single viewpoint can represent all of the types of information that decision makers use to warrant judgments about risk. Speakers need information conveyed in gesture to visualize dy-namic processes that cannot be articulated in speech.

I began my research with the hope that I could convince agencies that they could take advantage of rhetorical theory to improve technical docu-mentation in risky environments. My research has demonstrated that rheto-ricians can also benefit from the methods of others in order to understand how individuals communicate outside of written texts.

Given the complexity of mimetic and analytic viewpoints that speakers present in gesture, it is perhaps not surprising that researchers accustomed to the stability and authority of texts might not perceive the value of a speaker's frequent shifts in gesture, multiple and complex perspectives, or simultaneous viewpoints. Without modern video technology to record and analyze gesture, neither rhetoricians nor agencies could develop a method to capture and stabilize knowledge embodied in gesture. The iconographic portraits of gesture produced by 18th century elocutionary theorists like Bulwer (1644/1974) and Austin (1806/1966) show speakers performing ges-tural acts, but these crude portraits cannot represent the details that enable audiences to see the spatial and temporal complexity that speakers can rep-resent in idiosyncratic and unscripted gesture. Without a method to capture gesture, researchers could not see the rhetorical transformation of experi-ence that makes argument, persuasion, and—ultimately—communication with an audience possible.

With new technologies, rhetorical theorists can rethink the relationship between speech and gesture in the rhetorical transformation of experience. They can use this knowledge to help agencies integrate stakeholder knowl-edge into risk judgments and risk assessments throughout the entire Cycle of Technical Documentation in Large Regulatory Industries.

In chap. 9, we see how investigators interpret and transform information embodied in speech and gesture at the site of the investigation, but investigators are neither systematic nor consistent in their methods. As a result, their investigation does not allow us to examine how they draw conclusions or what components of the original representation are captured in the investigators' interpretation of (undocumented) gestures. As chap. 9 argues, analysis of gesture can and must contribute to that process if agencies are to improve the rhetorical transformation of experience at critical moments in the Cycle.

IV

TRANSFORMING EXPERIENCE

9

Capturing Experience: The Moment
of Transformation

Q: *In your opinion, who is responsible for health and safety at South-*
   mountain Number Three Mine?
A: *Pardon?*
Q: *In your opinion, who is responsible for health and safety at South-*
   mountain Number Three Mine?
A: *All the men that works there.*
Q: *All the men that works there?*
A: *All the men that works there.*[1]

—J. D. Cooke, (1993)

I'd like to say, I've run this through my mind. A lot of them people were my
friends. I don't know.[2]

—K. Brooks, (1993)

In January, 1993, MSHA conducted 31 interviews to determine the cause or
causes of the Southmountain disaster in Norton, Virginia.[3] All interviews

---

[1]Cooke, 1993, p. 28.
[2]Brooks, 1993, p. 72.
[3]These include statements by McKenney, 1993; Adams, 1993; Brooks, J. P., 1993; Brooks, K., 1993; Combs, R, 1993; Cooke, J.D., 1993; Cox, 1993; Davis, B.R., 1993; Davis, J.E., 1993; Davis, M., 1993; Duncan, C.E., 1993; France, 1993; Goode, C.A., 1993; Goode, D.E., 1993; Goode, D.L., 1993; Meade, D., 1993; Mullins, D., 1993; Mullins, D.A., 1993; Mullins, J.E., 1993; Mullins, L.C., 1993; Ramey, P, 1993; Shane, 1993; Short, D., 1993; Short, G.P., 1993; Silcox, 1993; Thompson, T., 1993. Depositions were eliminated if they bore no relevance to the investigation.

were recorded, transcribed, and made available to interested parties. In these interviews, MSHA investigators explore a wide range of questions concerning liability, responsibility, policy, and practice. What kinds of decisions led to the disaster? To what extent did economic pressure influence decisions? Who made decisions to change ventilation? What were the lines of communication within the mine? What kinds of policies and procedures can prevent similar disasters in the future?

For writers, the MSHA interviews provide a glimpse of the rhetorical strategies that investigators employ as they attempt to capture miners' embodied experience in the written record. In these interviews, we hear the voices of miners as they describe their experience of risk, and we see MSHA investigators as they attempt to make sense of these voices for the official record. The Southmountain interviews thus provide a complicated study in risk assessment at the moment when embodied knowledge moves into the realm of science.

In these interviews, individual investigators can see events only in and through the embodied representations and situated viewpoints of individuals. In theory, investigators never transmit expert information to lay audiences. Instead, they must draw upon their own expertise to elicit sufficient evidence to construct a logical and coherent narrative.[4] Within the cycle, the agency observer's expertise provides a framework for analysis, a set of technical questions and issues that must be answered in the investigation, and conditions of logical probability (both scientific and common sense) by which observers can judge the validity of individual conclusions. In this process, local knowledge and experience move into the domain of science and engineering, closing the cycle of information in the process of risk management and assessment depicted in Fig. 9.1.

The Southmountain interviews show that agency investigators play a more active role in the rhetorical transformation of experience than we might expect. MSHA investigators are not acting as rhetoricians as they question witnesses, rephrase their questions, and translate miners' narratives into the language of engineers. But their questions suggest that they employ a variety of rhetorical strategies to reconcile the diverse and contradictory accounts of the disaster—interpreting miners' gestures, translating embodied experience into an analytic framework that supports their find-

---

[4]The interviews demonstrate the degree to which the agency's investigation as a whole makes visible the collective viewpoints of individuals as they are mapped and tested against the material realities of a risky environment. But these interviews also draw attention to the power differentials and legal liability of each individual within the institutional architecture of a mine.

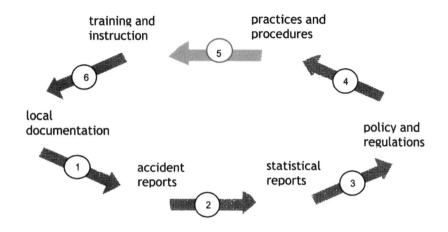

FIG. 9.1. Six critical moments of transformation.

ings and conclusions, and helping miners recall critical details about conditions and practices in the mine prior to the accident. To construct a coherent account of the investigation, they must weigh each individual's testimony against the physical findings of the investigation underground following the accident and the logical possibilities that those physical findings imply. In the process, they create new representations of the accident in the legal record.

Ideally, this legal record provides a complete and accurate account of miners' testimony. As the following analysis suggests, however, the written transcript provides only a partial record of the viewpoints that miners assume in speech and gesture. Miners and engineers speak different languages. They use different indices to gauge the magnitude of risk. And they recognize different levels of uncertainty. In some cases, transcribers indicate in the text that miners used hand motions or gestures to represent non-verbal actions. In other cases, deictic language in the text (such as when miners said, I moved "here" and then "there") provides evidence that miners were using their hand and bodies to represent information not conveyed in speech.

These interviews show how agency investigators work to transform miners' experience at the scene of the interview without the benefit of the kinds of data gathering, analysis, and interpretation that rhetoricians (or anthropologists) might perform in order to capture a full account of miners' narratives in speech and gesture. This observation suggests that much of the analysis and interpretation that should take place after the investigation is occurring at the scene of the investigation, where investigators have neither the data-gathering techniques nor tools of analysis to create an adequate reconstruction of miners' embodied experience.

As chap. 7 demonstrates, the viewpoints that miners create in speech and gesture are complementary, but they are not identical. Viewpoint affects the semantic content and meaning. When speakers shift viewpoint, they change the relationship between audience and speaker. New viewpoints can help them discover patterns, relationships, and discontinuities that would not be apparent in other modalities (cf. chap. 7). Gestures also serve important rhetorical functions in the negotiation and production of knowledge as speakers use gesture to explore, query, assert, and construct, both collaboratively and individually, a new understanding of risk. When agencies transform miners' experience in writing at the site of the investigation, the archival record does not provide an adequate basis for the kind of considered analysis and assessment that would improve policymaking within agencies.

When agencies fail to document the full range of viewpoints that miners represent in speech and gesture, they create public relations problems as well as regulatory indecision throughout the Cycle of Technical Documentation in Large Regulatory Industries. Unions complain that the agency's technical focus ignores miners' input; agency writers complain that the historical record provides an inadequate account of events, conditions, and decisions prior to the accident; and agencies are paralyzed by a process of revision and review because written accounts of the disaster do not provide the information they need to warrant judgments about risk.[5] Any perceived inadequacy in the agency's narrative affects the rhetorical force of risk representations throughout the entire Cycle.

Agencies would be horrified, we imagine, if investigators recorded their second-hand conclusions and impressions before they collected physical evidence from the accident site. But agencies have institutionalized the rhetorical transformation of experience in the post-accident interview.[6] Agencies use the methods of science to help them collect physical evidence at the site,

---

[5]In March, 1997, NIOSH investigated the larger problems of ground control in retreat mining in a series of seminars in five locations, including Norton, Virginia—the site of the Southmountain disaster. In these studies, researchers acknowledge the complexity of ground control conditions in arguing for a case-history approach based upon MSHA's large database of actual mining case histories. In relying solely on MSHA accounts, NIOSH's data reflects (and propagates) the uncertainties in MSHA's narratives of risk. Because NIOSH's databases reflect the limited technical viewpoint of MSHA's Investigative Reports, their findings cannot provide a complete account of risk (Mark, 1988).

[6]Cf. McAteer, 1995.

but they have not used the methods of rhetoricians, linguists, and social scientists to collect the rhetorical evidence they need to understand the disaster.[7] Without a consistent and adequate record, scientific or rhetorical, it becomes impossible to determine trends and patterns in the industry.[8]

There are also strong theoretical arguments in favor of using available digital technologies and linguistic analysis to capture the full range of viewpoints that individuals represent in speech and gesture before these viewpoints are transformed into writing. If rhetoric is the art of finding out the available means of persuasion, it is also about evaluating the best and most effective means of persuasion for particular audiences. If "all the men that works there" (Cooke, 1993) are responsible for risk, the knowledge they communicate in these investigations must be represented in policies and procedures designed to protect them.

This chapter describes the legal and rhetorical uncertainties that prevent agencies from capturing a "complete and forthright" (McAteer, 1995) account of miners' testimony in the post-accident investigation.[9] The Southmountain interviews help us locate—quite literally—the rhetorical interface between agencies and the world of experience. In these interviews, we hear the voices of miners as they describe their experiences of risk, and we see MSHA investigators as they attempt to make sense of these voices for the official record. But we cannot see their visible and embodied understanding at one critical moment of transformation.

The following section describes the controversy and uncertainty that followed the Southmountain disaster.

---

[7]The Southmountain interviews thus raise interesting questions about the nature of conventional (agency) representations of risk and their alignment with three-dimensional, uncertain environments. A discussion of the development of these genres is outside the scope of this project. The following discussion suggests that researchers must understand how agencies capture—or fail to capture—embodied experience in writing before agency writers begin the processes of transformation we have described in previous chapters. Only then can we attempt to recapture the knowledge that is rendered invisible in the transformation.

[8]South Africa's Leon Commission (Leon et al., 1994) confronted a similar problem when they attempted to assess the country's history in regard to silicosis and other dust-related diseases. The Commission noted that "the absence of a systematic approach to the control of respiratory disease reflects the absence of appropriate analysis of available data and the long standing fragmentation of services between distinct government departments" (Leon 1994, p. 51). The Commission recognized that outdated conclusions were useless in assessing trends; without the body of evidence that analysts used to derive their conclusions, the Commission could not reasonably make assumptions on the basis of the available evidence (pp. 48–52).

[9]McAteer, 1995.

CONTROVERSY AND UNCERTAINTY
FOLLOWING THE SOUTHMOUNTAIN
DISASTER

On December 7, 1992, an explosion in Southmountain, Va., killed eight miners and injured another. As in other mine disasters, excessive levels of methane contributed to the disaster. This time, however, MSHA blamed miners for igniting the explosion. In the abstract of its investigative report, MSHA concluded: "The methane was ignited on the 1 Left section in the No. 2 crosscut between the Nos. 1 and 2 entries by an open flame from a butane lighter."[10] The report explained:

> Persons were smoking in the mine and the operator's Smoking Search Program was not effective. One cigarette pack, containing nine unsmoked cigarettes, was found on the victim located at the point of origin. The butts of ten smoked cigarettes were also found in the victim's pockets. Some of the butts were in the container used to transport the victim to the medical examiner's office. A functional butane cigarette lighter was found on the mine floor at the rear of the scoop located at the point of origin.
>   The investigators believe that the open flame from this butane cigarette lighter was the ignition source for the December 7, 1992, explosion.[11]

MSHA's report of investigation cites eight violations that contributed to the disaster. According to MSHA, weekly examinations were not conducted; the smoking search program was "neither adequate nor conducted in its entirety"; the preshift exam on December 6, 1992, was not conducted in its entirety; ventilation control devices were not placed or maintained to direct and provide adequate ventilation; the incombustible content of mine dust samples showed that the percentage of incombustible content was inadequate; the approved ventilation plan was not being followed in the bleeder system; the volume and velocity of air was not sufficient to "dilute, render harmless, and carry away flammable and explosive gases that were liberated"; and the weekly exams were not conducted.[12]

Union officials (UMWA) criticized MSHA's conclusions both publicly and privately. According to Moore, MSHA broke with protocol in announcing that they had discovered smoking materials at the site. In a special edition of the *UMWA Journal*, Moore (1993) argued that the agency had set up a

---

[10]Thompson et al., 1993, p. 1
[11]Thompson et al., 1993, p. 43.
[12]Thompson et al., 1993, pp. 44–5.

"smokescreen" to cover its own failure to discover methane in previous inspections at the mine.[13] According to the UMWA, cigarettes could not produce enough heat to ignite methane. The union claimed instead that MSHA's new ventilation regulations had weakened miners' protection and were thus a contributing factor, particularly during the pillaring phase of room-and-pillar mining.[14]

The public (and highly crafted) rhetorical face of this debate reflects in part the tension between the agency's technical focus in the investigation and UMWA's demand for more expansive changes in the agency's policies toward small (and frequently non-union) mines. MSHA investigators (Thompson et al., 1993) had limited the scope of the agency's investigation in order to determine "the role, source, and location of explosive methane gas, the role, if any, of coal dust participation in the explosion, and the ignition source of the explosion." The union wanted MSHA to target dangerous sectors of the coal industry (e.g., non-union mines); protect whistle-blowers; restore critical health and safety regulations "gutted" by previous administrations; reevaluate agency procedures; and establish an independent investigations branch to investigate mine disasters, "including any MSHA wrongdoing."[15]

In the public debate, MSHA's focus on technical issues also seemed to ignore the human dimensions of the disaster, blaming victims for failures of enforcement and regulation. MSHA's investigative report, for example, listed miners' names in an appendix.[16] The UMWA, on the other hand, presented

---

[13]Moore, 1993, p. 5. The union based its conclusions on MSHA's investigation of the Pyro disaster (discussed in chap. 3). The union concluded: "MSHA had also found cigarette butts and matches on the site of the Pyro disaster, but MSHA had later rejected smoking as a cause of the accident, because the temperature of a burning cigarette is below the 1,100 degrees F needed to ignite methane" (Moore, 1993, p. 5).

[14]Moore, 1993, p. 7. The union linked the Southmountain disaster to a series of four serious accidents that occurred "back to back" and asked MSHA to "issue a national alert on the hazards of room-and-pillar mining" (Moore, 1993, p. 7). The union argued that pillar mining was under way at Southmountain at the time of the accident and during a nonfatal methane explosion at the Ammonite No. 31 mine. UMWA safety experts were investigating faulty or failed ventilation systems that may have caused these accidents. Moore (1993) explains the problem:

In room-and-pillar mining, the retreat or "pillaring" phase occurs when miners remove coal blocks originally used for support as the working section withdraws. The method requires stringent ventilation and roof control safeguards (p. 7).

[15]Moore, 1993, p. 8.
[16]Thompson et al., 1993, Appendix A, p. 1.

the victims as a close-knit team of dedicated workers and family men. According to the UMWA, James "Gar" Mullins was "putting his paychecks toward restoring his family's homestead."[17] Mike Mullins, "a devoted father, was planning Christmas surprises for his close-knit family."[18] " 'We were as close as any men could be,' " recalled Tony Parigan, who was off work that night due to his grandfather's death."[19]

In this chapter, we go behind the scenes of MSHA's investigation in order to examine the ways that individuals' speech and gestures are represented in the written record as they describe their perception of events, decisions, and conditions prior to the accident. To set the scene, we look first at the legal uncertainties that clouded the agency's investigation.

LEGAL UNCERTAINTIES CLOUD AGENCY INVESTIGATIONS

According to the Federal Mine Act, management is ultimately responsible for decisions and practices underground. But the Southmountain disaster raised questions about the locus of liability when both miners and management engaged in unsafe practices. (Ultimately, Southmountain Coal Company and several Southmountain officials pleaded guilty to criminal violations.)[20] Thus, while investigators assured miners that the interviews were voluntary and that they could not be discriminated against for their participation in the interviews, they also warned each respondent that they could be face civil and criminal sanction under federal and Virginia law. In addition, investigators recognized that economic pressures could influence safety decisions and their outcomes at every of level of authority within the mine—from the supervisor, Norman Deatherage, to the foremen and ordinary miners.

Preventing Self-Incrimination

Because of the threat of litigation, miners and management were cautious about their answers—particularly concerning smoking underground.

Several miners refused to answer questions about smoking practices. One miner twice told investigators that he'd "like to take the fifth amendment"

---

[17]Moore, 1993, p. 4.
[18]Moore, 1993, p. 4.
[19]Moore, 1993, p. 4.
[20]McAteer, 1995, p. 6, sec. 1.

when asked if he had observed anyone carrying matches underground.[21] Another "nodded negatively" but gave no "audible" response. The transcript reads:

Q: Have you ever observed anyone smoking underground?
A: (No audible response). (Witness nodded negatively).[22]

Paul Ramsey, a foreman, claimed to have no idea about the cause of the disaster. Deatherage, the supervisor, couldn't remember the details of coal production. In the testimony, investigators are skeptical about his faulty memory. His testimony reveals both the uncertainty and complexity of the institutional hierarchy at Southmountain:

A: I don't know. I have no idea. I mean, you know—
Q: You're the superintendent?
A: We just run what we can, but, I mean—
Q: You're the superintendent?
A: Right
Q: Who would know that? Everybody I've talked to, almost every management personnel, and nobody can answer that. I'm just asking for, is there more or less than usual?
A: Well, I—
Q: Would Ridley Elkins know?
A: Yeah, he probably would, I guess. I mean, I know I've probably got it written down. I mean, I've got it written down at the mines, but I mean, I don't keep track of—
Q: You don't keep track?
A: —Of whether we run more this month or last month because we just run what we can.[23]

Unraveling Institutional Authority and Liability

The complexities of institutional authority at Southmountain created a second source of uncertainty that unraveled as the investigation progressed. On paper, Ridley Elkins served as an engineering consultant to the mines, but MSHA ultimately determined that Elkins was more intimately involved in

[21]Silcox, 1993, pp. 38–39.
[22]Steele, 1993, p. 29.
[23]Deatherage, 1993, pp. 19–20.

engineering decisions than his role as consultant might suggest. Elkins also had potential conflicts of interest. Elkins, Meade, and Davis owned a mining supply company called EMD that provided supplies to the mine.[24]

Questions about conflict of interest pervaded many aspects of the investigation. Miners had a right to have a representative present at each interview, but unions and nonunion miners disagreed about who should represent miners' interests. Unions wanted union representatives present at the interviews. They questioned the loyalties of the two miners selected to serve as miners' representatives, Jessee Darrell Cooke and Donnie Short, Senior. As general superintendent of operations, Short had close ties to management. His close ties to Elkins also created potential conflict of interest. MSHA investigators wondered why he was put in charge of the mine when he had not been there for seven months. His own testimony suggested that his position allowed him to serve as a cover for Elkins, who was the real decision maker in the mines. The following excerpt raised questions about his ability to protect miners' interests in his role as representative of the miners:

Q: Ridley Elkins said that you were in charge for the company?

A: Yeah, he said I was in charge at that time for the company.

Q: Do you know why he would put you in charge when the superintendent was there and you hadn't been there in seven months?

A: I guess because of my past experience with MSHA. I have worked three or four of these mine explosions, ignitions. I was on the rescue team at Clinchfield back years ago when I was a lot younger, before, with MSHA. And I guess because of my experience.

Q: But Mr. Elkins was at the mine at this time?

A: Yes.

Q: And, would he not be the senior Southmountain official at that point?

A: Uh, evidently not. He put me in charge and he's my boss. *I do what I'm told.*[25]

Locating Responsible Agents

The Mine Act specifies that management must keep track of methane levels, record safety checks, and document pre-shift exams. The Act places the burden of compliance on management even when miners fail to follow safe practices—either deliberately or inadvertently—on the grounds that management

---

[24]Deatherage, 1993, pp. 19–20.
[25]D. Short, 1993, pp. 39–40, Italics added.

must insure that miners follow regulations.[26] Management is responsible even when miners illegally smoke underground, because management is responsible for checking for smoking materials before miners go underground. In addition, if monitoring does not work, management is ultimately responsible.

In their physical investigation of the mine following the disaster, MSHA investigators had discovered a methane monitor covered with a greasy rag. This physical evidence constituted a clear violation of the Mine Act, but investigators needed to determine who had ordered the monitor to be covered and who had put it into practice.

When methane monitors are working properly, they send a signal that cuts power to the mining machine when methane levels exceed safety standards set by the Mine Act. Miners might shortcut the safety switch (by covering the monitor) to keep production going if they felt the monitor was giving a false reading. They might also shortcut the safety switch if the methane levels were close to the safe standard, but the monitor was interrupting production too frequently. Any barrier—like the greasy rag—would prevent the monitor from taking a correct reading. Such actions are both illegal and unwarrantable. In a gassy mine, miners who worked with an inoperable monitor would be toying with disaster.

The MSHA interviews demonstrate the high level of uncertainty that clouded all testimony. In each of the 31 interviews, investigators asked what miners and management knew about the rag covering one methane monitor— and when they knew it. Altogether, 30 individuals testified that they knew nothing about the rag on the methane monitor. The mine's superintendent Deatherage claimed not to know about rag in methane monitor until after the explosion.[27] Cooke denied that he ever had any problems with the methane monitor, though he worked on the same shift as the sabotaged monitor:

Q: Did you ever have any problems with the methane monitor on your shift?
A: No. if I—Uh, I've had times that the lights would go out on it. The numbers go out. And, I'd send for a mechanic and he'd come up there and he' fix it right back. We never operate without that methane monitor.[28]

Another miner, Dane Meade, testified that he knew about the rag, but only after investigators recast the question five times. He claimed that he didn't know about the rag till after the investigation:

---

[26]See chap. 1.
[27]Deatherage, 1993, p. 67.
[28]J. D. Cooke, 1993, p. 49.

Q: Have you ever seen a cloth or rag placed in the methane monitor sniffer?

A: No sir.

Q: Do you know if anyone else had done this before?

A: No, I don't.

Q: Did you ever give anyone instructions to install a cloth into the sensor?

A: No sir.

Q: When was the first time you found out about the cloth in the sensor?

A: I can't remember whether it was in the, somebody at the mine told me, I
     believe.

Q: Was it after the, during the investigation or prior to the investigation?

A: It was after they'd took it off.

Q: So that would have been during the investigation.[29]

Thus, it comes almost as a complete surprise when one miner, Donnie
Mullins, explained very matter-of-factly why he put a water repellent rag on
the methane monitor. According to Mullins, water mist from the spray on
the continuous mining machine had caused the monitor to produce false
readings that would shut down the mining machine. Mullins described how
he tested the monitor after putting the oily rag on the "sniffer." The detailed
reasoning and honest enthusiasm in his testimony makes it obvious that others would have to have noticed the presence of the rag.[30]

Mullins' testimony raised questions about the locus of authority for decision making in the mine. What might it mean when miners claimed that they
had never suspected the presence of methane in the mines? What might it
mean when Billy Ray Davis asked two miners' representatives to leave the
room when he testifies? How should investigators' interpret Silcox's invoca-
tion of the Fifth Amendment? How should investigators interpret Billy Ray
Davis's enthusiasm for Southmountain Coal in the following passage?

---

[29]D. Meade, 1993, p. 34.

[30]Mullins, D. (1993) testified:

The reason I put that rag in there, that spray there in that pan, when you cut down that
water mist that cutter frame and, which they ain't that much can get in there in that, no
bigger hole than that was, I'd say two, two and half inch diameter hole. But two or three
drops on that sniffer will mess that methane monitor up. It cause it to, you know, to start
reading. And just like I told them up there at West Virginia when I went up there and
watched it, every test that they run on it. When he saturated that rag and screwed it back
down on that sniffer, just as soon as that water hit it, that thing went to reading. And it,
it'd read back and forth and it'd finally settled on eight (8) tenths percent. But the rag
was oily, it was water repellent, you know. And like I said, when I put it in there, I went
and got my bottle, I checked it, it read out, it gassed off. It read at nine (9) tenths. The
low light come on at one point nine (1.9). It kicked the power out and it read to two
point four (2.4). (pp. 28–29)

Yes sir, you know I was, a lot of people asked me, you know, just like my parents, you know. I, I'd rather work at Southmountain Coal, boys, as anywhere I've worked. I sure had. And, I like to work over there. And, one, one of the reasons, people has asked me this and I'll tell you all. They was good to keep the equipment up. They sure was. And, that's one thing I can say, because I like to run those buggies over there. They kept them in working order.[31]

Corroborating Factual Findings

Whatever miners say, investigators must weigh their findings in the interview against the material evidence they discover during the underground investigation.

As Mullins's testimony suggests, testimony does not always correspond to the evidence underground. Some miners and management lie to protect themselves and their colleagues; some honestly cannot recollect events; all miners stand in a different relation to risk, see events differently, and represent these events in different combinations of viewpoint in speech and gesture.

Investigators could test some claims against the physical evidence in the mine: the location of ventilation doors, curtains, and brattices (a set of curtains that helps direct air flow underground); the rag on the methane monitor; or a weakened pillar that shows evidence that miners had taken an illegal (extra) sixth cut. Some representations can be tested against an external reality because that reality is shaped by human agency. Someone placed a rag on the methane monitor; someone took a sixth cut. Thus, when the supervisor Deatherage could not remember whether he'd ever seen a sixth cut, investigators prompted his memory by asking a series of questions that tested his recollection against the evidence at the scene:

Q: Have you ever seen a sixth cut mined out of that block on the backside?
A: Not that I recollect.
Q: You haven't seen one?
A: Not that I know of.
Q: Have you been in the explosion area?
A: Yeah.
Q: That block that—
A: I mean I—
Q: The block that the continuous miner is sitting in now, are you aware of any sixth cut being mined out of that block?

---

[31]B. R. Davis. 1993, pp. 13–14.

A: I didn't look at it real close.

Q: You haven't looked at it.

A: Not real close.

Q: Do you know whether it's a practice to mine a sixth cut out of that out of the back end of the block?

A: No. We have done it before, but it's not a practice, no.

Q: Who would have done it?

A: I don't know?

     . . .

Q: Would a foreman make that decision?

A: Evidently so. I mean, I, you know?

Q: You say you have seen it done?

A: Yeah, I've seen it done.[32]

Investigators know that evidence at the scene indicates that human agents acted to shape the material environment. Other evidence is less certain, however, particularly when miners and investigators stand in different positions—rhetorically and geographically—in relation to the accident.

UNCERTAINTIES IN THE PROCESS OF RHETORICAL TRANSFORMATION

Multiple fatal accidents like the explosion at Southmountain account for only a small proportion of the accidents in risky environments. Agencies also investigate dozens of single fatalities and nonfatal but single accidents. At MSHA, many investigators are former miners; they have worked with miners; and they have participated in many investigations at all levels within the agency. We would thus expect investigators and miners to share similar representations of risk, use similar vocabularies, and understand each others' questions. We would not expect investigators to use language that confuses miners, nor would we expect miners to use terms that require lengthy negotiation and elaboration to determine their meanings.

In the Southmountain interviews, however, miners and investigators frequently seem to misunderstand each other. They do not seem to speak a common language, and they do not seem to share the same referents. In these interviews, investigators must work hard to transform miners' embodied understanding into the discourses of science and engineering at the site of the

---

[32]Deatherage, 1993, pp 41–42.

investigation. In the process, they create rhetorical and linguistic uncertainty that cannot be reconstructed after the interviews. If investigators take note of knowledge embodied in speech and gesture, that knowledge is retained only in individual memories. Because speech does not easily accommodate the temporal and spatial relationships that become visible in miners' gestures, miners' spoken testimony often seems inarticulate and disjointed. This problem is exacerbated when speaker and audience do not share a common vocabulary or shared set of assumptions and experience.

If policy makers believe that stakeholders need technical assistance to participate in the regulatory process, it is perhaps because they have not understood how the rhetorical transformation of experience in writing has shaped what policy makers see in the official record.[33] In the following examples, we see traces of that lost knowledge, but we can only draw on our previous analyses of miners' narratives to speculate about what that knowledge might look like or how that knowledge might help agencies understand risk.

Differences in Vocabulary Reflect More Fundamental Differences in the Relationship to Risk

Appalachian miners and professional engineers speak different languages. To communicate with miners, investigators must translate engineering jargon into terms that miners understand. In some cases, investigators must literally give miners a language to describe events, conditions, and locations in the mine in terms that engineers can apply to assess risk. As the following examples suggest, these differences in vocabulary reflect more fundamental differences in the ways that Appalachian miners and engineers represent their relationship to risk.

In the following exchange, Willis (1993) seems to resist the investigator's attempt to transform his embodied understanding of the environment into the language of engineers. He tells investigators to "speak English" when they ask a particularly tortured question:

Q: Did you ever have occasion to post the number one entry of the first left section, which is the bleeder entry?

A: What do you mean? Speak English. The post?

Q: Timber. Have you ever timbered the number one entry?

A: You talking the bleeder line?

---

[33]Cf. chap. 2 for a complete discussion of this topic.

Q: Yes.
A: Yeah, we timbered the bleeder line a lot.[34]

The apparently superficial differences in vocabulary in this exchange may reflect deeper assumptions about language and safety. Miners locate themselves in relation to physical objects, events, and conditions. Investigators locate these same objects, events, and conditions as two-dimensional coordinates on a mine map. Miners describe themselves in relation to the "bleeder line."[35] Investigators call the section "the number one entry of the first left section." Miners say "timber." Investigators say "post." These differences reflect underlying differences in viewpoint that reflect deeper assumptions about risk and safety. The "bleeder" describes a critical component of the mine's ventilation system. The "first left section" describes the section in relation to the mine's plan of development. The number of the section reveals little about the ventilation system. As a result, it tells us little about the potential hazard or benefit of a particular location in a crisis. In a fire, for example, miners could use the direction of air flow in the bleeder to help them locate an exit.

As the history of mining suggests, miners have developed their own terminology to express their embodied understanding of the environments in which they work.[36] These terms are often incomprehensible to outsiders, but they help miners locate themselves in relation to risk. In chap. 4, we saw how miners located themselves inby and outby particular events and locations. When miners describe chandeliers, gobs, bumps, faces, props, and timber, these terms have special meaning in the physical context of mining. They are difficult to define outside of the embodied sensory experience that gives them meaning.

Both engineering jargon and miners' vernacular expressions serve different purposes as warrants for risk decisions. The point is not that miners' vernacular expressions are better or more accurate representations of risk. Rather, we must pay attention to the relationships and concepts that are lost in the rhetorical transformation from one set of terms to another.

---

[34]Willis, 1993, p. 17.

[35]Willis, 1993, p. 17.

[36]Agricola's first work on mining, the *Bermannus, sive de re metallica* (1530) attempted to reconcile differences in meaning in German, classical, and medieval texts on mining. Agricola's work was the first in a long tradition of mining dictionaries which attempted to collect and translate this difficult discourse. When the Welsh opened their tin mines in the early 1800's, they borrowed freely from early German sources. These Welsh miners brought their knowledge of mining to the U.S. with the opening of the Appalachian coalfields and the copper mines in the Northern Peninsula of Michigan. See Agricola, 1556/1950.

Differences in Language Reflect Underlying Differences in the Ways That Miners and Investigators Warrant Judgments About Risk

Investigators look for indices (indications) of risk; miners hear bumps. When McKinney asks whether there was "any indication of a problem any time before the explosion," he gets no answer. When he asks, "Were you aware of bumps, any major bumps?" Davis answers: "Yes, you, you could hear them, you know, off and on, on that pillar work."[37] Bumps ground risk assessments in embodied experience; indices ground risk assessments in the accumulated evidence that constitutes engineering experience.

The word "bump" never appears in MSHA's report of investigation. Bumps describe the sound that a mine roof makes when there is a release of pressure within the strata. Bumps have a specific location and direction. Bumps are local events that cannot be heard throughout the mine. When bumps occur, they create concern. Good bumps can scare miners enough to chase them out of the section. Engineers, by contrast, know how the air is supposed to be controlled and thus describe general ventilation plans rather than local events like bumps. When accidents occur, investigators document these embodied experiences so that they can locate events underground.[38]

Because miners' embodied experiences provide local information about the magnitude of risk prior to the accident, investigators take seriously a miner's description of bumps. Goode, for example, describes one particularly big bump (a "pretty good bump") that scared him enough to chase him out of the section. Investigators ask about the size, location, and duration of the bump as they attempt to confirm the event and characterize conditions in the mine prior to the explosion.[39]

---

[37]Davis, 1993, p. 14, Italics added.

[38]Donnie Short (1993) testified:

Q: What direction did the air flow in the belt entry at the Number Three Mine?
A: In this mine we have a natural ventilation problem. The ventilation intake air was pulled to the tailpiece and really had a split of air there, one going into the face, ventilating one and coming back down the belt to the fifth or sixth crosscut inside and then regulated into the return. (p. 18)

[39]D. L. Goode (1993) testified that he heard bumps in the mine:

Q: When you were servicing the miner.
A: Yeah, at that time Gleason and I both was, he was pumping the grease gun and I was in on top of the miner holding the, holding the grease end of the, the hose into a plug. And, *it bumped enough to chase us both out of, out of there* [Italics added].

The intensity of miners' emotional reactions helps investigators assess the magnitude of events prior to the accident. Silcox experienced the bump "right down in them old works." He describes his fear—"the first time I've been scared in the mine in years"—in order to define the magnitude of the event.[40] Silcox tells investigators that miners have a "sense of survival" that tells them "danger is there." They use their "sense of survival" to warrant judgments about imminent danger:

> No, it was just that, that bump and that rip. That rip of that top. And, it really scared me. I mean, you have, you have your sense of survival if you, I'd say you've worked in the mines and you know how, I mean, bang, you get out of there. You know danger is there. . . .[41]

Miners' embodied experience also provides important information about the location and magnitude of risk. Cooke, for example, could not tell where the bumps were coming from, but he notes "these was pretty close."[42] Cooke speculates that the bumps were pressure bumps that indicated the sandstone layer of strata was breaking up: "That's what it sounded like . . . Pressure, pressure bumps."[43] The foreman, Kenneth Brooks, was too far away to hear the

---

Q: Was it a pretty good bump?
A: Yes.
Q: Do you, could you tell where it was coming from?
A: It was overhead. It sounded like it was over top of us.
Q: Were there any more bumps or did it settle down right away?
A: It settled instantly.
Q: That was it, just one big one.
A: Yes sir.
Q: There wasn't any—
A: No—
Q: Any before that or after?
A: None prior or after. (pp. 23–24)

[40]Silcox (1993) testified:

A: Well, during the time, I'd like to tell you something that did happen while we was servicing that miner. . . . The first time I've been scared in the mine in years. They was a heck of a bump. It shook the whole area and it just Pow!, just right down, right down toward them old works, you know, and, I run out of there. I was really scared. . . .
Q: Was there any falls?
A: No, it was just a big bump and shook. I mean, it sounded like that whole mountain broke. (p. 12–13)

[41]Silcox, 1993, p. 25.
[42]J. D. Cooke, 1993, p. 10.
[43]J. D. Cooke, 1993, p 19.

noise. He never heard the bumps because "of course, I was up at the end," but he qualifies his testimony. He's hard of hearing, and there were other noises in the mine.[44] Although he was too far away to judge conditions for himself, Brooks was persuaded by miners' representations of experience because "both of those men were certified people." Brooks interprets their testimony as an indication that the top "had to have been breaking up"—a sign of a weak roof:

> A: And those guys telling me and that did concern me, that while they was servicing this miner and both of those men were certified people. They said it bumped so hard in there that they ran. Once or twice they ran away from that miner. Now, where that they never, they never said where the bump come from, but the top had to have been breaking.[45]

The agency's final report of investigation shows how miners' testimony helps the agency gauge the force and extent of the explosion. In the interviews, Fleming testified that he felt a lot of pressure during the explosion—so much that his ears popped and he momentarily lost some hearing. Investigators later tell him, "We know what kind of pressure it takes to, to do that."[46] The final report reframes Fleming's experience in terms of the pressure "a person" might experience in an explosion:

> Tests by researchers on the physiological effects of blast pressure have shown that a peak overpressure of about one psi will knock a person down, five psi is the minimum pressure that will cause damage to the eardrums, and 15 psi results in lung damage. The dynamic pressure of 12 psi in the entries and a static pressure of six psi in the crosscuts outby and including the No. 4 crosscut would correlate with the condition of Fleming after the explosion.[47]

---

[44]K. Brooks, 1993, p. 17.

[45]K. Brooks, 1993, p. 13.

[46]Fleming, 1993, pp. 23-4. Fleming (1993) describes the force of the explosion:

Q: Kevin, let me just ask you one more time. It means a lot to us as to what you might have, at the time of the explosion, you said that you first saw dust, and I believe you said you felt heat, and your hard hat got knocked off. And, as you bent down then some greater force hit you.

A: Uh-huh (yes).

Q: Do you have anything else you can tell us about what you saw or felt?

A: I just felt, I felt a lot of pressure right about the time that the dust got there, because I know right there in that drive, I thought my ears was sort of popping. (pp. 23–24)

[47]Fleming, 1993, pp. 36–37.

The Southmountain interviews thus allow us to examine one aspect of the rhetorical transformation of experience when local knowledge moves into the domain of science where it is captured and transformed in writing. But these interviews also demonstrate how easily the absence of any sort of negative feedback can create a sense of complacency that can lead to disaster. In fact, one of the questions that is never explicitly stated is why miners who heard bumps did not communicate with management or persuade management to take a different course of action.[48] When Shane describes how the rock dust "always kept going toward the outside," investigators ask if this was normal.[49] They are particularly concerned about the direction of dust flow in the mine, which seems to indicate problems with the ventilation system prior to the accident.[50] But that concern did not move individuals to action.

In the darkness of a coal mine, miners experience a constant stream of embodied sensory information. To prevent disasters in the future, investigators must learn what makes certain forms of embodied knowledge more persuasive than others. How can they teach individuals to translate embodied experience into warrants that can guide future action?

---

[48]The transcripts show that investigators are confused when miners' answers do not meet their expectations of what is normal in a coal mine. Silcox (1993), for example, testified that the air "thinned out" as they approached the inside of the mine. His testimony implies that the dust was heavier at the mine entrance where (presumably) the air should be fresher:

A: [Silcox] You know, how it, how it kind of thinned out and dissipate as it goes further, you know.
Q: So the further to the outside, the worse it got or, I don't quite understand?
A: Well, we were leaving the section, all right? And, I don't remember exactly how far we got up there. We kind of run into that thin dust and it gets a little thicker and thicker and then as you go on past, it starts getting thinner and thinner.
Q: So you think the air was flowing in which direction?
A: It was coming down the belt.
Q: Toward s the inside of the mine?
A: Yeah.
Q: Is that normal? (pp. 47–48)

[49]Shane, 1993, p. 14.
[50]D. L. Goode (1993) noticed that the dust hit his face as he was heading the outside. The flow was strong enough that he "had to go inbye another five breaks just to let it settle down enough for me to go out" (p. 19). Silcox (1993) remembered the direction of the air flow because when they came in there with a scoop, "it would blow dust out, you can tell, you know, coming this way." Silcox later told investigators that "there was so much air" coming through the curtains,"it would send them flopping" (pp. 46–7). The last night before the explosion, Silcox testified, the dust in the mine got thicker as they went out.

THE TRANSCRIPTS INDICATE THAT MINERS GESTURE, BUT INTERVIEWERS DO NOT EMPLOY CONVENTIONS OF CODING THAT WOULD ALLOW US TO SEE WHAT INDIVIDUALS EXPRESS WITH THEIR HANDS

The written transcripts provide surprising evidence that investigators indicate miners' gestures in the written record in order to understand events, practices, and conditions prior to the accident. Transcribers have conventions for entering these gestures into the legal record, but they do not employ conventions of coding that would enable us to see (and thus interpret) the viewpoint and embodied understanding that these gestures represent.

When Ramey testifies, investigators ask him to "take the pen and show us how you just described with your fingers."[51] When Cooke testifies, investigators note that "he's pointing to the first and second right" as he speaks.[52] In the following exchange, Cooke's spoken testimony would be incomplete without an understanding of his gesture:

A: No, not in a while. I've went with him on gas checks. I believe it was up in here where we was down in here, walk these old places up in here, and he wanted somebody to go with him.

Q: Okay, he's pointing to first right and second right.[53]

In some cases, miners use deictic expressions (here, there, etc.) that point to invisible spaces. In these cases, we can only speculate that miners are using gesture to indicate spatial relationships, air flow, curtains, and structural support (timbers) in their testimony:

A: Well, normally *there would come up into these two areas.* Come up *into here* and they've got the miner here so, there would have been a set of eight breaker timbers here. It went *all the way across* and then which, since this is your last one here this would have been breakered off, breakered off. It would have been curtains *in front of all these breakers.* The air would have came let's see, they would have left this one open. *It would have came in here and,* and more or less swept the area going this way in this direction, but mostly into this direction, taking the air from the miner operator into the gob area. (Italics added).

---

[51]J. D. Cooke, 1993, p. 10.
[52]J. D. Cooke, 1993, p. 38.
[53]J. D. Cooke, 1993, p. 38.

The written transcript suggests that miners might use gesture to answer questions that are better expressed through demonstration than speech. In the following exchange, we can only speculate that Davis employs a spacing gesture similar to the ones we saw in chap. 7. But the record does not tell us what this gesture might look like:

Q: And, how was it blocked open.
A: I didn't really notice.
Q: How far was it open? Six inches? Twelve inches? Three foot?
A: *Like that, I guess?*
Q: Approximately a foot?
A: Maybe not a foot.[54] (Italics added)

The meanings conveyed in gesture are not always obvious; investigators must often work to interpret miners' meanings in gestures. The following exchange suggests that this rhetorical process is both explicit and self-conscious. Investigators address their comments to the transcriptionist (rather than to miners or investigators) in order to explain for the record what Steele is indicating with his hands:

A: I can't remember if it was that day or if I looked at it, but I know all the time we would, it wouldn't be, a lot of times there might be a little opening right there. I can't remember if it was this side or this side. But, we direct some air up that way and—
Q: Mr. Steele is indicating an opening on the curtain up your number one entry between C2 and C3. And, Mr. Steele, I just have to follow up what with your, what you do on the map so, we'll have it on the record.
A: Okay. But, I know most of the time they would have, you know, most of the air directed down this way. And, they'd have some going, you know, through there, too.[55]

Investigators pay particular attention to gesture when miners and management describe smoking checks at the mine, but the written transcript tells us little about the nature of the searches or how they were conducted. If investigators had captured knowledge embodied in gesture, we might be able to determine whether management had followed proper procedures. But the record tells us nothing about the spatial or temporal dimensions of the practices or procedures we see indicated in the transcript.

---

[54]M. Davis, 1993, p. 22.
[55]Steele, 1993, pp. 14–15.

In the following two examples, the written transcript hints at the presence of gesture but, again, provides no detail. In both cases, witnesses apparently use their hands to describe how searches were conducted, but miners' speech provides no information about the process. (In both cases, miners answer "just.") In both cases, investigators fill gaps in the record by naming the process that miners presumably describe with their hands. In neither case does the archival record allow us to confirm investigators' conclusions. In both cases, it would be useful to know whether miners represented themselves as they were patted down (in a mimetic viewpoint gesture) or whether they saw others being patted down (in an analytic viewpoint gesture). The record unfortunately does not indicate whether these miners were patted down all over their bodies, at their hips, or at the location of their pockets:

Q: How were the searches conducted?
A: Just (WITNESS INDICATING).
Q: You're patted down?
A: Patted down, yeah.[56]

Q: When were they held?
A: In the mornings before you go in.
Q: And how were they conducted?
A: Well, let's see, they just (interviewee indicating).
Q: They pat you down?[57]

Supervisors and management were also asked how miners were searched, but the transcript fails to indicate whether foremen and supervisors gestured when they described smoking checks. When investigators asked Deatherage how searches were conducted, for example, he responded, "pat down."[58] Deatherage ordered the searches but did not participate. Brooks, the foreman, was responsible for patting miners down. He tells investigators, "The only thing you can do is just pat the man down. If you don't feel any, he's not got any."[59] Charles Ernest Duncan conducted searches at the mine entrance. His viewpoint reflects the physicality of the experience: "I'd stop the mantrip and I'd go to each end of the mantrip and feel over their bodies, you know, their shirt and their pants pockets. And, that was normally the way we did it."[60]

---

[56]J. D. Cooke, 1993, p. 28.
[57]Willis, 1993, p. 14.
[58]Deatherage, 1993, p. 29.
[59]K. Brooks, 1993, p. 38–39.
[60]C. E. Duncan, 1993, p. 12.

Fleming (the sole survivor) was never searched but testifies that he saw others being searched:

A: I've seen them pat a couple of them down.
Q: Were you ever searched?
A: No.
Q: Did he ever ask you if you were carrying cigarettes?
A: Once.[61]

If we could be certain that investigators indicated all cases of gesture in the record, we might assume that the absence of any such indication would provide evidence that the speaker did not gesture. But we know that miners gesture frequently, and they assume many different viewpoints as they describe their experience of risk. It would be difficult to imagine that the number of gestures indicated in the transcript represents an accurate account of the gestures that miners actually produced in their testimony.

If we assume that investigators marked only a small number of gestures, we might assume that these indications might point to important or highly suggestive gestures that would provide clues to the practices and procedures that precipitated the accident. But the record is silent about the nature of the gestures and their role in investigators' interpretations of the evidence.

Without a video record of the interaction, it is also difficult to determine whether investigators made correct assumptions about the gestures indicated in the transcript. When Mullins testifies, for example, the transcript reads: "(No audible response). (Interviewee nodded affirmatively)."[62] Mullins apparently nods again when investigators ask him whether he understands that he can ask representatives of the coal company to leave if he objects to their presence during his testimony.[63] The transcript notes that the "interviewee nodded affirmatively"—presumably indicating that he understands. But there is no indication that he agreed to (or objected to) the presence of coal company representatives. Given the controversy that surrounded questions of representation, miners may have hesitated to respond—and simply nodded to indicate that they understood the question:

Q: Representatives of the Southmountain Coal Company of the miners are
     now present. If you have any objection to their being here, those repre-
     sentatives will not be present during your statement and your statement

---

[61]Fleming, 1993, p. 16.
[62]D. Mullins, 1993, p. 4.
[63]D. Mullins, 1993, p. 4.

will be treated as a confidential communication pursuant to law. Do you understand this?

A: (No audible response). (Interviewee nodded affirmatively).[64]

Ultimately, the transcripts confirm both the limitations of individual experience and the explicit transformation that must take place in order for embodied knowledge to enter the written record. Transcribers record those moments in the testimony when gaps in the text fail to represent a miners' experience, but they do not record those occasions when gesture accompanied speech, enhanced speech, or presented different and perhaps conflicting viewpoints in speech and gesture. We do not see miners creating complex narratives that incorporate multiple viewpoints, and we cannot see the interaction of embodied knowledge and analysis in miners' gestures.

The absence of gesture creates uncertainties in the written record, as if we are receiving only one channel of a stereo recording or—perhaps more analogously—text without accompanying visuals.

TWO-DIMENSIONAL MAPS CANNOT REPRESENT THE SPATIAL AND TEMPORAL COMPLEXITY THAT IS POSSIBLE WHEN SPEAKERS COORDINATE SPEECH AND GESTURE

To locate events in the disaster, investigators create an individual map of each miner's experiences. At these moments, investigators must work to translate miners' dynamic sense of institutional and geographic space into the two-dimensional surface of a conventional mine map.

The transcripts show that miners remember events as dynamic narratives that resist representation as static coordinate positions on a map. Shane remembers his position as an activity that moves from section to section. His narrative begins at number two drive, but moves to the other drives and back again. The uncertainty of his memory reflects the uncertainty of his position at any single moment underground:

A: I started at number two drive and met that other boy down at number three drive and we'd go on, on down to the other drives together and then we'd come back and shovel the belt.

Q: Okay, so you got number two drive, he got number three drive?

A: Yeah.

---

[64]D. Mullins, 1993, p. 4.

Q: And, you went together to?

A: To number four and did it. Then we shoveled, started shoveling on number, finished shoveling number three belt.

Q: Three belt is, do you remember the cross cut numbers?

A: Let's see, seventy-, seventy-two (72) to thirty-nine (39), if I'm not mistaken.[65]

Steele remembers curtains because he passed through them, but he can not recall "where we were at."[66] Silcox confesses his uncertainty when confronted with a map. Investigators must show him familiar landmarks to orient him to space in this new perspective.[67]

Miners are surprisingly uncertain about their location despite the fact that the miners I have interviewed claim that they maintain an inner sense of their location underground. In some cases, the problem reflects a lack of direct experience. Shane claims he does not know the direction of the air flow because he "very seldom dusted that drive up there, because that other boy did it, so I wouldn't—."[68] Willis may be pretending to be "simple" to avoid more difficult questions about the disaster. He tells investigators, "I don't read maps . . . I'm just a simple mining roof bolt man."[69] But Shane has real

---

[65]Shane, 1993, p. 9.

[66]Steele, 1993, p. 12.

[67]Silcox, 1993, p. 28. In the following lengthy exchange, Silcox (1993) attempts to locate a check curtain on the map but he can only remember the location because he had to move the "scoop" (a small vehicle) and charge the battery:

Q: Wayne [one of the investigators], can you show him where that is? Between number four and five entries, the power center. Do you remember whether there was a braddice of—

A: I think there was a charger there with a check curtain.

Q: Okay, up the charger you say there was a check curtain?

A: I, I believe now. I believe that's where it was at

Q: Okay,

A: It was either in front of it or behind it. *I don't exactly remember where.*

Q: *Okay, how about one block down in the number five entry, do you recall what was there?*

A: We walked down number five.

Q: Where Wayne is point to?

A: No, I don't, *I don't recall what was there.*

Q: Okay, but you think there was a check curtain there?

A: But, the reason I remember, you know, is the check curtain where the charger was at is because see, we had to change scoops out, you know, charge it and stuff. (p. 28, Italics added)

[68]Shane, 1993, p. 19.

[69]Shane, 1993, p. 10.

difficulty reading maps. He is not able to find the Number Three entry on the map. He says, "I am not familiar with reading these things."[70]

Cooke needs to look at the map to locate the spaces he is talking about, but these spaces are not the same coordinate spaces that the map depicts. Though he admits he can "hardly tell where it's at," Cooke nevertheless describes both the top of the mine and the rider seam above him. He frequently shifts to a viewpoint within the mine as if he were explaining the geography before him: "See," he says, "that's the reason we pull off here and then come back down here. This top over in here. The rider seam, you just couldn't hold it if you tried. Even with super bolts we tried to hold it."[71] The two-dimensional map does not correspond to the viewpoint represented in his speech (he would need a cutaway diagram to represent events inside the roof of the mine), but the map prompts him to recollect the experience as if he were present inside the mine. Although Cooke's viewpoint does not correspond to the bird's-eye view of the map, he describes problems as he sees them present in his memory.

Investigators' comments suggest that they were aware of the need to enter particular types of invisible information into the record.[72] When Deatherage tells investigators that he traveled the bleeder "up through all the way to the back of it and I don't know exactly how far, but it's somewhere right around in right here," investigators note for the record that he's "indicating the top end of first left about midway back towards second right."[73] When Duncan tells investigators "we were down in this area here," investigators explain, "and, he's referring to exhibit 31-b, which we refer that as second right."[74] When miners are unsure, investigators help them reconstruct events and conditions at the time of the accident.[75]

---

[70]Shane, 1993, p. 18.

[71]Cooke, 1993, p. 13.

[72]Cf. Deatherage, 1993, p. 79.

[73]Deatherage, 1993, p. 79.

[74]C. E. Duncan, 1993, p. 8.

[75]Adams, 1993, pp. 13–14. Adams (1993) knows where the power center is, but he can't transfer this knowledge to a map—even when Mr. Davis points out the position. In this example, investigators instruct him to draw a line to indicate his position "up by the feeder." They construct a complete narrative of his actions near the power center, then tell him how to locate that narrative on map:

A: Where's the power center?
By MR. DAVIS: There's the power center.
A [Adams]:   I ain't really for sure.
Q: Did you travel up by the feeder?
A: Yeah, we went up by the feeder.

Both directly and indirectly, investigators seem to recognize that the testimony must be elaborated to make sense. They often work together to negotiate differences of opinion that arise when the spoken testimony is inarticulate or incomplete.[76] When miners use deictic pronouns without a visible referent, investigators work together collaboratively to reconstruct a consistent narrative account in the written record.[77]

As the investigation proceeded, investigators discovered that Southmountain's mine maps bore little correspondence to changing conditions under-

---

Q: Okay, just draw you a line right up by the feeder then. Then when you went to the power center did you go to the power center, the back of the power center, the part that's near the outside of the mine or the part that's near the inside?
A: Uh—I ain't really for sure. (pp. 13–14)

[76]In the following exchange, Jessee Darrell Cooke (1993) has difficulty locating the continuous mining machine. Investigators offer to help him remember, finishing his sentence in one case and directing him to a different location on the map. Cooke becomes the questioner in this exchange, as investigators themselves answer questions for the record:

Q: Can you go to exhibit 2-a, which is the larger scale map that shows the working section, and with that black pen, mark up where you mined on that shift?
A: This is where they, that's where—
Q: That's where the—
Q: The miner was at the time of the explosion.
A: Okay. I'm trying to think which one we was in. This one?
Q: Okay, if we could help you out a little bit. The two blocks to the right of the continuous miner were probably—Now go in by one block?
A: Here?
Q: No, go in by—
A: Here?
Q: Yeah. Those two blocks were probably the most recent ones mined, if the proper sequence is being followed. (pp.

[77]In the following passage, Cooke tells investigators that he developed "this. . . ." The pronoun creates a high level of uncertainty in the written record. Both investigators prompt Cooke to elaborate. McKinney leaves the phrase open; Thompson supplies the missing referent—a term that miners could not use because the sections are not numbered underground. In this example, the transcriptionist interprets Cooke's guttural response as a positive "yes" (transcribed in parentheses). Cooke himself never actually voices the words of the investigators; he merely responds, "uh-hun (sic)." The official record records the four alternative voices [Cooke, McKinney, Thompson, and the transcriptionist (in parentheses)] as they negotiate the official version of events prior to the accident:

Q: Where did you develop next?
A: This . . .
BY MR. MCKINNEY: Which is referred to—

ground. No one in charge conducted surveys to determine the actual positions of mining operations underground. Deatherage argues that these maps contain the approved ventilation plan, but they make no attempt to represent "real" conditions underground.[78] Burns testifies that such surveying is "not really feasible" because of the rapid changes as mining progresses. As a result, maps are full of uncertainties and discrepancies:

Q: (McKinney) and, so you're saying that blocks that you'd draw indicate pillared five cut plans or full recovery that those are not exact, they're, they're—

A: They're not surveyed.

Q: They're not surveyed. So they're not in there for each block, so they go by the mine foreman's map?

A: Yes, that's a good description. It's not really feasible with the amount of quickness and moving time that they do

Q: How are uncertainties noted? If any discrepancies were to be found by the surveyors or mine inspectors, what would take place? What would happen?

A: Uh, I, I guess it would depend on what the discrepancy was and to what extent it was, was noted.[79]

Deatherage (1993) tells investigators that that maps are not always accurate. When bosses draw mining practices, "They get it a little bit wrong."[80] Thus, he is not surprised when the official mine map seems to contain mistakes that are theoretically impossible given the material conditions in the mines. For Deatherage, the logic of geography outweighs any representation on a map: "Sometimes I make mistakes, I guess . . . "but they ain't no way you can do that [mine in the direction indicated on the map]."[81]

Neither engineers nor management at Southmountain seem to believe that the discrepancies between maps and mines constituted a problem of

---

BY MR. THOMPSON: Referred to as first right.

A: Right.

BY MR. MCKINNEY: First right.

A: uh-huh (yes). (p. 30).

[78]Deatherage, 1993, p. 35.
[79]Burns, 1993, p. 31.
[80]Deatherage, 1993, p. 44.
[81]Deatherage, 1993, p. 44.

safety.[82] Maps are representations of institutional decisions and plans, past and future; they do not necessarily correspond to real spaces or practices. Burns appeals to the uncertainty of maps in a rapidly changing environment to justify his failure to map rapidly changing (and potentially hazardous) conditions. He does not seem concerned by the discrepancy between the map and worsening conditions underground even when (very real) water blocked critical ventilation to the face.[83]

Ultimately, the problem of mapping draws attention to the tensions between embodied and agency representations of risk. Although contemporary maps frequently depict bolting machines, continuous miners, and temporary and permanent roof support, they rarely include (living) miners, even in the case of accidents—though they may mark the location of victims' bodies. After an accident, maps can capture static viewpoints. They serve as findings of fact in the investigation, and they can be used rhetorically to reconcile the relative positions of observers in relation to the accident—as John Nagy located bodies and machines on a two-dimensional grid in order to reconcile differing accounts of the disaster.

Conventional mine maps show cutaway views of the mine from many different angles. These two-dimensional representations enable mining engineers to map large spaces, develop mine shafts and haulageways, and plan roof support and mine development. Investigators use these accounts to determine how practices and procedures (as miners remember them) reflect plans and projections. When there is a poor fit between plans and practices, investigators can begin to assess liability. Were miners poorly instructed? Did they fail to follow safe practices and procedures? Did management press miners to push production? Ultimately, investigators can use maps as a rhetorical

---

[82]Evidence in the hearings suggests that engineers at Southmountain were particularly careless about documenting mining practices. Maps are incomplete and inaccurate. Short testifies: "[T]he advancing dates are shown on the map" but "the retreat dates are not on here and there's no way I'd tell you what dates they are." Management failed to conduct preshift exams on December 6, 1992, between the hours of 9:30 p.m. and 10:30 p.m. when methane may have been present in the mine. MSHA cites this failure as a contributing factor in the accident (30 CFR 75. 364).

[83]Burns (1993) testifies:

Q: Were you aware that water had accumulated in the face of west mains?
A: Back to a certain point, yes.
Q: And how was that controlled since the air, the water could have stopped ventilation based on the water elevation shown on the map?
A: The water line was ventilated straight across the water line, and it was roofed in by it, so it left no voyage for anything to accumulate.

tool that enables them to speculate about the differences between theory (expressed in a mine's MSHA-approved plans) and practice.

Unlike ordinary miners, mining engineers seem more comfortable with maps, but their viewpoint does not necessarily reflect their experience underground. Burns oversaw the mapping, drafting, and some of the outside surveying projects. He conducted mine feasibility studies and handled general mine engineering and civil engineering projects.[84] But he did not visit the mine directly for "probably at least a year."[85] He also cannot recall when employees assigned to surveying were last in the mine, and he has no records of the survey.[86] The mine surveyors also spent most of their time on the surface. As a result, their maps reflect a prospective view—a plan for development rather than a picture of daily mining operations underground. Maps serve a variety of purposes, including recovery work, as Burns tells investigators: "We're looking to make sure that the, that the reserves are being recovered and that there's not a lot of coal being lost, things of that nature."[87] Surveyors are not at all concerned with "any kind of safety inspection."[88]

As the testimony reveals, maps are useful in planning and development, but they tell little about actual conditions in a mine. When engineers and management view mine space solely through the perspective of static two-dimensional representations, they may inadvertently turn their attention away from the embodied experiences and changing conditions that signal risk in a hazardous environment.

THE COMPLETE AND FORTHRIGHT
TESTIMONY OF WITNESSES

On May 10, 1995, MSHA announced that the agency would conduct a review of its mine accident investigation policy, particularly with respect to witness interviews. According to McAteer (1995), then Assistant Secretary of Labor for Mine Safety and Health, "the complete and forthright testimony of any witnesses to the events leading up to an accident [is] essential in determining the cause of the accident and preventing similar accidents in the future."[89]

---

[84]Burns, 1993, p. 5.
[85]Burns, 1993, p. 8.
[86]Burns, 1993, p. 9.
[87]Burns, 1993, p. 28.
[88]Burns, 1993, p. 28.
[89]McAteer, 1995.

According to McAteer, MSHA's accident investigation procedures are designed to establish "all relevant facts about a mining accident in an orderly manner." The investigation includes three phases: physical inspection of the mine, complete analysis and testing of mining equipment that may have been involved in the accident, and interviews of persons who may have relevant information about the conditions which led up to the accident.

The agency emphasized the critical importance of witness testimony in the interview phase of the investigation. McAteer (1995) acknowledges that fatal accidents in the past have raised questions about the presence of mine operators and other interested parties during the interview: "Ideally, we would be able to obtain candid testimony from all witnesses without negating the information needs of mine operators, miners' representatives, and the accident victims' families." The agency was particularly concerned that the presence of particular parties may have discouraged witnesses from telling the truth.

Analysis of the Southmountain interviews suggests that both ethical and methodological arguments for understanding how the agency's rhetorical practices affect the kinds of information the agency obtains from its investigation. The rhetorical and political uncertainties we have described in this chapter reflect deeper assumptions about the political and technical value of engineering experience and embodied experience as warrants for future risk judgments. These assumptions may inadvertently silence or exclude the knowledge that agencies need to assess and manage risk.

When agency investigators help miners rephrase their answers in the language of science and engineering, they assist in the rhetorical transformation of experience, but they also privilege engineering experience and scientific knowledge at the expense of miners' embodied experience. Warrants grounded in engineering experience and scientific knowledge have communicative value inside of agencies and in public hearings, but they tell us little about the conditions and practices that precipitated the accident. As we have seen in chap. 4, the agency's technical focus in the investigation helps us locate the technical source of the disaster, but it tells us little about why some miners recognized danger in time to escape.

The examples in this chapter show how the agency's rhetorical practices create uncertainty in the written transcript. But they also suggest that agencies can draw upon the methods of rhetoricians to create more "complete and forthright" representations of what individuals know. The examples in this chapter show that written testimony cannot accommodate the dynamic uncertainty that miners represent in speech and gesture, but it also suggests that miners' testimony does help agencies create a more complete understanding of risk.

Within the Cycle of Technical Documentation in Large Regulatory Industries, rhetorical choices can be strategic or habitual. As we have shown in previous chapters, interviewers can prompt miners to assume new viewpoints. Agencies can use these techniques to create the rhetorical conditions that evoke elaboration, demonstration, and redefinition. They can use gesture to explore, query, and co-construct a more adequate representation of risk.

The interviews in this chapter remind us that agencies must learn to account for the ways that habitual and strategic rhetorical practices create uncertainty in the transcript even as they learn to use rhetorical strategies to create better representations of risk. In the final chapter, we explore the implications of our research more generally for rhetorical theorists.

10

Conclusion: The Last Canary?

Written documentation provides a stable object that can be analyzed and interpreted long after the ephemeral features of speech and gesture have vanished in real-time communication. Myers (1990) argues that written texts provide an important site for research because texts "hold still" and provide "portable" objects that are amenable to analysis.[1] Olson (1994) suggests that one of the "virtues of print" is that it allows researchers to examine how societies change as they come to depend more and more on writing.[2] But print culture has not entirely replaced oral culture; nor can print alone provide an adequate modality for representing the dynamic uncertainty of the material world. "What writing represents" and "what it does not" depend upon conceptions of literacy, epistemological notions of how we come to know the world, systems of record keeping that preserve some literate practices and not others, and systems of interpretation that enable us to understand and make sense of "what writing means" in a culture.[3]

Both the stability of print and its portability created a revolution in early modern Europe that, Eisenstein argues (1993), made possible the very notion of modern science.[4] The print revolution enabled "communication at a distance" (Kaufer and Carley, 1993) and increased the dissemination of knowledge at a rate not possible in oral culture.[5] It enabled institutions to maintain

---

[1]Myers, 1990.
[2]Olson, 1996, p. 17.
[3]Olson, 1996.
[4]Eisenstein, 1993.
[5]Kaufer & Carley, 1993.

and produce a "continuous accumulation of fixed records" (Olson, 1994) they could draw on to understand the past and predict events in the future.[6] Print facilitated the emergence of large bureaucracies and created "distinctive modes of thought that are conveyed through systematic education."[7]

For early modern scientists, printed texts created "immutable mobiles" (Latour, 1990) that could be studied, compared, and synthesized to produce new knowledge.[8] Print offered certainty because it was repeatable, reproducible, and verifiable. Printed texts helped scientists organize and make sense of the scattered and fragmented incunabula that had previously documented the material and institutional histories of late medieval Europe. Latour (1990) writes: "No 'new man' suddenly emerged sometime in the sixteenth century, and there are no mutants with larger brains working inside modern laboratories who can think differently from the rest of us."[9] Instead, particle physics "must be" radically different from folk biology because print enables scientists to create stable images that maintain [the] internal consistency of the object regardless of perspective.[10]

With the invention of three-dimensional perspective, print made possible the kinds of rhetorical practices (such as mapping) that were necessary for technological expansion in Europe.[11] A new interest in truth became possible because printed texts increased the dissemination of knowledge and revealed the contradictions, disagreements, errors and difference that were not visible in individual (and widely dispersed) manuscripts.[12] It was easy to extend this research to draw parallels between the emergence of print culture and the emergence of new media at the dawn of the new millennium.

More recently, however, historians have questioned the simple dichotomy between oral and literate practices in science. Shapin (1994), for example, articulates the tensions between the "epistemic virtues of individual experience and individual reason" and an emerging scientific method.[13] If early

---

[6]Olson, 1996, p. 87.

[7]Olson, 1996, p. 17. Olson's analysis is useful for understanding the relationship between print culture and the conventions of language and "literacy" in oral and print cultures

[8]Latour, 1990, p. 19.

[9]Latour, 1990, p. 26.

[10]Latour 1990, p. 26.

[11]Cf. Ivins, 1973.

[12]Latour, 1990, p. 33. See also Shapin, 1994.

[13]Shapin, 1994, p. 201. Shapin (1994) demonstrates how print revealed both the "insufficiency of authoritative texts" and the "problematic status of testimony" (p. 202):

English empiricist individualism . . . was a rhetoric which insisted that no source of factual information possessed greater reliability or inspired greater confidence than the direct experience of an individual. The legitimate springs of empirical knowledge were located in the individual's sensory confrontation with the world. (p. 202)

modern scientists valued experience, they also recognized that new knowledge must build on a foundation of prior knowledge—communicated in and through both written and oral testimony. If only individual experience counted, Shapin argues, scientists would have to recreate all knowledge within a culture with each new generation.[14]

Unfortunately, not all texts are equally reliable; not all testimony can be trusted. As a result, early modern scientists needed a method for evaluating the truth-value of competing claims. Their questions sound surprisingly contemporary: Are craftsmen more reliable than those with a material or intellectual interest in the subject? Can scientists develop a set of principles—independent of competing claims—to adjudicate differences? Must scientists distrust every fact that cannot be verified through personal observation and experience? What kinds of knowledge are necessary to warrant claims in science? Whose knowledge counts? How can we represent that knowledge in writing?[15]

Ultimately, Shapin argues, early modern scientists valued the testimony of individuals who showed integrity and disinterestedness because such individuals seemed not to be motivated by either passion or self-interest. If it was impossible to demand "absolute certainty" of knowledge, seventeenth century scientists nonetheless demanded a "moral certainty" and "objectivity" that could both legitimate and set proper limits upon the role of testimony as a warrant for claims in science.[16]

Bazerman shows how the genre and activity of the experimental article emerged as one solution to the problem of how to evaluate testimony in the absence of "ocular proof."[17] Because the entire scientific community could not travel en masse to remote destinations, scientists chose "designated competent witnesses" to view experiments. Eventually, Bazerman demonstrates, the community developed documentation practices that could "stand in" for physical demonstration.

If the genre and activity of the experimental article never fully represented the complex interactions and negotiations that constitute science in the laboratory, the scientific method would come to dominate documentation prac-

---

[14]According to Shapin (1994), early modern scientists acknowledged that there was a "proper, valuable, and ineradicable role for testimony and trust within legitimate empirical practices" (p. 202).

[15]As Shapin (1994) argues, many of these methods (both formal and informal) were often counter-intuitive. Thus, despite appeals to the value of direct knowledge, scientists often rejected commonsense knowledge, craft knowledge, and the reports of the "vulgar" (p. 231–232).

[16]Shapin, 1994, p. 210. Cf. Shapiro, 1983.

[17]Bazerman, 1988, p. 74. Without a witness, the investigator must create a narrative account that provides "sufficiently good cause" for the conclusions the author reports (p. 74).

tices and intellectual activity in many disciplines. But contemporary writers still wrestle with the uncertainty of personal observation and experience when they create regulations and procedures to protect the health and safety of workers.

To achieve objectivity,[18] observers must share a set of rational standards and methods of collecting data that enable them to replicate experience interchangeably. The rational observer—the postulated positivist in social constructivist accounts of science—attempts to view the situation "from nowhere," maintains a rational and distant perspective, and avoids "implication in the phenomenon under study."[19]

The counter argument—social construction—argues that knowledge can never be known except through representation. Individual subjects construct their own representations of reality, which must be negotiated in social interactions to produce knowledge. Ultimately, however, knowledge must always be relative—a function of the situated perspectives of individual knowers who create facts when they reach consensus with other knowers in social and institutional interactions. Reality is inseparable from representation.

These crudely reductive versions of two opposing epistemological positions in the social and rhetorical studies of science distort their more subtle, and instructive, arguments about the relationship between material reality and the discursive practices that shape documentation in large regulatory industries. But the tensions between these two positions reveal a dilemma at the heart of rhetorical inquiry in the context of risk: *How can we reconcile the radical differences in individual accounts of experience with the need to create generalizable policies and procedures that can be applied across diverse situations?*[20] Harding writes:

> From the perspective of the conventional notion of objectivity—sometimes referred to as "objectivism"—it has appeared that if one gives up this concept, the only alternative is not just a cultural relativism (the sociological assertion that what is thought to be a reasonable claim in one society or sub-culture is not thought to be so in another) but, worse, a judgmental or epistemological relativism that denies the possibility of any reasonable standards for adjudicating between competing claims.[21]

---

[18]I have yielded to my editor's desire to eliminate unnecessary quotation marks in using this term without its customary framing in ironic quotes. This usage does not imply that I believe that scientific objectivity can be achieved in ordinary scientific practice.

[19]Nagel, cited in Petraglia, 1997, p. 196.

[20]When I speak of individual accounts, I include multi-authored, institutional and collaborative accounts which present a single argument or narrative account of risk.

[21]Harding, 1991, p. 139.

In *The Rhetoric of Risk*, I have argued that the problem of documentation is particularly difficult in hazardous environments—in part because the stakes are so high. Because no single individual has access to all aspects of the environment, each individual necessarily observes conditions from a highly local but situated view of the whole.[22] In these environments, all documents are admittedly incomplete reconstructions. No single document provides a complete account of events and conditions underground, but agencies must nonetheless maintain a system of documentation that can help them evaluate complex and difficult geographic and institutional situations to prevent accidents in the future.

In focusing on two specific moments of rhetorical transformation in large regulatory industries, my own inquiry raises many questions about the relationship between writing and experience in uncertain and dynamic material sites. What is the value of experience? What's at stake in its transformation? How does this research help us understand historically specific rhetorical activity? How does it answer questions not answered in other theoretical approaches? Are the practices described in this research specific to one highly idiosyncratic industry or can they be applied, more generally, to documentation practices to other sites? What are the implications for workplace discourse? For our understanding of literacy and orality in a global context?

No single volume can answer all of these questions, but I hope I have answered some questions and mapped a new territory for future work investigating documentation practices throughout the Cycle of Technical Documentation in Large Regulatory Industries.

In the following discussion, I suggest how *The Rhetoric of Risk* can be extended and applied in other sites, how this work helps us rethink notions of instruction and workplace training, and, more generally, what this work might contribute to a more robust theory of rhetoric. I hope that others will join me in examining the function of written documentation and its transformation in other institutional and geographic contexts.

THE FUNCTION OF WRITTEN
DOCUMENTATION IN HAZARDOUS
ENVIRONMENTS

The stability of written documentation serves an important function in large regulatory industries. Written documentation carries the weight of scientific authority as a warrant for experience. It represents the agency's legal author-

---

[22]Cf. Code, 1991; Haraway, 1991.

lay knowledge and its relation to expert discourses. Research in risk decision making can help us understand the assumptions that guide individuals when they make judgments about risk.

Olson's analysis of the relationship between language and writing provides a foundation for the kinds of work I am proposing. Olson (1994) has developed a set of principles intended to overturn previous assumptions about the relationship between language and writing.[23] These principles suggest that documentation in writing involves a transformation of understanding that is difficult, if not impossible to unthink, once we have become accustomed to understanding language in writing. Recognizing that no writing system can bring all aspects of language into awareness, Olson argues that an important function of literacy "derives from the attempt to compensate for what was lost in transcription."[24] Before we can understand what is lost, this work suggests, we must reexamine those moments when other forms of language are transformed in writing.

In *The Rhetoric of Risk*, I begin the project of understanding how the practice of documentation can be extended to include the knowledge embodied in speech and gesture. Ultimately, my work suggests, we must reexamine what it means to document experience in script, video, or multi-media construction. To understand how these new forms of documentation might func-

---

[23]Olson (1996) describes eight principles that overturn previous assumptions about the relation between language and writing:

(1) Writing was responsible for bringing aspects of spoken language into consciousness (p. 258);

(2) No writing system, including the alphabet, brings all aspects of what is said into awareness (p. 260);

(3) What the script as model does not represent is difficult, if not impossible, to bring into consciousness (p. 261);

(4) Once the script-as-model has been assimilated it is extremely difficult to unthink the model and see how someone not familiar with that model would perceive language (p. 262);

(5) The expressive and reflective powers of speech and writing are complementary, rather than similar (p. 264);

(6) An important implication of literacy derives from attempt to compensate for what was lost in the act of transcription (p. 265);

(7) Once texts are read in a certain new way, nature is "read" in an analogous new way (p. 268); and

(8) Once the illocutionary force of a text as recognized as the expression of a personal, private intentionality, the concepts for representing how the text is to be taken provide just the concepts necessary for representation of the mind (p. 270).

[24]Olson, 1996, p. 265.

tion, we must first expand our notion of what it means to be literate in a multi-modal, visual-verbal culture.

TOWARD A BETTER UNDERSTANDING OF WORKPLACE DISCOURSE

Just as the genre and activity of the scientific article emerged out of the conventions and practices of 17th century (English) society, so the genres and activities of technical writing have emerged out of the conventions and practices of an emerging industrial economy. Although writers recognize the power differentials implicit in the imperative mode of conventional instructions, few but the most radical theorists have sought to dismantle either the apparatus of workplace instruction (as a means of controlling workers) or the legal framework that legitimizes conventional practices of instruction as a measure of the quality of instructional discourse. These practices are reinforced in technical writing textbooks that provide models of instruction for imitation and practice.[25]

The Rhetoric of Risk challenges conventional notions of workplace discourse grounded in a notion of instruction as a set of hierarchical and highly specified procedures that will necessarily produce reproducible—and by implication—safe results if workers understand the instructions and follow them precisely:

- Conventional instructions assume that workers or machinery can operate according to well-defined tasks and procedures to produce controlled and predictable results. In hazardous environments, workers must observe and evaluate a constantly changing stream of information so that they can adjust practices as conditions warrant in highly specific problems in the mine.
- Conventional instructions provide step-by-step procedure. In these instructions, the order of events may be critical to the productions of safe and successful outcomes. In hazardous environments, a list of instruc-

---

[25]See, for example, Burnett, 1997, pp. 413–447. Burnett challenges the reader to ask critical questions that might lead readers to revise their notion of instruction. Unfortunately, most readers do not have the time or resources to conduct the kinds of research that might help them understand complex systems. Nor do they necessarily have the rhetorical skill to rethink previous practices. As a result, most writers use sample documents as models that reinforce the notion of instructions as a set of hierarchical procedures. Writers may also hesitate to change their conventional practices because they fear they may be held liable if changes in documentation are perceived to affect safety.

tions may not reflect the same assumptions about order. In hazardous environments, workers must attend to many different aspects at once; the order of tasks is not necessarily important because each event is equally critical to the safety and health of workers.

In arguing for a more reflexive understanding technology that draws on local experience, researchers in science and technology studies (STS) may forget the value of simple heuristics in a crisis. In uncertain and hazardous environments, individuals may face unexpected conditions for which they have no instruction manuals and no prior experience. Without an instruction manual, workers become dependent upon management to make decisions for them. Or, they must draw upon whatever limited resources they have available to reduce and manage risk. Workers need theoretical (science-based) knowledge to apply the lessons learned from experience. When individuals express this theoretical knowledge in gesture, researchers may not appreciate the degree to which lay audiences construct complex narratives that integrate theory and practice.

As we have seen in previous chapters, local knowledge is critical for effective risk assessment, but local experience may be so highly specific (site specific) that the lessons learned cannot be translated to new environments. Generalized instructions may require adjustment (as when carpenters file down a new door to fit an old door frame). In a crisis, individuals may not have time to articulate all of the uncertainties in the decision process. They need numerical risk assessments to help them assess whether they can bring conditions into compliance with standards in a timely manner. They know that any risk estimate is an uncertain approximation of material hazards underground, but they need some kind of approximation so that they can estimate risk and adjust their practices accordingly.[26]

As I argued in chap. 5, individuals draw on many types of knowledge to warrant judgments about risk. Having more than one basis for assessing risk also creates a healthy skepticism that can alert individuals to signs of trouble—when the methanometer always produces the same reading, for example, or when the drill seems to hit resistance in the rock.

Ultimately, however, the burden will fall upon writers (trained and untrained) to develop new documentation practices that can balance the tension between liability and flexibility, formal standards and the realities of day-to-day uncertainty in local sites. Before we can create new forms of documentation, we must reexamine the fit between current documentation practices and the uncertainties of work in hazardous environments.

---

[26]South African miners are, interestingly, the strongest advocates for standards they can use to legitimate their own concerns about risk and safety in the workplace.

INDUSTRIAL STRENGTH DOCUMENTATION
IN AN ELECTRONIC CULTURE

The title of this chapter refers to my own experience retiring the last canary in a British mine. As I describe in the preface, this canary was made redundant (retired) by British legislation in 1994 when I was visiting the mine. The bird no longer went underground, but it served as a symbol of the authority of experience in a coal mine. When the bird was finally captured (after no small struggle), the manager and I placed it in a small brown box, and he turned to take the bird to its new home. "We should say something," I suggested. He then pronounced (in a thick Yorkshire accent), "Thus ends two hundred years of industrial history."

The last canary symbolizes the profound changes in Britain's industrial landscape in the 10 years since I first began this project. In northern England, thousands of miners have also been made redundant in the wake of mine closures and the denationalization of the British mining industry following the 1994 strike. This strike divided families and broke the NUM's power as a voice for Labor in the north. It also affected safety. Because the safety practices that had made British mines some of the safest in the world were also the most expensive, British mines could no longer compete with cheap coal from South Africa and Eastern Europe. When the British government denationalized the mines in the wake of the strike, the 1994 Boyd Commission recommended mine closures on the basis of two criteria: the stability of the labor force and the willingness of miners to accept new cost-saving technologies like roof bolting.[27] Mines that resisted change or had a history of labor problems were closed, and many miners were made redundant.

The problems of documentation that I have described necessary reflect the changing history of labor-management relations in U.S. and British mines, the declining economics of the mining industry, and the special character of manual labor in a heavily regulated, but dying industry. As the following examples suggest, however, the perceived difference between old commerce and new may reflect differences of scale and focus rather than differences of kind.

Expert Systems Are Constructed on a Foundation
of (Frequently Unarticulated) Tacit Knowledge

To construct expert systems, researchers in artificial intelligence must develop methods for mining (extracting) experience. I first encountered the

---

[27]See John T. Boyd Company, Department of Trade and Industry, 1993.

notion of mining experience in the work of researchers in artificial intelli-gence.[28] Data mining has replaced coal as the object of this attempt to extract objects of value in a capital market, but the notion reminds us that mining experience has been a critical component of risk management and assessment since Agricola sought to codify his own observations and experience at the beginning of the 16th century.

As Collins recognized, the principles and practices that constitute expert knowledge in a discipline are built on a foundation of (frequently unartic-ulated) tacit knowledge. Experts attempt to codify this knowledge as a set of principles, but what experts know frequently resists codification in writing.

The Rhetoric of Risk suggests that much of what constitutes tacit knowledge may be invisible because we do not have the means to interpret and capture this knowledge in writing. Throughout the Cycle, documentation practices shape our understanding of the worlds we seek to understand and regulate. These practices emerge as solutions to historically specific problems and situ-ations, but the solutions to these problems affect knowledge production long after the problems are solved. Once we have developed systems for extracting experience it becomes difficult (in Olson's terms) to unthink habitual prac-tices. But it is not impossible.

When we unthink the primacy of written communication, we begin to no-tice how often speakers deploy multiple modalities—including speech and gesture—when they describe the shape of a robotic arm or structure of infor-mation in a data warehouse. In the machine shop at the Sarnoff laboratory in Princeton, New Jersey, for example, an engineer used his hand and arm to demonstrate how a robotic arm should move. Another engineer used gestures to explain how a closely milled joint would operate. The machine shop tech-nicians read these gestures, translating them into three-dimensional images and then into objects.

As we observe how individuals gesture in other contexts, we also discover the limitations of gesture if speakers do not align their viewpoint with the viewpoint of their audiences, if speaker and audience do not share a common interpretative framework, if the meanings conveyed in gesture conflict with meanings conveyed in speech, or—most importantly—if speakers deploy no gesture at all.

As we learn more about the function of other modalities in video and digi-tal technologies, we must extend our notion of literacy education to help in-dividuals read and interpret meanings embodied in tone and gesture. We know, for example, that actors can imitate gesture deliberately and strategi-cally. We also know that individuals can learn to assume appropriate view-

---

[28]Collins, 1990.

points in gesture, strategically, as they learn other rhetorical strategies (like Power point presentations and graphics)—once they are convinced of the rhetorical function of these modalities in their communication practices. Although audiences can recognize the difference between these actors' gestures and the more naturalistic, idiosyncratic gestures that are the focus of this inquiry (Chawla & Krauss, 1994), the fact that audiences can discern the difference does not reduce the communicative function of these gestures in real-time interactions.[29]

The Uncertainties in New Technologies Are Less Visible, but the Problem Is a Question of Scale, Not Kind

On the surface, computer software seems to share little with the material and geographic uncertainty of hazardous environments. Yet these virtual architectures are by no means certain. Changes in one part of the system may produce unexpected changes, errors and bugs in other parts of the system. Beta versions are dynamic and flexible systems; each version attempts to correct errors in previous versions, but differences across multiple platforms, new systems software, and unpredicted changes in hardware can create an uncertain and unpredictable environment. Software engineers work against the clock to meet product release deadlines; they know that the software contains many bugs that will only be recognized when audience use the product for real-time work. Because writers must revise their documentation each time that software engineers revise the system, software documentation practices are also highly uncertain. Ultimately, the cycle of documentation and revision is much smaller than in large regulatory industries, but the cycle is similar in kind.

In most cases, the problems that precipitate change remain undetected because they do not affect how users work. Sometimes, however, problems affect the public in ways that require regulatory review and regulation. The Y2K problem (at the turn of the millennium) and the Eliza virus remind us that seemingly small problems can produce large and unpredictable changes across global networks. When problems reach a critical magnitude, federal and state agencies must take action to control and manage risk.

Uncertain documentation practices (including both software documentation and coding practices developed on the fly in the early years of the computer industry) contribute to the magnitude and cost of the problem. In the

---

[29]Chawla & Krauss, 1994. I am grateful to Martha Alibali for her suggestion in this matter.

case of the Y2K problem, stakeholders debated the imminence of the problem, its magnitude, and its predicted effects throughout the system. In the end, most agencies and institutions developed appropriate procedures to achieve Y2K compliance. But not all companies were equally successful in finding errors in the system.

The Move Off Shore Has Relocated Many of the Problems of Transformation to a More Difficult Rhetorical Context

Many U.S. industries have moved off shore in order to take advantage of cheap labor and a more favorable regulatory climate in developing countries. In these countries, workers face many of the problems of safety and enforcement that miners faced prior to the promulgation of regulations under the 1969 Mine Act. Researchers in rhetoric have investigated the problems of cross-cultural communication in nations that share some affinity with U.S. culture, but they have not yet investigated the problems of instruction and safety in offshore manufacturing. Differences in language and culture as well as the controversial subject matter will make such investigations difficult, if not impossible, in the near future. As a result, we lose the opportunity to understand how documentation functions in these sites and how changes in documentation practices might affect change in policy and procedure.

Electronic Commerce Has Not Entirely Eliminated the Problems of Industrial Labor, Even Within Its Own Institutions

Institutions depend upon the labor of many individuals to clean, operate, maintain, and repair the buildings, facilities, and equipment that allow institutions to continue to function. At one large data storage facility, for example, 25 operators controlled the computer operations throughout the facility. The data storage system could retrieve information from hospitals and institutions throughout the world. But the physical labor that supports data storage—billing, mailing, cleaning, and moving boxes—had not been entirely eliminated. As I walked through the hallways, workers lugged rolls of printer paper as large as bales of cotton. At night, cleaning personnel cleaned under each floor tile to keep the facility as sterile and dust-free as possible.

Although much of this labor is invisible even to many individuals in the company, these workers nevertheless fall within OSHA regulatory policy in regard to issues of workplace safety like ergonomic injury.

IMPLICATIONS FOR A THEORY OF RHETORIC

The Rhetoric of Risk provides a rhetorical framework that builds on the known and answers questions not answered by other approaches. It can help rhetors interpret and produce new knowledge in collaborative interactions. It has application in other domains, and it enables us to discriminate historically specific activities, discourse structures, and genres. The framework is both modular and scalable, and it opens a new area of inquiry that can guide future research in the relationship between speech, writing, and gesture. The framework is both epistemic and descriptive. It can hopefully stir new research in the construction of knowledge in organizations and, more pragmatically, it can help writers understand how new documents emerge from prior documentation and experience in large regulatory agencies. Most importantly, it allows us to see rhetoric, once again, as a full-bodied art in which gesture complements speech as a mode of delivery—even when we focus our attention on texts.

This final argument has two consequences for rhetorical theory. First, it means that we do not need to construct a new definition of delivery to accommodate texts;[30] instead, we can see written knowledge as built upon, complementary to, and tightly integrated with oral and gestural modalities. Secondly, it implies that rhetoric is no longer what Gross and Keith (1996) call a "thin" art grounded in the interpretation of neo-Aristotelian categories of analysis. Instead, rhetoric can emerge as a robust public art that influences public policy and decision making through the art-full production and reproduction of knowledge in specific rhetorical contexts.[31] Rhetoric in this sense is both backward looking and forward looking. To engage fully in the understanding and production of texts, rhetors must analyze the past in order to understand how the material and discursive world might be different in the future. Rhetors can engage in that project as "civil epistemologists" (Fuller, 1991).[32] Or they can engage in that project as members of the discursive and institutional communities charged to regulate the public health and well-being.

Ultimately, *The Rhetoric of Risk* argues that rhetoricians must develop, study, and adopt documentation practices that will have the same rhetorical force and authority that we now associate with the written word. The rhetoric I imagine will not see gesture as additive, but it will help us understand how writers and speakers integrate many different forms of knowledge—ana-

---

[30]Reynolds, 1993.
[31]Gross & Keith, 1996.
[32]Fuller, 1991.

lytic and experiential, scientific and embodied—in many different modalities at many different sites of rhetorical production.

This inquiry will require thick description and careful analysis of language in real situations and in experimental studies that test these results across many different sites of rhetorical production. In my own work, I have begun to explore the implications of this work in training in one particularly difficult cross-cultural situation in the new South Africa. I hope that others will join me in this project.

References

Adams, R. (1993). Investigation. In Re: *Southmountain Coal Co. Inc. No. 3 Mine. Wise County. Norton, VA. Explosion of Dec. 7, 1992* (Deposition). Southmountain, VA: U.S. Mine Safety and Health Administration/Virginia Dept. of Mines.

Agricola, G. (1950). *De re metallica* (H. Hoover & L. H. Hoover, Trans.). New York: Dover. (Original work published in 1556)

Alibali, M., Bassok, M., Olseth, K. L., Syc, S., & Goldin-Meadow, S. (1995). *Gestures reveal mental models of discrete and continuous change.* Paper presented at the 17th Annual Conference of the Cognitive Science Society, Hillsdale, NJ.

Alibali, M., & Goldin-Meadow, S. (1993). Gesture-speech mismatch and mechanisms of learning: What the hands reveal about a child's state of mind. *Cognitive Psychology, 25,* 468–523.

Allen, J. (1991). Gender issues in technical communication studies: An overview of the implications for the profession, research, and pedagogy. *JBTC, 5*(4), 371–92.

American National Standards Institute, National Safety Council. Association of Casualty and Surety Companies (1962), *American National Standard method of recording basic facts relating to the nature and occurrence of work injuries.* (Report reaffirmed 1969. Revision of Z16.2–1941).

Anon. (British manager, transcript of training session), near York, England, July, 1995.

Anon. (Interview with British miner), August, 1994.

Anon. (Interview with British novice, N1), Doncaster, England, January, 1996.

Anon. (Interview with British novice, N2), Doncaster, England, January, 1996.

Anon. (Interview with miner, U.S.), Osage, WV, February, 1999.

Anon. (Interview with female miner, U.S.), NIOSH Conference, Morgantown, West Virginia, March, 1997.

Anon. (Interview with former inspector, U.S.). April, 1992.

Anon. (Interview with mine manager, UK), January, 1995.

Anon. (Interview with miner, UK), January, 1995.

Anon. (Interview with miner, U.S.), April, 1992.

Anon. (Interview with MSHA official), April, 1992.

Anon. (Interview with roof bolting engineer, U.K.), January, 1995.

Anon. (Interview with safety manager, UK), January, 1995.

Anon. (Transcript of training session, Canterbury Coal Co.), April, 1997.

Anon. Trainer, Canterbury Coal Company (Transcript of training session), April, 1997.

Anon. (Interview with miners, Canterbury Coal Company), April, 1997.

Anon. (Interview with British mechanical engineer), January, 1995.

Anon. (Interview with former NUM official), January, 1995.

Anon. (Interview at Health and Safety Executive, Bootle, England), August, 1994.

Anon. (Interview with Welsh miner), August, 1995.

Aristotle (1991). *Aristotle on rhetoric: A theory of civil discourse* (G. A. Kennedy, Trans.). New York: Oxford University Press.

Austin, G. (1806/1966). *Chironomia: Or a treatise on rhetorical delivery.* (Ed. M. M. Robb & L. Thonssen), Carbondale, IL: Southern Illinois University Press.

Bardach, E., & Kagan, R. (1982). *Going by the book: The problem of regulatory unreasonableness.* Philadelphia: Temple University Press.

Barrett, E. A., & Kowalwalski, G. (1995). *Effective hazard recognition using a latent-image, three-dimensional slide simulation exercise* (Report of Investigation/1995 RI/9527). Arlington, VA: U.S. Department of the Interior, Bureau of Mines.

Bazerman, C. (1988). *Shaping written knowledge: The genre and activity of the experimental article in science.* Madison, WI: Wisconsin University Press.

Bazerman, C. (1999). *The languages of Edison's light.* Cambridge, MA: The M.I.T. Press.

Bazerman, C., & Paradis, J. (1991). *Textual dynamics of the professions.* Madison, WI: Wisconsin University Press.

Beason, R. L. (1987). Memorandum for: Alan C. McMillan, Acting Assistant Secretary, Arlington, VA, Re: Gassy Standard Changes Dated June 12, 1986. In U.S. House of Representatives, 100th Cong., 1st Sess., Committee on Labor and Human Resources, *Oversight of the Mine Safety and Health Administration. Hearings before the Committee on Labor and Human Resources . . . on examining activities of the Mine Safety and Health Administration* (p. 474). Washington, DC: U.S. Government Printing Office.

Beason, R. L. (1987). Memorandum: To Gassy Standard Committee members, Re: Reply to Gassy Mine Standards, dated June 12, 1986. With attachment from Roy L. Bernard. In U.S. House of Representatives, 100th Cong., 1st Sess., Committee on Labor and Human Resources, *Oversight of the Mine Safety and Health Administration. Hearings before the Committee on Labor and Human Resources . . . on examining activities of the Mine Safety and Health Administration* (p. 477). Washington, DC: U.S. Government Printing Office.

Beck, U. (1992). *Risk society: Towards a new modernity.* Newberry Park, CA: Sage.

Behavioral Research Aspects of Safety and Health Group (BRASH), Institute for Mining and Minerals Research (IMMR). (1987). *Roof fall entrapment exercise* (Instructor's copy). Lexington, KY: University of Kentucky.

Bereiter, C., & Scardamalia, M. (1987). *The psychology of written composition.* Hillsdale, NJ: Lawrence Erlbaum Associates.

Bernard, R. L. (1987). Memorandum for: Ronald L. Beason, Metal and Nonmetal Mine Safety and Health Inspector Re: Changes to Gassy Mine Standards dated June 12, 1986. In U.S. House of Representatives, 100th Cong., 1st Sess., Committee on Labor and Human Resources, *Oversight of the Mine Safety and Health Administration. Hearings before the Committee on Labor and Human Resources . . . on examining activities of the Mine Safety and Health Administration* (pp. 478–491). Washington, DC: U.S. Government Printing Office.

Bertin, J. (1981). *Graphics and graphic information processing* (W. I. Berg and P. Scott, Trans.). Berlin: DeGruyter.

Blake, K. (1985/1992). Handwritten report. In D. W. Huntley, R. J. Painter, J. K. Oakes, D. R. Cavanaugh, & W. G. Denning (Eds.), *Report of investigation: Underground coal mine fire. Wilberg Mine. I.D. No. 42-00080. Emery Mining Corporation. Orangeville, Emery County, Utah. December 19, 1984.* Appendix F. Arlington, VA: U.S. Department of Labor, Mine Safety and Health Administration.

Blau, P. (1969). *The dynamics of bureaucracy: A study of interpersonal relations in two government agencies.* Chicago: University of Chicago Press.

Bloor, D. (1976). *Knowledge and social imagery.* London: Routledge & Kegan Paul.

Bradshaw, G. A., & Borchers, J. G. (2000). Using scientific uncertainty to shape environmental policy. An expanded version of Bradshaw, G. A., & Borchers, J. G. (2000). Uncertainty as Information: Narrowing the science-policy gap. Conservation Ecology 4(1):7. [On-line]. Available: http://www.consecol.org/vol4/iss1/art7/.

Brasseur, L. E. (1993). Contesting the objectivist paradigm: Gender issues in the technical and professional communication curriculum. *IEEE Transactions on Professional Communication, 36*(3), 114–123.

Bremmer, J. (1991). Walking, standing, and sitting in ancient Greek culture. In J. Bremmer & H. Roodenburg (Eds.), *A cultural history of gesture* (pp. 15–35). Ithaca, NY: Cornell University Press.

British Coal, Nottinghamshire Area, Lound Hall Training Centre. (n.d.). *Roof bolting.* Eastwood Hall, England: British Coal.

British Coal Corporation. (n.d.). *Rockbolting in mines* (Operational instructions code of practice CP/30/MR; Notes of guidance NG/30/MR; Code of practice CP/30/SF; Notes of guidance NG/30/SF). Eastwood Hall, England: British Coal Corporation.

Brooks, J. P. (1993). Investigation. In Re: *Southmountain Coal Co. Inc. No. 3 Mine. Wise County. Norton, VA. Explosion of Dec. 7, 1992* (Deposition). Southmountain, VA: U.S. Mine Safety and Health Administration/Virginia Dept. of Mines.

Brooks, K. (1993). Investigation. In Re: *Southmountain Coal Co. Inc. No. 3 Mine. Wise County. Norton, Va. Explosion of Dec. 7, 1992* (Deposition). Southmountain, VA: U.S. Mine Safety and Health Administration.

Bruce, W. E. (1987). Memorandum for: Madison McCulloch, Director, Technical Support . . . through A. Z. Dimitroff, Chief, Safety and Health Technology Transfer Center. Re: Draft revision of Gassy Mine Standards for Metal and Nonmetal Mines. In U.S. House of Representatives, 100th Cong., 1st Sess., Committee on Labor and Human Resources, *Oversight of the Mine Safety and Health Administration. Hearings before the Committee on Labor and Human Resources . . . on examining activities of the Mine Safety and Health Administration* (pp. 392–458). Washington, DC: U.S. Government Printing Office.

Bryner, G. C. (1987). *Bureaucratic discretion: Law and policy in federal regulatory agencies.* New York: Pergamon Press.

Bulwer, J. (1644/1974). *Chirologia: Or the natural language of the hand* (J. Cleary, Ed.). Carbondale: Southern Illinois University Press.

Burnett, R. E. (1997). *Technical communication.* 4th ed. Belmont, CA: Wadsworth Publishing Company.

Burns, R. E. (1993). Investigation. In Re: *Southmountain Coal Co. Inc. No. 3 Mine. Wise County. Norton, VA. Explosion of Dec. 7, 1992* (Deposition). Southmountain, VA: U.S. Mine Safety and Health Administration/Virginia Dept. of Mines.

Carroll, J., 1976. Comment. In U.S. Congress. House of Representatives, Committee on Education and Labor, Subcommittee on Labor Standards. (1977). *Oversight hearings on the Coal Mine Health and Safety Act of 1969 (Excluding Title IV)* (pp. 4–32). 95th Cong., 1st Sess. (Hearings). Washington, DC: U.S. Government Printing Office.

Chawla, P., & Krauss, R. M. (1994). Gesture and speech in spontaneous and rehearsed narratives. *Journal of Experimental and Social Psychology, 30*, 580–601.

Childers, M. S., Elam, R., Eslinger, M. O., Kattenbraker, S. R., Phillips, R. L., & Ritchie, E. B. (1990). *Report of investigation: Underground coal mine explosion. Pyro No. 9 Slope, William Station Mine—I.D. No. 15-13881. Sullivan, Union County, Kentucky. September 13, 1989* (Accident investigation report). Arlington, VA: U.S. Department of Labor, Mine Safety and Health Administration.

[Cicero]. (1989). *Ad G. Herennium: De ratione dicendi (Rhetorica ad Herennium)* (H. Caplan, Trans.). Cambridge, MA.: Loeb Classical Library.

Cicero, M. T. (1942). *De oratore: Book III.* In G. P. Goold (Ed.), *De oratore, De fato, Paradoxa stoicorum, De partitione oratoria* (Vol. 4). (H. Rackam, Tr.). Cambridge, MA: Harvard University Press.

Cleveland, M. J., & Turner, E. (1977). *Compilation of judicial decisions which have an impact on coal mine inspection* (MESA informational report/1977 1060). Arlington, VA: U.S. Department of Labor, Mining Enforcement and Safety Administration.

Code, L. (1991). *What can she know? Feminist theory and the construction of knowledge.* Ithaca, NY: Cornell University Press.

Cohn, C. (1987). Sex and death in the rational world of defense intellectuals. *Signs: Journal of Women in Culture and Society, 12*(4), 687–718.

Collins, H. (1990). *Artificial experts: Social knowledge and intelligent machines.* Cambridge, MA: MIT Press.

Combs, R. (1993). Investigation. In Re: *Southmountain Coal Co. Inc. No. 3 Mine. Wise County. Norton, VA. Explosion of Dec. 7, 1992* (Deposition). Southmountain, VA: U.S. Mine Safety and Health Administration/Virginia Department of Mines.

Cooke, J. D. (1993). Investigation. In Re: *Southmountain Coal Co. Inc. No. 3 Mine. Wise County. Norton, VA. Explosion of Dec. 7, 1992* (Deposition). Southmountain, VA: U.S. Mine Safety and Health Administration.

Cox, D. B. (1993). Investigation. In Re: *Southmountain Coal Co. Inc. No. 3 Mine. Wise County. Norton, VA. Explosion of Dec. 7, 1992* (Deposition). Southmountain, VA: U.S. Mine Safety and Health Administration/Virginia Dept. of Mines.

Davis, A. (1991). *Extended cuts: An accident analysis and inventory of approvals.* Arlington, VA: U.S. Mine Safety and Health Administration, Division of Safety, Coal Mine Safety and Health.

Davis, B. R. (1993). Investigation. In Re: *Southmountain Coal Co. Inc. No. 3 Mine. Wise County. Norton, VA. Explosion of Dec. 7, 1992* (Deposition). Southmountain, VA: U.S. Mine Safety and Health Administration/Virginia Dept. of Mines.

Davis, J. E. (1993). Investigation. In Re: *Southmountain Coal Co. Inc. No. 3 Mine. Wise County. Norton, VA. Explosion of Dec. 7, 1992* (Deposition). Southmountain, VA: U.S. Mine Safety and Health Administration/Virginia Dept. of Mines.

Davis, M. (1993). Investigation. In Re: *Southmountain Coal Co. Inc. No. 3 Mine. Wise County. Norton, VA. Explosion of Dec. 7, 1992* (Deposition). Southmountain, VA: U.S. Mine Safety and Health Administration/Virginia Dept. of Mines.

Dawes, R. M. (1988). *Rational choice in an uncertain world.* Harcourt Brace.

Dawes, R. M. (2001). *Irrationality in everyday life: Professional arrogance and outright lunacy.* Unpublished manuscript.

Dear, P. (1991). *The literary structure of scientific argument: Historical studies.* Philadelphia: Pennsylvania University Press.

Deatherage, F. (1993). Investigation. In Re: *Southmountain Coal Co. Inc. No. 3 Mine. Wise County. Norton, VA. Explosion of Dec. 7, 1992* (Deposition). Southmountain, VA: U.S. Mine Safety and Health Administration.

DeKock, A., Kononov, V. A., & Oberholzer, J. (n.d.). *Remote control as a factor in increasing safety and productivity in the coal mining industry* (Report). Auckland Park, Johannesburg, SA: Miningtek (CSIR).

de Saussure, F. (1959). *Course in general linguistics* (W. Baskin, Trans.). New York: Philosophical Library. (Original work published 1916). In Goldin-Meadow, McNeill, and Singleton, 1996.

Dombroski, P. M. (1991). The lessons of the challenger investigations. *IEEE Transactions on Professional Communication, 34*(4), 211–216.

Dombroski, P. M. (1992). A comment on "The construction of knowledge in organizations: Asking the right questions about the *Challenger*." *Journal of Business and Technical Communication, 6*(1), 123–7.

Dowlatabadi, H. (1999). Integrated Assessment: Implications of Uncertainty. In *Encyclopedia of life support systems*. Oxford: Oxford University Press. [On-line]. Available: http://www.hdgc.epp.cmu.edu/publications/abstracts/integrated3.htm.

Drew, P., & Heritage, J. (1992). Analyzing talk at work: An introduction. In P. Drew, and John Heritage (Ed.), *Talk at work: Interaction in institutional settings* (pp. 3–65). Cambridge, England: Cambridge University Press.

Driessen, H. (1991). Gestured masculinity: Body and sociability in rural Andalusia. In J. Bremmer & H. Roodenburg (Eds.), *A cultural history of gesture* (pp. 237–252). Ithaca, NY: Cornell University Press.

Duncan, C. E. (1993). Investigation. In Re: *Southmountain Coal Co. Inc. No. 3 Mine. Wise County. Norton, VA. Explosion of Dec. 7, 1992* (Deposition). Southmountain, VA: U.S. Mine Safety and Health Administration/Virginia Dept. of Mines.

Duranti, A., & Goodwin, C. (Eds.). (1992). *Rethinking context: Language as an interactive phenomenon*. Cambridge, England: Cambridge University Press.

Efron, D. (1941). *Gesture, race, and culture*. The Hague, Netherlands: Mouton.

Eisenstein, E. L. (1993). *The printing revolution in early modern Europe*. Cambridge, England: Cambridge University Press.

Engeström, Y. (1996). Developmental studies of work as a test bench of activity theory: The case of primary care medical practice. In S. Chaiklin & J. Lave (Eds.), *Understanding practice: Perspectives on activity and context* (pp. 64–103). Cambridge, England: Cambridge University Press.

Fahnestock, J. (1996). Series reasoning in scientific argument: "Incrementum" and "graditio" and the case of Darwin. *Rhetorical Society Quarterly, 26*, 13–40.

The Federal Mine Safety and Health Act of 1977, Public Law 91-173 § 104(d)(1), as amended by Public Law 95-164 [On-line]. Available: http://www.msha.gov/REGS/ACT/ACT1.HTM#13act.

Fischhoff, B. (1980). For those condemned to study the past: Reflections on historical judgment. *New directions for methodology of social and behavioral research* (Nov. 4, 1980), 79–93.

Fischhoff, B. (1988). Judgment and decision making in the psychology of human thought. In R. J. Sternberg & E. E. Smith (Eds.), *The psychology of human thought* (pp. 152–187). Cambridge, England: Cambridge University Press.

Fischhoff, B. (1991). Values elicitation: Is there anything there? *American Psychologist, 46*(8), 835–847.

Fischhoff, B., Watson, S. R., & Hope, C. (1984). Defining risk. *Policy Sciences, 17*, 123–139.

Fleming, R. K. (1993). Investigation. In Re: *Southmountain Coal Co. Inc. No. 3 Mine. Wise County. Norton, VA. Explosion of Dec. 7, 1992* (Deposition). Southmountain, VA: U.S. Mine Safety and Health Administration.

Flower, L. (1994). *The construction of negotiated meaning: A social cognitive theory of writing*. Carbondale, IL: Southern Illinois University Press.

France, C. (1993). Investigation. In Re: *Southmountain Coal Co. Inc. No. 3 Mine. Wise County. Norton, VA. Explosion of Dec. 7, 1992* (Deposition). Southmountain, VA: U.S. Mine Safety and Health Administration/Virginia Dept. of Mines.

Frank, F. W., & Treichler, P. A. (1989). *Language, gender and professional writing: Theoretical approach and guidelines for non-sexist usage*. New York: MLA.

Fuller, S. (1991). *Social epistemology*. Bloomington, IN: Indiana University Press.

Garry, D. (1987). In United States. Department of Labor. Bureau of Mines, *Roof fall entrapment videotape: Dave Garry's account*, (MSHA video). Pittsburgh: United States. Department of Labor. Bureau of Mines.

Gaventa, J. (1980). *Power and powerlessness: Quiescence and rebellion in an Appalachian valley*. Urbana, IL: Illinois University Press.

Geertz, C. (1973). *The interpretation of cultures: Selected essays*. New York: Basic Books.

Gershung, H. L., McConnell-Ginet, S., & Wolfe, S. J. (1989). *Language, gender and professional writing: Theoretical approaches and guidelines for non-sexist usage*. New York: MLA.

Glaser, R. (1984). In C. Bereiter & M. Scardamalia (Eds.), *The psychology of written composition* (pp.). Hillsdale, NJ: Lawrence Erlbaum Associates.

Goldin-Meadow, S., & Alibali, M. W. (1995). Mechanisms of transition: Learning with a helping hand. *The psychology of learning and motivation, 33*, 115–157. New York: Academic Press.

Goldin-Meadow, S., McNeill, D., & Singleton, J. (1996). Silence is liberating: Removing the handcuffs on grammatical expression in the manual modality. *Psychological Review, 103*(1), 34–55.

Goode, C. A. (1977). Testimony. In U.S. Congress. House of Representatives, Committee on Education and Labor, Subcommittee on Labor Standards, *Oversight hearings on the Coal Mine Health and Safety Act of 1969 (Excluding Title IV)* (pp. 354–373), 95th Cong., 1st Sess. (Hearings). Washington, DC: U.S. Government Printing Office.

Goode, D. L. (1993). Investigation. In Re: *Southmountain Coal Co. Inc. No. 3 Mine. Wise County. Norton, VA. Explosion of Dec. 7, 1992* (Deposition). Southmountain, VA: U.S. Mine Safety and Health Administration/Virginia Dept. of Mines.

Goodman, S. N. (1999). Toward evidence-based medical statistics. 2. Annals of internal medicine. 130. 1005–1013. [On-line]. Available: http://www.annals.org/issues/v130n12/full/199906150-00009.html#R1-9.

Goodwin, C. (1999, May). *Pointed as situated practice*. Paper presented at the LISO-CLIC Conference, University of California, Santa Barbara.

Graf, F. (1991). Gestures and conventions: The gestures of Roman actors and orators. In J. Bremmer & H. Roodenburg (Eds.), *A cultural history of gesture* (pp. 36–53). Ithaca, NY: Cornell University Press.

Greenblatt, S. (1984). *Renaissance self-fashioning: From More to Shakespeare*. Chicago: University of Chicago.

Gross, A. (1990). *Rhetoric of science*. Cambridge, MA: Harvard University Press.

Gross, A., & Keith, W. M. (Eds.). (1996). *Rhetorical hermeneutics: Invention and interpretation in the age of science*. Albany, NY: State University of New York Press.

Hacker, S. (1990). Doing it the hard way. In D. E. Smith & S. M. Turner (Eds.), *Investigations of gender and technology*. Boston: Unwin Hyman.

Hanna, K., Conover, D., & Haramy, K. (1991). *Coal mine intersection behavior study*. (BOM technical report RI 9337). Pittsburgh, PA: U.S. Bureau of Mines.

Haraway, D. (1991). *Simians, cyborgs, and women: The reinvention of nature*. London: Routledge.

Harding, S. (1991). *Whose science? Whose knowledge?* Ithaca, NY: Cornell University Press.

Herbert, A. (1989). *Practical risk assessment in coal mines: Open learning module.* Edwinstowe, Nottinghamshire, UK: Safety/Environmental Group, British Coal Nottinghamshire Group.

Herndl, C., Fennell, B. A., & Miller, C. (1991). Understanding failures in organizational discourse: The accident at Three Mile Island and the *Shuttle Challenger* disaster. In C. Bazerman & J. Paradis (Eds.), *Textual dynamics of the professions* (pp. 279–305). Madison, WI: Wisconsin University Press.

Hill, J. R. M. (1993). Intelligent drilling system for geological sensing. In *Mills* (pp. 495–501). Yokahama, Japan.

Horton, W. (1991). *Illustrating computer documentation: The art of presenting information graphically on paper and online.* New York: Wiley.

Huntley, D. W., Painter, R. J., Oakes, J. K., Cavanaugh, D. R., & Denning, W. G. (1992). *Report of Investigation: Underground coal mine fire. Wilberg Mine. I.D. No. 42-00080. Emery Mining Corporation. Orangeville, Emery County, Utah. December 19, 1984.* Arlington, VA: U.S. Department of Labor, Mine Safety and Health Administration.

Hurst, E. (1977). *Statement of the Association of Bituminous Contractors on the Coal Mine Health and Safety Act,* 95th Cong., 1st Sess., Subcommittee on Labor Standards of the Committee on Education and Labor (Ed.), *Oversight hearings on the Coal Mine Health and Safety Act of 1969 (Excluding Title IV).* Washington, DC: U.S. Government Printing Office.

Hutchins, E. (1996). Learning to navigate. In S. Chaiklin & J. Lave (Eds.), *Understanding practice: Perspectives on activity and context* (pp. 35–65). Cambridge, England: Cambridge University Press. (testimony of E. Hurst).

Irwin, A. (1995). *Citizen science: A study of people, expertise, and sustainable development* (Environment and society). London: Routledge & Kegan Paul.

Irwin, A., Dale, A., & Smith, D. (1996). Science and hell's kitchen: The local understanding of hazard issues. In A. Irwin & B. Wynne (Eds.), *Misunderstanding science: The Public reconstruction of science and technology* (pp. 47–64). Trowbridge, England: Cambridge University Press.

Irwin, A., & Wynne, B. (1996). *Misunderstanding science: The public reconstruction of science and technology.* Trowbridge, England: Cambridge University Press.

Ivins, W. M. (1973). *On the rationalization of sight, with an examination of three Renaissance texts on perspective [by] William M. Ivins, Jr. De artificiali perspectiva [by] Viator, reproducing both the 1st ed. (Toul, 1505) and the 2d ed. (Toul, 1509).* New York: Da Capo Press.

Jackendoff, R. (1996). The architecture of the linguistic-spatial interface. In P. Bloom, M. A. Peterson, & M. F. Garrett (Ed.), *Language and space* (pp. 1–30). Cambridge, MA: MIT Press.

Jasanoff, S. (2001, April). *Technologies of uncertainty: Reconstructing politics through rhetorics of risk, conference.* Cornell University, Ithaca, NY.

John T. Boyd Company, Department of Trade and Industry. (1993). *Independent review: 10 collieries under consultation. British Coal Corporation* (U.K. report 2265.7; March 1994). London: Her Majesty's Stationer's Office.

John T. Boyd Company, Pittsburgh, PA. (1993). *Executive summary: Independent analysis. 21 closure review collieries. British Coal Corporation. United Kingdom.* London: Department of Trade and Industry, Coal Review Team.

Johnstone, B., & Bean, J. M. (1997). Self-expression and linguistic variation. *Language in Society, 26,* 221–246.

Kaufer, personal communication with the author, March, 1996.

Kaufer, D., & Butler, B. (1996). *Rhetoric and the arts of design.* Mahwah, NJ: Lawrence Erlbaum Associates.

Kaufer, D. S., & Carley, K. M. (1993). *Communication at a distance: The influence of print on sociocultural organization and change.* Hillsdale, NJ: Lawrence Erlbaum Associates.

Keller, E. F. (1985). *Reflections on gender and science.* New Haven, CT: Yale University Press.

Kelly, S. D., & Church, R. B. (1997). Can children detect conceptual information conveyed through other children's nonverbal behaviors? *Cognition and Instruction, 15*(1), 107–134.

Kelly, S. D., & Church, R. B. (1998). A comparison between children's and adults' ability to detect conceptual information conveyed through representational gestures. *Child Development, 69*(1), 85–93.

Kennedy, E. (1987). Opening Statement of Senator Kennedy. In U.S. Congress, House of Representatives, Committee on Labor and Human Resources, *Oversight of the Mine Safety and Health Administration. Hearings before the Committee on Labor and Human Resources . . . on examining activities of the Mine Safety and Health Administration* (pp. 1–2). 100th Cong., 1st Sess. (Hearings). Washington, DC: U.S. Government Printing Office.

Kentucky Deep Mining Safety Commission. (1976). *Final report.* Frankfurt, KY: Author.

Killingsworth, J., & Palmer, J. (1992). *Ecospeak: Rhetoric and environmental politics in America.* Carbondale, IL: Southern Illinois University Press.

Klishis, M. J., Althouse, R. C., Layne, L. A., & Lies, G. M. (1992). Increasing roof bolter operator awareness to risks of falling roof material during the bolter cycle. In U.S. Department of the Interior, Bureau of Mines, *Preventing coal mine groundfall accidents: How to identify and respond to geologic hazards and prevent unsafe worker behavior* (Information circular/1992 ed., pp. 69–78). Washington, DC: U.S. Government Printing Office.

Knorr, K. (1977). Producing and reproducing knowledge: descriptive or constructive? Toward a model of research production. *Soc. sci. inform, 16,* 669–696.

Kunreuther, H., Slovic, P., & Macgregor, D. (1996). Risk perception and trust: Challenges for facility siting. *Risk: Health, safety, and environment, 7*(109) [On-line]. Available: http://www.Fplc.Edu/risk/vol7/ spring/kunreuth.Htm

Labor Research Department. (1989). *The hazards of coal mining.* London: Labor Research Department Publication/ RAP Ltd.

Langdon, B. (1994). *Extensive fall of roof at Bilsthorpe Colliery: A report of the HSE's investigation into the extensive fall of roof at Bilsthorpe Colliery, Nottinghamshire, on 18 August, 1993.* Sudbury, Suffolk, England: Health and Safety Executive Books.

Larkin, J. H., & Simon, H. (1995). Chapter 3: Why a diagram is (sometimes) worth ten thousand words. In B. Chandrasekaran, J. Glasgow, N. H. Narayanan (Eds.), *Diagrammatic reasoning: Cognitive and computational perspectives* (pp. 69–110). Cambridge, MA: MIT Press.

Latour, B. (1990). Drawing things together. In M. Lynch & S. Woolgar (Eds.), *Representation in scientific practice* (pp. 19–68). Cambridge, MA: MIT Press.

Latour, B., Woolgar, S., & Salk, J. (1986). *Laboratory life.* Princeton, NJ: Princeton University Press.

Lay, M. M., Gurak, L. J., Gravon, C., & Myntti, C. (Eds.). (2000). *Body talk: Rhetoric, technology, reproduction.* Madison, WI: Wisconsin University Press.

Leon, R. N., Davies, Q. W., Salamon, M. D. G., & Davies, J. C. A. (1994). *Report of the Commission of Inquiry into Safety and Health in the Mining Industry* (Vol. 1). Braamfontein, SA: Commission of Inquiry into Safety and Health in the Mining Industry.

Levelt, W. J. M. (1996). Perspective taking and ellipsis in spatial descriptions. In P. Bloom, M. A. Peterson, & M. F. Garrett (Eds.), *Language and space* (pp. 76–107). Cambridge, MA: MIT Press.

Levinson, C. (1996). Frames of reference and Molyneux's question: Crosslinguistic evidence. In P. Bloom, M. A Peterson, & M. F. Garrett (Eds.), *Language and space* (pp.). Cambridge, MA: MIT Press.

Levinson, S. C. (1992). Activity types and language. In P. Drew, and John Heritage (Ed.), *Talk at work: Interaction in institutional settings* (pp. 3–65). Cambridge, England: Cambridge University Press.

Lewis, J. L. (1947). *Testimony of John L. Lewis. before the House of Representatives, Subcommittee on Miners' Welfare of the Committee on Education and Labor. April 3, 1947, and Subcommittee of the Senate Committee on Public Lands to investigate the Centralia mine explosion. April 17, 1947.* Washington, DC: Labor's Non-Partisan League.

Libby (1999). (Interview with female miner, U.S.), Osage, West Virginia Workshop, February 27, 1999.

Libby. (Interview with female miner, U.S.), NIOSH Conference, Morgantown, West Virginia. March, 1997.

Littleton, E. B., & Alibali, M. (1997, March). *What children's hand gestures reveal about learning to write.* Paper presented at the Conference on College Composition and Communication, Phoenix, AZ.

Lukes, S. (1989). *Power: A radical view.* New York: Macmillan.

Lynch, M. (1985). *Art and artifact in laboratory science: A study of shop work and shop talk in a research laboratory.* London: Routledge & Kegan Paul.

Lynch, M. (1990). The externalized retina: Selection and mathematization in the visual documentation of objects in the life sciences. In M. Lynch & S. Woolgar (Eds.), *Representation in scientific practice* (pp. 153–186). Cambridge, MA: MIT Press.

Mallett, L., Vaught, C., & Peters, R. H. (1992). *Training that encourages miners to avoid unsupported roof.* In U.S. Department of the Interior, Bureau of Mines, *Preventing coal mine groundfall accidents: How to identify and respond to geologic hazards and prevent unsafe worker behavior* (Information circular/1992, pp. 32–45). Washington, DC: U.S. Government Printing Office.

Mark, C. (1998). *Comparison of ground conditions and ground control practices in the United States and Australia.* Paper presented at the 17th International Conference on Ground Control in Mining, Morgantown, WV.

Mark, C., Chase, F. E., & Iannacchione, A. (1991). Longwall mining: Geologic considerations for better performance. *E&MJ,* 16c–16k.

Mark, C., & DeMarco, M. J. (1993). Longwalling under difficult conditions in U.S. coal mines. *CIM Bulletin* (April), 31–38.

Markley, R. (1993). *Fallen languages: Crises of representation in Newtonian England, 1660–1740.* Ithaca, NY: Cornell University Press.

McAtteer, D. (1995). In U.S. Department of Labor. MSHA. MSHA to Conduct Review of Accident Investigation Policy. (MSHA News Release No. 95-018). May 10, 1995. [On-line]. Available: http://www.msha.gov/media/press/1995/nr950510.htm.

McAteer, J. D. (1995). Statement of J. Davitt McAteer, Assistant Secretary of Labor for Mine Safety and Health, Submitted to the Subcommittee on Workforce Protections of the Economic and Educational Opportunities Committee of the United States House of Representatives on H.R. 1834, the Safety and Health Improvement and Regulatory Reform Act of 1995. [On-line] (p. 6, sec. 1). Available: http://www.msha.gov/MEDIA/CONGRESS/CT950901.HTM.

McCloskey, D. D. (1985). *Rhetoric of economics.* Madison, WI: University of Wisconsin Press.

McCloskey, D. D. (1992). *Rhetoric of economics.* Madison, WI: University of Wisconsin Press.

McConnell-Ginet, S. (1989). The sexual (re)production of meaning: A discourse-based theory. In F. W. Frank & P. A. Treichler (Eds.), *Language, gender and professional writing: Theoretical approach and guidelines for non-sexist usage* (pp. 35–50). New York: MLA.

McKinney, O. (1993). Investigation. In Re: *Southmountain Coal Co. Inc. No. 3 Mine. Wise County. Norton, VA. Explosion of Dec. 7, 1992* (Deposition). Southmountain, VA: U.S. Mine Safety and Health Administration/Virginia Dept. of Mines.

McNeill, D. (1998). Personal communication with the author. Data session. University of Chicago Dept. of Psychology and Linguistics, April, 1998.

McNeill, D. (1992). *Hand and mind: What gestures reveal about thought.* Chicago: Chicago University Press.

McNeill, D., & Duncan, S. D. (2000). Growth points in thinking-for-speaking. In D. McNeill (Ed.), *Language and gesture: Window into thought and action* (pp. 141–161). Cambridge, England: Cambridge University Press.

Meade, D. (1993). Investigation. In Re: *Southmountain Coal Co. Inc. No. 3 Mine. Wise County. Norton, VA. Explosion of Dec. 7, 1992* (Deposition). Southmountain, VA: U.S. Mine Safety and Health Administration/Virginia Dept. of Mines.

Merchant, C. (1983). *The death of nature.* San Francisco: HarperSanFrancisco.

Miller, C. (1994). Opportunity, opportunism, and progress: *Kairos* in the rhetoric of technology. *Argumentation, 8*(1), 81–96.

Miller, C. (1998). Learning from history: World War II and the culture of high technology. *Journal of Business and Technical Communication, 12*(3), 288–315.

Miller, C. (1990). The Rhetoric of decision science, or, Herbert A. Simon says. In H. W. Simons (Ed.), *The rhetorical turn: Invention and persuasion in the conduct of inquiry* (pp. 162–184). Chicago: University of Chicago Press.

Miller, C., Jasanoff, S., Long, M., Clark, W., Dickson, N., Iles, A., & Parris, T. (1997). Shaping knowledge, defining uncertainty: The dynamic role of assessments (Working Group 2-Assessment as a communications process. Chapter 3). [On-line]. Available: http://grads.iges.org/geaproject1997/gea4.html.

Minick, N. (1997). The early history of the Vygotskian school: The relationship between mind and activity. In M. Cole, Y. Engeström, & O. Vasquez (Eds.), *Mind, culture and activity: Seminal papers from the laboratory of comparative cognition* (pp. 117–127). Cambridge, England: Cambridge University Press.

Molinda, G. M. (1992). *Recognition and control of roof falls caused by faults* (Information circular/1992 IC 9332). Washington, DC: U.S. Department of the Interior.

Molinda, G. M., & Mark, C. (1994). *Coal mine roof rating (CMRR): A practical rock mass classification for coal mines.* U.S.MB IC 9387 (Bureau of Mines information circular/1994 IC 9387). Washington, DC: U.S. Department of the Interior, Bureau of Mines.

Molinda, G. M., & Mark, C. (1996). *Rating the strength of coal mine roof rocks* (Bureau of Mines information circular 9444/1996). Washington, DC: U.S. Department of the Interior, Bureau of Mines.

Molinda, G., & Mark, C. (Eds.). (1996). *Rating the strength of coal mine roof rocks* (Information circular/1996. IC: 9444 (2–19). Arlington, VA: U.S. Department of the Interior, Bureau of Mines.

Moore, M. (1993). Danger stalks small non-union mines. *UMWA Journal* (Special edition: Coal's killing fields), *104*(2), 4–8.

Morgan, M. G. (1998). Commentary: Uncertainty analysis in risk assessment. *Human and Ecological Risk Assessment, 4*(1), 25–39.

Morgan, M. G., Fischhoff, B., Bostrom, A., & Atman, C. J. (2002). *Risk communication: A mental models approach.* Cambridge, England: Cambridge University Press.

Morgan, M. G., Fischhoff, B., Bostrom, A., Lave, L., & Atman, C. (1992). Communicating risk to the public: First, learn what people know and believe. *Environ. sci. technol., 26*(11), 2048–2056.

Morgan, M. G., & Henrion, M. (1992). *Uncertainty: A guide to dealing with uncertainty in quantitative risk and policy analysis.* Cambridge, England: Cambridge University Press.

Morone, D. (1987). In United States. Department of Labor. Bureau of Mines, *Roof fall entrapment videotape: Dave Garry's account*, (MSHA video). Pittsburgh: United States. Department of Labor. Bureau of Mines.

Morrel-Samuels, P., & Krauss, R. M. (1992). Word familiarity predicts temporal asynchrony of hand gestures and speech. *Journal of Experimental Psychology: Learning, Memory, and Cognition, 18*(3), 615–622.

Mullins, D. (1993). Investigation. In Re: *Southmountain Coal Co. Inc. No. 3 Mine. Wise County. Norton, VA. Explosion of Dec. 7, 1992* (Deposition). Southmountain, VA: U.S. Mine Safety and Health Administration/Virginia Dept. of Mines.

Mullins, D. A. (1993). Investigation. In Re: *Southmountain Coal Co. Inc. No. 3 Mine. Wise County. Norton, VA. Explosion of Dec. 7, 1992* (Deposition). Southmountain, VA: U.S. Mine Safety and Health Administration/Virginia Dept. of Mines.

Mullins, J. E. (1993). Investigation. In Re: *Southmountain Coal Co. Inc. No. 3 Mine. Wise County. Norton, VA. Explosion of Dec. 7, 1992* (Deposition). Southmountain, VA: U.S. Mine Safety and Health Administration.

Mullins, L. C. (1993). Investigation. In Re: *Southmountain Coal Co. Inc. No. 3 Mine. Wise County. Norton, VA. Explosion of Dec. 7, 1992* (Deposition). Southmountain, VA: U.S. Mine Safety and Health Administration.

Myers, G. (1990). *Writing biology: Texts in the social construction of knowledge.* Madison, WI: Wisconsin University Press.

Petraglia, J. (1997). *Reality by design: The rhetoric and technology of authenticity in education.* Mahwah, NJ: Lawrence Erlbaum Associates.

Nagy, J. (1987). Wilberg Mine Fire: Cause, location and initial development. In D. W. Huntley, R. J. Painter, J. K. Oakes, D. R. Cavanaugh, & W. G. Denning (Eds.), *Report of investigation: Underground coal mine fire. Wilberg Mine. I.D. No. 42-00080. Emery Mining Corporation. Orangeville, Emery County, Utah. December 19, 1984* (Appendix II). Arlington, VA: U.S. Department of Labor, Mine Safety and Health Administration.

National Coal Association, Bituminous Coal Operators' Association. (1977). Statement of the Association of Bituminous Coal Operators' Association. In U.S. House of Representatives, 95th Cong., 1st Sess., Subcommittee on Labor Standards of the Committee on Education and Labor, *Oversight hearings on the Coal Mine Health and Safety Act of 1969 (Excluding Title IV).* Washington, DC: U.S. Government Printing Office.

Nelkin, D., & Brown, M. S. (Eds.). (1984). *Workers at risk: Voices from the workplace.* Chicago: University of Chicago Press.

Nelson, J., Megill, A., & McCloskey, D. (Eds.). (1987). *The rhetoric of the human sciences: Language, argument, and scholarship in public affairs.* Madison, WI: Wisconsin University Press.

Nichols, M. (1992). Personal interview with author.

O'Keefe, B. J. (1988). The logic of message design: Individual differences in reasoning and communication. Communication Monographs, 55, 80–103, p. 88. In Van Eemeren, F. H., Grootendorst, R., & Henkemans, F. S. (1996). *Fundamentals of argumentation theory: A handbook of historical backgrounds and contemporary developments.* Mahwah, NJ: Lawrence Erlbaum, p. 202.

O'Keefe, J. (1996). The spatial prepositions in English, vector grammar, and the cognitive map theory. In P. Bloom, M. A. Peterson, & M. F. Garrett (Eds.), *Language and space* (pp. 1–30). Cambridge, MA: MIT Press.

Olson, D. R. (1996). *The world on paper: The conceptual and cognitive implications of writing and reading.* Cambridge, England: Cambridge University Press.

Omenn, G. S. (1997). Preface. In *Presidential/Congressional Commission on Risk Assessment and Risk Management. Framework for environmental health risk management. Final report.* Vol. 1. [On-line]. Available: http://www.riskworld.com/Nreports/1997/risk-rpt/html/epajanb.htm.

O'Sullivan, Tyrone. (Interview with the author). January, 1995.

Oversight hearings on the Coal Mine Health and Safety Act of 1969 (Excluding Title IV), 95th Cong., 1st Sess., Committee on Education and Labor. Subcommittee on Labor Standards (1977) (testimony of J. D. McAteer). Washington, DC: U.S. Government Printing Office.

Oversight hearings on the Coal Mine Health and Safety Act of 1969 (Excluding Title IV), 95th Cong., 1st Sess., by the Subcommittee on Labor Standards of the Committee on Education and Labor (1977) (pp. 332–339). Washington, DC: U.S. Government Printing Office (testimony of T. E. Boettger).

Oversight hearings on the Coal Mine Health and Safety Act of 1969 (Excluding Title IV), 95th Cong., 1st Sess., Subcommittee on Labor Standards of the Committee on Education and Labor (1977) (testimony of J. Davenport; pp. 445–457). Washington, DC: U.S. Government Printing Office.

Oversight hearings on the Coal Mine Health and Safety Act of 1969 (Excluding Title IV), Statement on behalf of the Council of the Southern Mountains, Inc., 95th Cong., 1st Sess., before the Subcommittee on Labor Standards of the Committee on Education and Labor, Subcommittee on Labor Standards of the Committee on Education and Labor (1977), Washington, DC: U.S. Government Printing Office. (testimony of L. T. Galloway & J. D. McAteer)

Oversight hearings on the Coal Mine Health and Safety Act of 1969 (Excluding Title IV), 95th Cong., 1st Sess., before the Subcommittee on Labor Standards of the Committee on Education and Labor (1977). Washington, DC: U.S. Government Printing Office.

Oversight hearings on the Coal Mine Health and Safety Act of 1969 (Excluding Title IV), 95th Cong., 1st Sess., Subcommittee on Labor Standards of the Committee on Education and Labor (1977) (testimony of C. A. Goode; pp. 354–373). Washington, DC: U.S. Government Printing Office.

Oversight hearings on the Coal Mine Health and Safety Act of 1969 (Excluding Title IV), 95th Cong., 1st Sess., Subcommittee on Labor Standards of the Committee on Education and Labor (1977) (testimony of E. Gibert; pp. 407–413). Washington, DC: U.S. Government Printing Office.

Oversight hearings on the Coal Mine Health and Safety Act of 1969 (Excluding Title IV), 95th Cong., 1st Sess., Subcommittee on Labor Standards of the Committee on Education and Labor (1977) (testimony of R. H. Querum; pp. 327–332). Washington, DC: U.S. Government Printing Office.

Oversight of the Mine Safety and Health Administration. Hearings before the Committee on Labor and Human Resources . . . on examining activities of the Mine Safety and Health Administration, 100th Cong., 1st Sess. Committee on Labor and Human Resources, 1982/1987. Abstract of investigation. Multiple fatal roof fall accident. report release date: Jul 29, 1982. Mine ID Number 15-06778. Washington, DC: U.S. Government Printing Office.

Oversight of the Mine Safety and Health Administration. Hearings before the Committee on Labor and Human Resources . . . on examining activities of the Mine Safety and Health Administration, 100th Cong., 1st Sess., Committee on Labor and Human Resources (1987) (testimony of P. Helton; p. 226). Washington, DC: U.S. Government Printing Office.

Oversight of the Mine Safety and Health Administration. Hearings before the Committee on Labor and Human Resources . . . on examining activities of the Mine Safety and Health Administration, 100th Cong., 1st Sess., to the Committee on Labor and Human Resources (pp. 1–2). Washington, DC: U.S. Government Printing Office. (Opening statement of E. Kennedy).

Paradis, J. (1990). Text and action: The operator's manual in context and in court. In C. Bazerman & J. Paradis (Eds.), *Textual dynamics of the professions: Historical and contemporary studies of writing in professional communities* (pp. 256–278). Madison, WI: Wisconsin University Press.

Perelman, C., & Olbrechts-Tyteca, L. (1969). *The new rhetoric: A treatise on argumentation* (J. Wilkinson & P. Weaver, Trans.). Notre Dame, IN: Notre Dame Press.

Perrow, C. (1984). *Normal accidents*. New York: Basic Books.

Peters, R. H. (1992). *Miners' views about how to prevent people from going under unsupported roof* (Information circular/1992 IC 9332). Washington, DC: U.S. Department of the Interior.

Peters, R. H., & Randolph, R. F. (1991). *Miners' views about why people go under unsupported roof and how to stop them* (Bureau of Mines information circular 9300). Washington, DC: Bureau of Mines.

Peters, R. H., & Wiehagen, W. J. (1988). *Human factors contributing to groundfall accidents in underground coal mines: Workers' views* (Bureau of Mines information circular/1988 IC 9185). Pittsburgh, PA: U.S. Department of the Interior.

Petroski, H. (1985). *To engineer is human: The role of failure in successful design*. New York: St. Martin's Press.

Pickering, M. (1995). *The mangle of practice: Time, agency, and science*. Chicago: Chicago University Press.

Pinch, T., & Bijker, W. E. (1984). The social construction of facts and artifacts: Or how the sociology of science and the sociology of technology might benefit each other. *Social Studies of Science, 14*(3), 399–441.

Polanyi, M. (1983). *Tacit dimension*. In H. Collins (Ed.), *Artificial experts: Social knowledge and intelligent machines* (pp. 108–9). Cambridge, MA: MIT Press.

Prelli, L. J. (1989). *A rhetoric of science: Inventing scientific discourse*. Columbia, SC: South Carolina University Press.

Presidential/Congressional Commission on Risk Assessment and Risk Management. (1997). *Framework for environmental health risk management. Final report*. Vol. 1. [On-line]. Available: http://www.riskworld.com/Nreports/1997/risk-rpt/html/epajanb.htm

Presidential/Congressional Commission on Risk Management and Assessment (1997). *Risk assessment and risk management in regulatory decision-making* [On-line]. (Volume 2). Washington, DC: U.S. Government Printing Office. Available: http://www.riskworld.com.

Ramey, P. (1993). Investigation. In Re: *Southmountain Coal Co. Inc. No. 3 Mine. Wise County. Norton, VA. Explosion of Dec. 7, 1992* (Deposition). Southmountain, VA: U.S. Mine Safety and Health Administration.

Reynolds, J. F. (1993). *Rhetorical memory and delivery: Classical concepts for contemporary composition and communication*. Hillsdale, NJ: Lawrence Erlbaum Associates.

RJB Mining (U.K.) Limited. (n.d.). Standards-based scheme of training for appreciation in longwall faces and narrow workings (20 days minimum). Part I. Foundation training: Skill specification (Draft copy).

Rodrigue, C. M. (1999). Social Construction Of Technological Hazard: Plutonium On Board The Cassini-Huygens Spacecraft (narrative for a proposal resubmitted to the Decision, Risk, and Management Science Program. National Science Foundation. January 14.

Roodenburg, H. (1991). The "hand of friendship": Shaking hands and other gestures in the Dutch republic. In J. Bremmer & H. Roodenburg (Eds.), *A cultural history of gesture* (pp. 152–189). Ithaca, NY: Cornell University Press.

Sames, G. P., & Moebs, N. N. (1992). Roof control near coalbed outcrops in small hilltop operations. In U.S. Department of the Interior, Bureau of Mines, *Preventing coal mine groundfall accidents: How to identify and respond to geologic hazards and prevent unsafe worker behavior* (IC 9332. Information circular), U.S. Bureau of Mines Technology Transfer seminar (pp. 15–24). Pittsburgh, PA: Bureau of Mines.

Sauer, B. (1992). The Engineer as rational man: Imminent danger in non-rational environments. *IEEE Transactions on Professional Communication, 35*(4), 242–49.

Sauer, B. (1993). Sense and sensibility in technical documentation and accident reports: How feminist interpretation strategies can save lives in the nation's mines. *Journal of Business and Technical Communication, 7*(1), 63–83.

Sauer, B. (1994a). Sexual dynamics of the professions: Articulating the *ecriture masculine* of science and technology. *Technical Communication Quarterly, 3*(3), 309–12.

Sauer, B. (1994b). The dynamics of disaster: A three-dimensional view of documentation in a tightly regulated industry. *Technical Communication Quarterly, 3*(4), 393–419.

Sauer, B. (1996). Communicating risk in a cross-cultural environment: A cross-cultural comparison of rhetorical and social understandings in U.S. and British mine safety training programs. *Journal of Business and Technical Communication, 10*(3), 306–329.

Sauer, B. (1998). Embodied knowledge: The textual representation of embodied sensory information in a dynamic and uncertain material environment. *Written Communication, 15*(2), 131–169.

Sauer, B. (1999). Embodied experience: Representing risk in speech and gesture. *Discourse Studies, 1*(3), 321–354.

Sauer, B., & Franz, L. (1997, March). *Authority and gesture.* Paper presented at the Conference on College Composition and Communication, Phoenix, AZ.

Sauer, B., & Palmer, T. (1999, March). *Invisible knowledge.* Paper presented at the Conference on College Composition and Communication, Atlanta, GA.

Scardamalia, M., & Bereiter, C. (1989). Intentional learning as a goal of instruction. In L. B. Resnick (Ed.), *Knowing, learning, and instruction: Essays in honor of Robert Glaser* (pp. 361–392). Hillsdale, NJ: Lawrence Erlbaum Associates.

Schachter, D. L. (1996). *Searching for memory: The brain, the mind, and the past.* New York: HarperCollins.

Schegloff, E. A. (1992). On talk and its institutional occasions. In P. Drew, and John Heritage (Ed.), *Talk at work: Interaction in institutional settings* (pp. 101–134). Cambridge, England: Cambridge University Press.

Schmitt, J.-C. (1991). The rationale of gestures in the west. In J. Bremmer & H. Roodenburg (Eds.), *A cultural history of gesture* (pp. 36–53). Ithaca, NY: Cornell University Press.

Schnakenberg, G. Telephone communication with the author, October, 2001.

Scribner, S. (1997). Mind, culture, and activity. In M. Cole, Y. Engeström, & O. Vasquez (Eds.), *Seminal papers from the laboratory of comparative human cognition* (pp. 354–368). Cambridge, England: Cambridge University Press.

Secretary of Labor, Mine Safety and Health Administration (MSHA) v. Emery Mining Corporation. (1987). Docket No. WEST 86-35-R, p. 3. (December 11, 1987). [On-line] Available: http://www.msha.gov/SOLICITOR/FMSHRC/FREQDECS/87121997.htm.

Secretary of Labor v. Emery Mining Corporation. http://www.msha.gov/solicitor/fmshrc/freqdecs/87121997.htm.

Secretary of Labor v. Consolidation Coal Company (Robinson Run No. 95). March 27, 2000. (WEVA 98-111). [On-line]. Available: http://www.msha.gov/solicitor/fmshrc/decision/0003xxx4.htm.

Selber, S. A. (1999). User-centered technology: A rhetorical theory for computers and other mundane artifacts. *JBTC, 13*(4), 457–470.

Selzer, J. (Ed.). (1993). *Understanding scientific prose.* Madison, WI: University of Wisconsin Press.

Shane. (1993). Investigation. In Re: *Southmountain Coal Co. Inc. No. 3 Mine. Wise County. Norton, VA. Explosion of Dec. 7, 1992* (Deposition). Southmountain, VA: U.S. Mine Safety and Health Administration.

Shapin, S. (1994). *A social history of truth: Civility and science in seventeenth-century England.* Chicago: Chicago University Press.

Short, D. (1993). Investigation. In Re: *Southmountain Coal Co. Inc. No. 3 Mine. Wise County. Norton, VA. Explosion of Dec. 7, 1992* (Deposition). Southmountain, VA: U.S. Mine Safety and Health Administration/Virginia Dept. of Mines.

Short, G. P. (1993). Investigation. In Re: *Southmountain Coal Co. Inc. No. 3 Mine. Wise County. Norton, VA. Explosion of Dec. 7, 1992* (Deposition). Southmountain, VA: U.S. Mine Safety and Health Administration/Virginia Dept. of Mines.

Shrader-Frechette, K. S. (1985). *Science policy, ethics, economic methodology.* Dordrecht, The Netherlands: D. Reidel Publishing Company.

Shrader-Frechette, K. (1990). Scientific method, anti-foundationalism and public decision-making. *Risk: Health, safety, and environment (Franklin Pierce Law Center) 1*(23). [Online]. Available: http://www.fplc.edu/risk/vol1/winter/Shrader.htm

Silcox, G. (1993). Investigation. In Re: *Southmountain Coal Co. Inc. No. 3 Mine. Wise County. Norton, VA. Explosion of Dec. 7, 1992* (Deposition). Southmountain, VA: U.S. Mine Safety and Health Administration.

Silvey, P. (1992). Personal interview with the author.

Simons, H. (1980). Are Scientists rhetors in disguise? An analysis of discursive practices within scientific communities. In E. E. White (Ed.), *Rhetoric in transition: Studies in the nature and uses of rhetoric* (pp. 115–130). University Park: Penn State University Press.

Sismondo, S. (1993). Some social constructions. *Social Studies of Science, 23,* 515–555.

Souder, W. E. (1988). *A Catastrophe-theory model for simulating behavioral accidents* (Bureau of Mines information circular/1988 IC 9178). Washington, DC: U.S. Department of the Interior, Bureau of Mines.

Steele, J. P. (1993). Investigation. In Re: *Southmountain Coal Co. Inc. No. 3 Mine. Wise County. Norton, VA. Explosion of Dec. 7, 1992* (Deposition). Southmountain, VA: U.S. Mine Safety and Health Administration/Virginia Dept. of Mines.

Stewart, W., & Feder, N. (1986). Professional practices among biomedical scientists: A study of a sample generated by an unusual event. In U.S. House of Representatives, Committee on Science and Technology, Task Force on Science Policy, *Science policy study* (Vol. 22: Research and Publication Practices; Second Session). Washington, DC: U.S. Government Printing Office.

Strayer, L. (1987). In United States, Department of Labor, Bureau of Mines, *Roof fall entrapment videotape: Larry Strayer's account* (MSHA video). Pittsburgh: United States. Department of Labor. Bureau of Mines.

Tannen, D. (1993). *Framing in discourse.* Oxford, England: Oxford University Press.

Tattersall, W. (1991). *Internal review of MSHA actions at William Station Mine (July 24, 1990)* (Internal review). Arlington, VA: U.S. Mine Safety and Health Administration.

Thompson, T. (1993). Investigation. In Re: *Southmountain Coal Co. Inc. No. 3 Mine. Wise County. Norton, VA. Explosion of Dec. 7, 1992* (Deposition). Southmountain, VA: U.S. Mine Safety and Health Administration/Virginia Dept. of Mines.

Thompson, T. J., Pyles, J. M., Morgan, E. R., Rhea, R. W., Vallina, J. O. (Jr.), Painter, R., Urosek, J. E., & Clete, R. S. (1993). *Report of investigation. Underground coal mine explosion. #3 mine - ID. No. 44-06594. Southmountain Coal Co. Inc., Norton, Wise County, Virginia. December 7, 1992.* Arlington, VA: U.S. Department of Labor, Mine Safety and Health Administration.

Tibbetts, P. (1992). Representation and the realist-constructivist controversy. In M. Lynch & S. Woolgar (Eds.), *Representation in scientific practice* (pp. 242–249). Cambridge, MA: MIT Press.

Totten, C. (1999). Interview with female miner, U.S.), Osage, West Virginia, February, 1999.

Toulmin, S. (1995). *The uses of argument.* Cambridge, England: Cambridge University Press.

Treichler, P. A. (1989). From discourse to dictionary: How sexist meanings are authorized. In F. W. Frank & P. A. Treichler (Eds.), *Language, gender and professional writing: Theoretical approach and guidelines for non-sexist usage* (pp. 50–79). New York: MLA.

Trento, J. (1987). *Prescription for disaster.* New York: Crown Publishers, Inc.

Tufte, Edward R. The Visual Display of Quantitative Information, Graphics Press, Cheshire, CT 1983.

Tufte, Edward R. Envisioning Information, Graphics Press, Cheshire, CT 1991.

U.K. Health and Safety Executive. (1983). *Health and safety annual.* London: Her Majesty's Stationery Office.

U.K. Her Majesty's Inspector of Mines and Quarries. (1958). *Reports of HM Inspectors of Mines and Quarries . . . for 1957* (T. A. Jones, Ed.). London: Her Majesty's Stationery Office.

U.K. Her Majesty's Inspector of Mines and Quarries. (1960). *Reports of HM Inspectors of Mines and Quarries . . . for 1959* (T. A. Jones, Ed.). London: Her Majesty's Stationery Office.

U.K. Her Majesty's Inspector of Mines and Quarries. (1963). *Reports of HM Inspectors of Mines and Quarries . . . for 1962* (C. Leigh, Ed.). London: Her Majesty's Stationery Office.

U.K. Her Majesty's Inspector of Mines and Quarries. (1964a). *Reports of HM Inspectors of Mines and Quarries . . . for 1963* (C. Leigh, Ed.). London: Her Majesty's Stationery Office.

U.K. Her Majesty's Inspector of Mines and Quarries. (1966). *Reports of HM Inspectors of Mines and Quarries . . . for 1965* (C. Leigh, Ed.). London: Her Majesty's Stationery Office.

U.K. Her Majesty's Inspector of Mines and Quarries. (1968). *Reports of HM Inspectors of Mines and Quarries . . . for 1967* (C. Leigh, Ed.). London: Her Majesty's Stationery Office.

U.K. Her Majesty's Inspector of Mines and Quarries. (1969). *Reports of HM Inspectors of Mines and Quarries . . . for 1968* (L. D. Rhydderch, Ed.). London: Her Majesty's Stationery Office.

U.K. Her Majesty's Inspector of Mines and Quarries. (1970). *Reports of HM Inspectors of Mines and Quarries . . . for 1969* (L. D. Rhydderch, Ed.). London: Her Majesty's Stationery Office.

U.K. Her Majesty's Inspector of Mines and Quarries. (1971). *Reports of HM Inspectors of Mines and Quarries . . . For 1970* (J. S. Marshall, Ed.). London: Her Majesty's Stationery Office.

U.K. Her Majesty's Inspector of Mines and Quarries. (1972). *Reports of HM Inspectors of Mines and Quarries . . . for 1971* (J. S. Marshall, Ed.). London: Her Majesty's Stationery Office.

U.K. Her Majesty's Inspector of Mines and Quarries. (1973). *Reports of HM Inspectors of Mines and Quarries . . . for 1972* (J. S. Marshall, Ed.). London: Her Majesty's Stationery Office.

United Mineworkers, Department of Occupational Health and Safety. (n.d.). *Report on the disaster at Wilberg Mine. Emery Mining Corp. Orangeville. Emery County, Utah. December 19, 1984.* Washington, DC: Author.

U.S. Bureau of Mines. (1992). *Errors and unexpected data encountered in MSHA's address/employment and accident/injury databases and miscellaneous suggestions for improving the usability of those databases* (Report). Arlington, VA: Mine Safety and Health Administration.

U.S. Congress, Committee on Education and Labor, Subcommittee on Health and Safety. (1982). *MSHA oversight hearings on coal mine explosions during December 1981 and January 1982, 97th Cong., 2nd Sess. February 23; March 2, 16, 23; and May 4, 1982* (Hearings). Washington, DC: U.S. Government Printing Office.

U.S. Congress, Committee on Education and Labor, Subcommittee on Labor Standards. (1977). *Oversight hearings on the Coal Mine Health and Safety Act of 1969 (Excluding Title IV). 95th Cong., 1st Sess. June 7, 9, 28, and 29, 1977* (Hearings). Washington, DC: U.S. Government Printing Office.

U.S. Congress, House of Representatives. (1969). *Report, together with minority, supplemental, and separate views [to accompany H.R. 13950].* [On-line] (Report No. 91-563), 91st Cong., 1st Sess. October 13, 1969. Available: http://www.msha.gov/SOLICITOR/COALACT/69hous.htm.

U.S. Congress, House of Representatives, Committee on Education and Labor, Subcommittee on Labor Standards. (1977). *To amend the Federal Coal Mine Health and Safety Act: H.R. 4287. March 29, 1977*. 95th Cong., 1st Sess. (Hearings). Washington, DC: U.S. Government Printing Office.

U.S. Congress, House of Representatives, Committee on Science and Technology, Task Force on Science Policy. (1986). *Science policy study* (Vol. 22: Research and Publication Practices; Second Session).

U.S. Congress, Senate. (1977). *Report, together with minority views. Federal Mine Safety and Health Act of 1977*. [On-line] (Report No. 95-181), 95th Cong., 1st Sess. May 17, 1977. Available: http://www.msha.gov/SOLICITOR/COALACT/leghist2.htm.

U.S. Congress, Senate, Committee on Commerce, Science, and Transportation, Subcommittee on Science, Technology, and Space, 100th Cong., 1st Sess. (1987). *Space shuttle oversight*. Washington, DC: U.S. Government Printing Office.

U.S. Department of Labor. (1978). *Coal mine injury and employment experience by occupation, 1972–7*. Washington, DC: U.S. Government Printing Office.

U.S. Department of Labor. (1991). *Accident investigation report. Underground multiple fatal. Roof fall accident. J & T Coal, Inc. No. 1 Mine. I.D. No. 44-05668. St. Charles, Lee County, Virginia. February 13, 1991*. Washington, DC: Mine Safety and Health Administration.

U.S. Department of Labor, Mine Safety and Health Academy. ( ). *Fire safety manual*. Arlington, VA: Mine Safety and Health Administration.

U.S. Department of Labor, Mine Safety and Health Academy. (1990). *Accident investigation*. Arlington, VA: Mine Safety and Health Administration.

U.S. Department of Labor, Mine Safety and Health Academy. (1991). *Accident investigation*. Arlington, VA: Mine Safety and Health Administration.

U.S. Department of Labor, Mine Safety and Health Administration, National Mine Health and Safety Academy. (1986). *Roof and rib control* (Programmed instruction workbook no. 3). Arlington, VA: Mine Safety and Health Administration.

U.S. Department of Labor, Mine Safety and Health Administration. CFR 30, Code of *Federal regulations, Part 75-Mandatory safety standards underground coal mines*. [On-line]. Available: http://www.msha.gov/regdata/msha/75.0.htm

U.S. Department of Labor, Bureau of Mines, Human Factors Group. (1994). Errors and unexpected data encountered in MSHA's address/employment and accident/injury databases and miscellaneous suggestions for improving the usability of those databases (Draft report). November 10 and 18, 1994.

U.S. Department of Labor, Mine Safety and Health Administration. (n.d.). *Gassy mine standards* (Draft document).

U.S. Department of Labor, Bureau of Mines. (1987). *Roof fall entrapment videotape: Dave Garry's account, (Instructor's guide)*. Pittsburgh: United States. Department of Labor. Bureau of Mines.

U.S. Department of Labor, Bureau of Mines. (1987). *Roof fall entrapment videotape: Larry Strayer's account, (Instructor's guide)*. Pittsburgh: United States. Department of Labor. Bureau of Mines.

U.S. Department of Labor, Mine Safety and Health Administration. (1990). *Mine Act* (Docket No. WEST 86-35-R).

U.S. Department of Labor, Mine Safety and Health Administration. (1978). *Mandatory safety and health training of miners. Provisions in the new federal act and a review of German, British, and Polish Programs* (Informational report IR 1982).Washington, DC: U.S. Government Printing Office.

U.S. Department of Labor. Bureau of Mines. (1954/1985). Explosions and fires in bituminous coal mines (Circular No. 50). March, 1954. In J. Nagy, Wilberg mine fire: Cause, location and initial development, (Appendix H). In Huntley, D. W., Painter, R. J., Oakes, J. K., Cavanaugh, D. R., & Denning, W. G., *Report of investigation: Underground coal mine fire. Wilberg Mine. I.D. No. 42-00080. Emery Mining Corporation. Orangeville, Emery County, Utah. December 19, 1984.* Arlington, VA: U.S. Department of Labor, Mine Safety and Health Administration.

U.S. Department of Labor, Mine Safety and Health Administration. (1985). *Training and retraining of miners* (MSHA administrative manual 30 CFR Part 48, July 1, 1985). Washington, DC: U.S. Government Printing Office.

U.S. Department of Labor, Mine Safety and Health Administration. (1988a). *Coal accident analysis and problem identification course.* Washington, DC: U.S. Government Printing Office.

U.S. Department of Labor, Mine Safety and Health Administration. (1988b). *Coal accident analysis and problem identification instruction guide* (Instruction Guide. MSHA IG 67). Arlington, VA: U.S. Government Printing Office.

U.S. Department of Labor, Mine Safety and Health Administration. (1991a). *Coal-underground fatalities. First half-1991.* Washington, DC: U.S. Government Printing Office.

U.S. Department of Labor, Mine Safety and Health Administration. (1991b). *Courses for MSHA and the mining industry. FY 1991.* Washington, DC: U.S. Government Printing Office.

U.S. Department of Labor, Mine Safety and Health Administration. (1991c). *Instructor's manual for mine rescue training* (Training module—Mine rescue). Washington, DC: U.S. Government Printing Office.

U.S. Department of Labor, Mine Safety and Health Administration. (1991d). *Internal review of MSHA's actions at the Granny Rose Coal Company, Barbourville, Knox County, Kentucky, March 3, 1991* (Internal review). Arlington, VA: Mine Safety and Health Administration.

U.S. Department of Labor, Mine Safety and Health Administration. (1991e). *Abstracts with illustrations. Analysis and suggested uses. Coal underground fatalities. Second half-1991. Fatalities by victim's experience. Underground coal. First Half 1991. Fatalities by victim's experience. Underground coal. Second Half 1991.* Arlington, VA: Mine Safety and Health Administration.

U.S. Department of Labor, Mine Safety and Health Administration. (1991f). *Fatalities by victim's experience. Underground coal. First Half 1991. Coal - Underground fatalities. Second half - 1991. Abstracts with illustrations: Analysis and suggested uses.* Arlington, VA: Mine Safety and Health Administration.

U.S. Department of Labor, Mine Safety and Health Administration. (1991g). *Fatalities by victim's experience. Underground coal. Second Half 1991. Coal - Underground fatalities. Second half - 1991. Abstracts with illustrations: Analysis and suggested uses.* Arlington, VA: Mine Safety and Health Administration.

U.S. Department of Labor, Mine Safety and Health Administration. (1993). *Coal-first half 1993. Surface and underground fatalities. Analysis, abstracts, and illustrations.* Arlington, VA: U.S. Government Printing Office.

U.S. Department of Labor, Mine Safety and Health Administration. (1995). *MSHA to conduct review of accident investigation policy* (MSHA news release, No. 95-018. Mine Safety and Health Administration. May 10, 1995. [On-line]. Available: http://www.msha.gov/media/press/1995/nr950510.htm.

U.S. Department of Labor, Mine Safety and Health Administration. (1996). Conducting inspections and investigations. In *MSHA policy manual, III. Special inspections* (pp. 21–24a). Arlington, VA: Mine Safety and Health Administration.

U.S. Department of Labor, Mine Safety and Health Administration. (1996). *Conducting inspections and investigations. MSHA Policy Manual, III. Special Inspections.* Arlington, VA: Mine Safety and Health Administration.

U.S. Department of Labor, Mine Safety and Health Administration. (1996). *Special inspections. Procedures for processing hazardous complaints. MSHA Policy Manual, III. Special Inspections.* Arlington, VA: Mine Safety and Health Administration.

U.S. Department of Labor, Mine Safety and Health Administration, Coal Mine Safety and Health. (n.d.). *Preliminary report of investigation. Underground coal mine fire. Wilberg Mine. I.D. No. 42-00080. Emery Mining Corporation. Orangeville. Emery County, Utah. December 19, 1984.* Arlington, VA: Office of the Administrator. Coal Mine Safety and Health.

U.S. Department of Labor, Mine Safety and Health Administration, Division of Safety. (1989). *Informational report of investigation: Underground coal mine explosion and fire. Consol No. 9 Mine. Mountaineer Coal Company. Farmington, Marion County, West Virginia. November 20, 1968.* Arlington, VA: Mine Safety and Health Administration.

U.S. Department of Labor, Mine Safety and Health Administration, Federal Mine Safety and Health Commission. (1997). Secretary of Labor, Mine Safety and Health Administration (MSHA) v. Emery Mining Corporation (Docket No. WEST 86-35-R). [On-line]. Available: http://www.msha.gov/SOLICITOR/FMSHRC/FREQDECS/87121997.htm

U.S. Department of Labor, Mine Safety and Health Administration, Federal Mine Safety and Health Commission. (1997). *Secretary of Labor, Mine Safety and Health Administration (MSHA) v. Consolidation Coal Co.(Robinson Run No. 95)* (FMHSHRC No. WEVA 98-111. ed.). [On-line]. Available: http://www.msha.gov/solicitor/fmshrc/decision/0003xxx4. htm.

U.S. Department of Labor, Mine Safety and Health Administration, Metal and Nonmetal Mine Safety and Health. (1983). *Coal mine inspection manual: Procedures, orders, citations, and inspection reports* (Vol. 1). Washington, DC: U.S. Government Printing Office.

U.S. Department of Labor, Mine Safety and Health Administration, Metal and Nonmetal Mine Safety and Health. (1987b). *Fatalgram. June 1987. A 37 year old miner, with 18 years mining experience, was fatally injured when a 2000 pound slab of unsupported roof fell on him. The victim was charging a drift round at the time of the accident. There were no witnesses.* Arlington, VA: U.S. Government Printing Office.

U.S. Department of Labor, Mine Safety and Health Administration, Metal and Nonmetal Mine Safety and Health. (1987a). *Fatalgram. June 1987. A 20-year-old kiln operator with less than 5 months mining experience suffocated after falling into a surge hopper which contained fine coal.* Arlington, VA: U.S. Government Printing Office.

U.S. Department of Labor, Mine Safety and Health Administration, Metal and Nonmetal Mine Safety and Health. (1987d). *Fatalgram. MSHA 8000-7, June 87. December 30, 1992.* Arlington, VA: Mine Safety and Health Administration.

U.S. Department of Labor, Mine Safety and Health Administration, Metal and Nonmetal Mine Safety and Health. (1987c). *Fatalgram. MSHA 8000-7. December 30, 1991. 91-045.* Arlington, VA: Mine Safety and Health Administration.

U.S. Department of Labor, Mine Safety and Health Administration, Metal and Nonmetal Mine Safety and Health. (1991a). Abstract of Investigation. Date: April 17, 1991. Slide No: 4 (Fatal case number: 4). Accident classification: Fall of roof. Type of mine: Potash-underground. Location: New Mexico. In *Abstracts with illustrations. Analysis and suggested uses. metal/nonmetal-underground fatalities. First half-1991* (pp.). Arlington, VA: U.S. Government Printing Office.

U.S. Department of Labor, Mine Safety and Health Administration, Metal and Nonmetal Mine Safety and Health. (1991b). Abstract of investigation: Fatal Case Number 26. Slide No. 8. May 28, 1991 (Died May 29, 1991). In *Abstracts with illustrations. Analysis and sug-*

gested uses. Metal–nonmetal-underground fatalities. First half-1991 (pp.). Arlington, VA: U.S. Government Printing Office.

U.S. Department of Labor, Mine Safety and Health Administration, Metal and Nonmetal Mine Safety and Health. (1991c). Abstract of investigation: Fatal Case Numbers 8–11. Slide No. 2. February 13, 1991. In *Abstracts with Illustrations. Analysis and suggested uses. Metal/nonmetal-underground fatalities. First half-1991* (pp.). Arlington, VA: U.S. Government Printing Office.

U.S. Department of Labor, Mine Safety and Health Administration, Metal and Nonmetal Mine Safety and Health. (1991d). *Abstracts with illustrations. Analysis and suggested uses. Metal/nonmetal-underground fatalities. First half-1991. Fatalities by victims' experience.* Arlington, VA: U.S. Government Printing Office.

U.S. Department of Labor, Mine Safety and Health Administration, Metal and Nonmetal Mine Safety and Health. (1991e). *Abstracts with illustrations. Analysis and suggested uses. Metal/nonmetal-underground fatalities. Second half-1991. Fatalities by victim's experience. Underground coal.* Arlington, VA: U.S. Government Printing Office.

U.S. Department of Labor, Mine Safety and Health Administration, Metal and Nonmetal Mine Safety and Health. (1991f). *Fatalgram. July 9, 1992. A 41 year old crusher operator with 11 months mining experience, 1 month as a crusher operator, was fatally injured when the load chain on a 1–12 ton come-along broke.* Arlington, VA: U.S. Government Printing Office.

U.S. Department of Labor, Mine Safety and Health Administration, National Mine Health and Safety Academy. (1989a). *Job safety analysis* (Safety manual No. 5). Washington, DC: U.S. Government Printing Office.

U.S. Department of Labor, Mine Safety and Health Administration, National Mine Health and Safety Academy. (1989b). *Roof and rib control* (MSHA IG 17). Arlington, VA: Mine Safety and Health Administration.

U.S. Department of Labor, Mine Safety and Health Administration, National Mine Health and Safety Academy. (1992). *Roof and rib control* (Draft Copy). Arlington, VA: Mine Safety and Health Administration.

U.S. Department of the Interior, Bureau of Mines. (1993). *Coal mine injury analysis: A model for reduction through training (Vol. VIII-Accident risk during the roof bolting cycle: Analysis of problems and potential solutions.)* August 1993. (Cooperative agreements C)167023 and C)178052; West Virginia University).

U.S. Department of the Interior, Mine Safety and Health Administration. (1977). *Manual for investigation of coal mining accidents and other occurrences relating to health and safety.* Washington, DC: U.S. Government Printing Office.

U.S. Department of Labor, Mine Safety and Health Administration, National Mine Academy. (1990). *Accident prevention* (Safety Manual No. 4; Revised 1990). Arlington: Mine Safety and Health Administration.

U.S. Environmental Protection Agency. (1990). *Advisory Committee Charter. Appendix A. 2. Mandate of the Commission on Risk Assessment and Risk Management.* http://www.riskworld.com/Nreports/1996/risk_rpt/html/nr6aa028.htm

U.S. Mine Enforcement and Safety Administration. (1977). Testimony. In U.S. House of Representatives, 95th Cong., 1st Sess., Subcommittee on Labor Standards of the Committee on Education and Labor, *Oversight hearings on the Coal Mine Health and Safety Act of 1969 (Excluding Title IV)* (pp. 458–525). Washington, DC: U.S. Government Printing Office.

VanDeMerwe, J. N. (1995). *Practical coal mining strata control: A guide for managers and supervisors.* Johannesburg, South Africa: Geoscientific Technology Development, Sasol Coal Division.

Van Eemeren, F. H., Grootendorst, R., & Henkemans, F. S. (1996). *Fundamentals of argumentation theory: A handbook of historical backgrounds and contemporary developments*. Mahwah, NJ: Lawrence Erlbaum Associates.

Vaught, C., Brnich, M. J., & Kellner, H. J. (1988). *Effect of training strategy on self-contained self-rescuer donning performance* (Bureau of Mines information circular/ 1988 IC 9185, pp. 2–14). Pittsburgh, PA: U.S. Department of the Interior.

White, E. (Ed.). (1980). *Rhetoric in transition: Studies in the nature and uses of rhetoric*. University Park, PA: Penn State University Press.

Willis, I. J. (1993). Investigation. In Re: *Southmountain Coal Co. Inc. No. 3 Mine. Wise County. Norton, VA. Explosion of Dec. 7, 1992* (Deposition). Southmountain, VA: U.S. Mine Safety and Health Administration/Virginia Dept. of Mines.

Winsor, D. (1990). The construction of knowledge in organizations: asking the right questions about the *Challenger*. *JBTC, 4*(2), 7–20.

Woolgar, S. (1992). Time and documents in researcher interaction: Some ways of making out what is happening in experimental science. In M. Lynch & S. Woolgar (Eds.), *Representation in scientific practice*. Cambridge, MA: MIT Press.

Wynne, B. (1992a). Misunderstood misunderstandings: Social identities and public uptake of science. *Public Understanding of Science, 1*(1), xx–xx.

Wynne, B. (1992b). Public understanding of science research: New horizons or hall of mirrors. *Public Understanding of Science, 1*(1), 37–43.

Author Index

A

Adams, R., 287, 313
Agricola, G., 28, 107, 108, 302
Alibali, M. W., 220, 226, 227, 245, 246, 247, 250, 273
Allen, J., 132
Althouse, R. C., 129, 148
Aristotle, 3, 5, 17, 58, 86, 98, 99, 261
Atman, C., 13, 14, 15, 66, 82, 83
Austin, G., 253, 282

B

Bardach, E., 40
Bassok, M., 220
Bazerman, C., 4, 5, 37, 322
Bean, J. M., 123
Beason, R. L., 47, 50
Beck, U., 65
Bereiter, C., 80
Bernard, R. L., 47, 49, 50
Bertin, J., 168
Bijker, W. E., 118
Blake, K., 139
Blau, P., 40
Bloor, D., 3, 181
Borchers, J. G., 101
Bostrom, A., 13, 14, 15, 66, 82, 83

Bradshaw, G. A., 101
Brasseur, L. E., 4, 132
Bremmer, J., 224
Brnich, M. J., 85, 97
Brooks, J. P., 287
Brooks, K., 287, 305, 309
Bruce, W. E., 1, 44, 45, 46
Bryner, G. C., 40
Bulwer, J., 253, 282
Burnett, R. E., 328
Burns, R. E., 315, 316, 317
Butler, B., 97, 123, 124

C

Carley, K. M., 320
Carroll, J., 27
Cavanaugh, D. R., 34
Chase, F. E., 81, 82
Chawla, P., 224, 253, 254, 280, 332
Childers, M. S., 34, 35, 36, 52, 53
Church, R. B., 257, 258, 272, 273
Cicero, 219, 271
Cleveland, M. J., 60, 63, 104, 105, 106
Code, L., 227, 324
Cohn, C., 4
Collins, H., 182, 204, 205, 331
Combs, R., 287
Conover, D., 192, 193

Cooke, J. D., 287, 297, 304, 307, 309, 313, 314
Cox, D. B., 287

D

Dale, A., 97
Davies, J. C. A., 88, 89, 90, 291
Davies, Q. W., 88, 89, 90, 291
Davis, A. , 208
Davis, B. R., 55, 287, 299, 303, 308
Davis, J. E., 287
Davis, M., 287, 308
Dawes, R. M., 85
de Saussure, F., 265
Dear, P., 37
Deatherage, F., 295, 296, 297, 300, 309, 313, 315
DeKock, A., 190, 215
DeMarco, M. J., 207, 208
Denning, W. G., 34
Dombroski, P. M., 4
Dowlatabadi, H., 101, 102
Dreissen, H., 224
Drew, P., 257, 259
Duncan, C. E., 287, 309, 313
Duncan, S. D., 220, 259, 262, 266, 268

E

Eisenstein, E. L., 320
Engeström, Y., 4
Efron, D., 223

F

Fahnestock, J., 4
Feder, N., 4
Fennell, B. A., 4
Fischhoff, B., 13, 14, 15, 16, 56, 66, 82, 83
Fleming, R. K., 305, 310
Flower, L., 16
France, C., 287
Frank, F. W., 4
Franz, L., 214, 220, 236, 245, 255
Fuller, S., 334

G

Garry, D., 146, 160, 161, 164

Gaventa, J., 2
Geertz, C., 272
Gershung, H. L., 4
Glaser, R., 80
Goldin-Meadow, S., 220, 226, 227, 245, 246, 247, 256, 258, 262, 263, 265, 266
Goode, C. A., 287
Goode, D. E., 287
Goode, D. L., 287, 306
Goodman, S. N., 100
Graf, F., 223
Greenblatt, S., 122
Grootendorst, R., 132
Gross, A., 4, 37, 334

H

Hacker, S., 4
Hanna, K., 192, 193
Haramy, K., 192, 193
Haraway, D., 132, 179, 227, 324
Harding, S., 179, 227, 323
Henkemans, F. S., 132
Henrion, M., 56, 101
Herbert, A., 82, 86, 112
Heritage, J., 257, 259
Herndl, C., 4
Hill, J. R. M., 203, 204
Hope, C., 66
Horton, W., 168
Huntley, D. W., 34, 130, 131, 139, 140
Hurst, E., 55
Hutchins, E., 4, 157, 272

I

Iannacchione, A., 81, 82
Irwin, A., 66, 97
Ivins, W. M., 321

J

Jackendoff, R., 134, 223
Jasanoff, S., 124
Johnstone, B., 123

K

Kagan, R., 40

Kahneman, P., 12
Kaufer, D. S., 97, 123, 124, 185, 320
Keith, W. M., 334
Keller, E. F., 4, 27, 179
Kellner, H. J., 85, 97
Kelly, S. D., 257, 258, 272, 273
Kennedy, E., 3
Killingsworth, J., 4
Klishis, M. J., 129, 148, 155
Knorr, K., 92
Kononov, V.A., 215
Krauss, R. M., 224, 253, 254, 280, 332
Kunreuther, H., 11, 66

L

Langdon, B., 108, 177, 195, 196, 197, 198, 199
Larkin, J, H., 80
Latour, B., 37, 79, 118, 193, 266, 321
Lave, L., 13, 66, 82
Layne, L. A., 129, 148
Leon, R. N., 88, 89, 90, 291
Levelt, W. J. M., 134, 223
Levinson, C., 223, 257, 259
Libby, 240, 241, 243, 247, 251, 256, 270, 273
Lies, G. M., 129, 148
Littleton, E. B., 227
Lukes, S., 2
Lynch, M., 6, 213

M

MacGregor, D., 11, 66
Mallett, L., 109, 154, 160, 161, 162
Mark, C., 81, 82, 94, 95, 191, 207, 208, 210, 290
Markley, R., 37
McAteer, J. D., 43, 110, 130, 291, 294, 317, 318
McCloskey, D. D., 5
McConnell-Ginet, S., 4
McKinney, O., 287
McNeill, D., 219, 220, 221, 222, 224, 225, 226, 236, 241, 246, 250, 256, 258, 259, 261, 262, 263, 265, 266, 268
Meade, D., 287, 298

Merchant, C., 4
Miller, C., 4, 101, 102, 127
Minick, N., 272
Moebs, N. N., 209
Molinda, G. M., 94, 95, 186, 191, 208, 210
Moore, M., 293, 294
Morgan, M. G., 13, 14, 15, 56, 57, 66, 82, 83, 101
Morone, D., 160
Morrel-Samuels, P., 254
Mullins, D., 287, 298, 310, 311
Mullins, D. A., 287
Mullins, J. E., 287
Mullins, L. C., 287
Myers, G., 4, 118, 266, 278, 320

N

Nagy, J., 140, 141, 143, 144, 145, 146, 147, 148, 149, 150, 151, 152
Nelkin, D., 99, 175, 190
Nichols, M., 67

O

Oakes, J. K., 34, 95, 130, 131, 139, 140
O'Keefe, J., 122, 134
Oberholzer, J., 190, 215
Olbrechts-Tyteca, L., 5
Olseth, K. L., 220
Olson, D. R., 320, 321, 327
Omenn, G. S., 68
O'Sullivan, T., 110

P

Painter, R. J., 34, 130, 131, 139, 140
Palmer, J., 4, 223
Paradis, J., 4, 175, 200
Perelman, C., 5
Perrow, C., 129, 130, 182
Peters, R. H., 109, 154, 158, 160, 161, 162, 181, 186, 190
Petraglia, J., 323
Petroski, H., 129, 130, 182, 190, 207
Pickering, M., 118
Pinch, T., 118
Polanyi, M., 204

Prelli, L. J., 4, 37

Q

Querum, R. H., 131

R

Ramey, P., 287
Randolph, R. F., 158, 181, 186, 190
Reynolds, J. F., 334
Rodrigue, C. M., 88
Roodenberg, H., 224

S

Salamon, M. D. G., 88, 89, 90, 291
Salk, J., 37, 79, 118
Sames, G. P., 209
Sauer, B., 2, 4, 72, 85, 176, 180, 185, 188,
        189, 190, 194, 212, 214, 220,
        223, 233, 236, 245, 255
Scardamalia, M., 80
Schachter, D. L., 225, 252
Schegloff, E. A., 257, 259
Schmitt, J. C., 224
Schnakenberg, G., 39
Scribner, S., 4
Selber, S. A., 14
Shane, 287, 306, 312, 313
Shapin, S., 321, 322
Short, D., 287, 296, 303
Short, G. P., 287
Shrader-Frechette, K. S., 12, 66
Silcox, G., 287, 295, 304, 306, 312
Silvey, P., 63
Simon, H., 80
Simons, H., 4

Singleton, J., 256, 258, 262, 263, 265, 266
Sismondo, S., 118
Slovic, P., 11, 66
Smith, D., 97
Souder, W. E., 154, 156, 157
Steele, J. P., 295, 308, 312
Stewart, W., 4
Strayer, L., 142, 160, 161, 162
Syc, S., 220

T

Tannen, D., 132
Tattersall, W., 54
Thompson, T. J., 34, 292, 293, 294
Thompson, T., 287
Tibbetts, P., 92
Totten, C., 271, 272
Toulmin, S., 58, 59
Treichler, P. A., 4
Trento, J., 129, 130, 182, 190
Tufte, E. R., 168
Turner, E., 60, 63, 104, 105, 106
Tversky, A., 12

V

VanDeMerwe, J. N., 57
Van Eemeren, F. H., 132
Vaught, C., 85, 97, 109, 154, 160, 161, 162

W

Watson, S. R., 66
Wiehagen, W. J., 158
Willis, I. J., 301, 302, 309
Winsor, D., 4
Wolfe, S. J., 4
Woolgar, S., 37, 79, 118, 166
Wynne, B., 66

Subject Index

A

Accident investigations, 287–291, 318, *see also* Cycle of Technical Documentation; Rhetorical transformation; Technical documentation
  corroborating factual findings in, 299–300
  implications for agencies, 319
  legal uncertainties of, 294–295
    and institutional authority, 295–296
    and liability, 294–296
    and responsible agents, 296–297
    self-incrimination, 294–295
  witness testimony in, *see* Witness testimony
Ambiguity, 31, 36, 43, 94, 104, 122, 253, 324–325, *see also* Interpretation
Amtrak train crash, *see* Disaster
Analytic viewpoint, *see* Viewpoints
Anomalous behavior, 19, 155–159, *see also* Irrational behavior
  attention or focus, 155, 157, 159, 161
  and habitual behavior, 159
Audience–speaker relationship
  and viewpoints, 260–262

B

Bilsthorpe, *see* Disaster
Boyd Report, 187
Bumps, 303
Bureau of Mines, 29–31, 186

C

Co-construction of knowledge, *see* Gesture
Compliance, 29–33, 50–57, 296–297, *see also* Regulation
Cycle of Technical Documentation in Large Regulatory Industries, 17–18, 66, 72–75, 83–85, 97–98, 152, 198, 325, *see also* Rhetorical transformation; Technical documentation
  accident reports, 17, 74, 140–153, *see also* Accident investigations
  local documentation, 74
  and new technologies, 199, 327–328
  policy and regulations, 74
  practices and procedures, 75
  Six Critical Moments of Rhetorical Transformation, *see* Rhetorical transformation

statistical reports, 74
training and instruction, 17, 75, *see also*
    Training

D

Decision making, *see* Risk decision making
Disaster, 1
    Amtrak train crash, 1
    Bilsthorpe, 9, 108, 176, 194, 198
    Farmington, 29–30
    Hyatt collapse, 1
    large-scale technological, 1, 130, 190
    narratives, 130
    Pyro, 33–36, 53
    Shuttle Challenger, 1
    Southmountain, 20, 34, 287–288,
        291–311
        controversy and uncertainty
            following, 292–294
    Three Mile Island, 1
    Wilberg, 19, 34, 129–133, 136–152
Documentation, technical *see* Technical
    Documentation

E

Electronic commerce, 333
Embodied experience, 77, 304
    embodied sensory experience, 112, 306
        see also Warrants
    representation in gesture, 219, *see also*
        Gesture
    representation in writing, 288, *see also*
        Technical documentation
Embodied knowledge, 112, *see also* Pit sense
    loss in written communication, 6,20,
        216
Enthymemes, 3
Expertise, 79–83, 175, 184, 192, 202, 271,
        see also Risk assessment
    and collaboration, 82, 98
Expert model, 13, 15, 16
    influence model, 83
    influence diagram, 13–14
Expert systems, 330–332

F

Farmington, *see* Disaster
FATALGRAMS, 166–175
Feminist perspective, 2, 4, 178–180
    silencing of alternative viewpoints, 6,
        15

G

Gender and power
Gesture, 6, 20, 224–228, 231–283, *see also*
        Embodied experience
    aid in understanding risky
        environments, 220–221,
        274–275
    and architectural landscape, 241
    co-construction of knowledge in,
        280–281
    in collaborative interactions, 279–281
    combined with speech, 219–224,
        255–256, 265–268
        compared to two-dimensional maps,
            311–317
    and elaboration of meaning, 259,
        266–268, 276–279
    in expert discourse, 244, 255
    and imitation, *see* Imitation
    interpretation of, *see* Interpretation
    limitations of, 331
    in representation of embodied
        knowledge, 219–220
    in representation of risk, 227–252,
        259–260
    in rhetorical construction of meaning,
        272–279
    and viewpoints, 221–226, 228–232, *see
        also* Viewpoints
    and writing, 220, 223, 257, 307–311
        leading to uncertainties, 311

H

Hazardous Environments
    decision making in, 14, 16
    inherence in mining, 27–28

uncertainty of, 14, 16, 19, 124, 156,
    182, 195, 202 *see also*
    Uncertainty of knowledge
HSE (Her Majesty's Safety Executive),
    194–197, 199
Hyatt collapse, *see* Disaster

I

Imitation, *see also* Viewpoints
    of geography, 238
    ironic imitation, 236–237
    of non-human others, 237–238
    of others, 234–236
    of self, 233
    of spaces, 233–234
Imminent danger, 55–58
Inby, 135–136
Influence diagram, *see* Expert model
Instruction, *see* Training
International mining community, 10–11
Interpretation, 106, *see also* Ambiguity
    of gesture and viewpoints, 252–254,
        258, 273
Interviews, *see* Methodology
Invention, *see* Rhetorical invention
Irrational behavior, 155, 157, 159, 163, *see
        also* Anomalous behavior

K

Knowledge representation, 16, 321–323
    and objectivity, 323
    and social construction, 323

L

Legislation, *see* Regulation
Local experience, 138–140, 178, 288, 306
    see also, Embodied knowledge,
        Embodied experience
Longwall mining, 10, 43, 54, 58

M

Maps

inadequacy and uncertainty of,
    311–317
Mental models approach, 13, 82, *see also*
    Expert model; Risk assessment
Methodology, 7–8, 10, 18, 254–255, 288
Mimetic viewpoint, *see* Viewpoints
Mine safety, 2, 108, 172
Monitors, 177–178, 198–199
    methane, 297
MSHA (Mine Safety and Health
    Administration)
    creation of, 31–32
    investigations, 130–132, 140–141,
        287–294
    regulation, 34, 35, 47, 62, 119–120
    training, 159–163, 166–175
        gaps in, 163–165

N

NIOSH (National Institute for
    Occupational Safety and
    Health), 10, 70, 255

O

Offshore industry, 333
Outby, 135–136

P

Perspective *see* Viewpoints
Persuasion, *see* Rhetorical force
Pit sense, 81, *see also* Embodied knowledge,
    Warrants
Premium data, *see also* Risk assessment,
    Uncertainty of knowledge
Presidential/Congressional Commission on
    Risk Assessment and Risk
    Management, 67–72
Proper perspective, *see* Viewpoints
Pyro, *see* Disaster

R

Regulation, 18, 34, *see also* Compliance

ambiguity and misinterpretation of,
    31–33, 60, *see also* Ambiguity
costs and benefits, 57
creation by regulatory agencies, 38–40
in "gassy" mines, 52
history of mining legislation, 29–33
inspection guidelines, 119–120
and judgment, 34, 37
revision and review, 39, 47–50
and standardizing experience, 37–38
and wording, 44
Rhetoric of science, 4–5
Rhetorical force, 7–11, 85–88
Rhetorical interface, 291
Rhetorical invention, 3–7, 83, 174, 324
Rhetorical knowledge, 16–18
Rhetorical practices, 2, 6
    effect on content and conduct of
        communication, 5, 15
Rhetorical theory, 3, 99–100
    implications for, 20, 213–216, 324–328,
        334–335
Rhetorical transformation, 17–18, 91–92,
    see also Technical
        documentation, Risk assessment
    in accident investigations, 92–93,
        129–133, 289–292, *see also*
        Accident investigations
    collective agreement, 147–148
    collective experience, 145–147
    embodied positions, 142–143
    outcome evaluation, 150–152
    previous experience, 148–150
    rhetorical positions, 144–145
    time reconstruction, 143–144
    and engineering knowledge, 209
    of local knowledge into writing, 306
    in regulatory agencies, 67, 91, 114
    in risk assessment and management,
        288
    Six Critical Moments of, 17, 18, 66–67,
        75–78, 119, 288–289 (*put
        this under cycle*
    in training, 93–97, 155, 166–175
    uncertainties in, 300–306
Risk assessment, 4, 16, 66 *see also*
    Expertise, Rhetorical

transformation, Stakeholder
    knowledge, Warrants
expert judgments, 11, 96, 303–304, 318
and focus of attention, 88–89
manipulation on paper, 19, 87–88,
    90–91
and preconceived notions, 88
Risk communication, 2–4, 112, 215, 223
Risk decision making, *see also* Warrants
complexity, 138
stakeholder involvement, 178, *see also*
    Stakeholder knowledge
theoretical framework development, 15
Roof falls, 108–109, 154, 162, 167, 169, 194
Roof support, 184–189, 193–197
faults, 186–187
full metal arch support, 187
roof bolts (British "rock bolts"), 9,
    176–178, 200–201, 269–270
timbers, 187
unpredictability of, 195–196

S

Scientific expertise, 206, *see also* Warrants
Situated knowledge, 133
    structure of in hazardous worksites,
        133–138
Six Critical Moments of Rhetorical
    Transformation, *see* Rhetorical
        transformation
Southmountain, *see* Disaster
Stakeholder knowledge, 11–14, 17, 67–72,
    291, *see also* Risk assessment,
        Risk decision making
Standards, *see* Regulation
Structure of mine, 133–135

T

Tacit knowledge, 204–207, 330–331
    loss of in writing, 331
Technical documentation, 9, *see also*
        Accident investigations,
        Rhetorical transformation,
        Written documentation
adequacy and timeliness of, 59–61, 65

cycle of, *see* Cycle of Documentation in
    Large Regulatory Agencies
embodied experience in, *see also*
    Embodied knowledge,
    Embodied experience
  loss of, 159
  in hazardous environments, 2, 64, 324
  of local experience, 138–140
  practices, 324, 329
    future development, 334
Technology
  uncertainties of, 332–333
Training, 82, 86, 154–159, 213, 324 *see*
    also Rhetorical transformation,
    Technical documentation
  effects of narrow focus, 20
  implications for, 329
  old apprenticeship system, 110–111
  new training programs, 111
  retransformation of embodied
    experience, 159
  speculation in, 163–175
Transformation, *see* Rhetorical
    transformation

U

Uncertainty of knowledge, 97–98,
    100–107, *see also* Hazardous
    environments
  in citations, 105–106
  benefits of articulation of, 101–102
  of documentation in hazardous
    worksites, 136–138, 324
  and imminent danger, 103–104
  of premium data, 112–116
    and classification, 113–116
  in social structure and organization,
    116–122
  sources of in technical documentation
    dynamic uncertainty of hazardous
      environments, 107–110
    variability and unreliability of
      human performance,
      110–112
Unwarrantable failure, 61–62
Uptake, 272–273

V

Viewpoints, 20, 134, 221–225, *see also*
    Gesture
  aid in integration of theory and
    practice, 268–270
  analytic, 221, 228–230, 238–244
    and emotional detachment,
      239–240
  character, 225
  construction of agency's perspective
    ("proper perspective"), 19,
    140–142, 155–156, 159–163,
    197
    gaps in, 163–165
  dual, 226
  effects on semantic content, 262–266
  and elaboration of meaning in speech,
    266–268
  effect on semantic context, 262–266
  incompleteness of single viewpoints,
    122–125, 133, 138, 324
  intentional shifting of, 20, 259–260,
    265
  loss in documentation, 290
  mimetic, 221, 228–229, 232, *see also*
    Imitation
  missing, 230, 244–249
    and lack of embodied experience,
      244–246
  multiple, 112, 122–123, 140, 152–153,
    230–231
    management of, 19, 155–156,
      175–178, 180
  observer, 225
  and representation of risk, 238–244,
    259–272
  recovery of lost viewpoints and
    rhetorical analysis, 179–180
  sequential, 231, 249–251
  simultaneous, 231, 243–244
  and transformation of understanding,
    271–272
Vocabulary
  differences related to different
    viewpoints, 301–302
  of embodied understanding, 302

W

Warrants, 20, 181–189, *see also* Risk
     assessment; Risk decision making
     effects on risk decisions and outcomes,
          193–199
     embodied sensory knowledge (pit
          sense), 182–185, 189–191
     engineering experience, 182–185, 191,
          207–210
     scientific (invisible) knowledge,
          182–185, 191–193, 210–213
Wilberg, *see* Disaster
Witness testimony

completeness forthrightness of,
     317–319
and use of rhetorical methods, 318
Workplace discourse, 6, 11, 15, 259–260,
     324, 328–329
Written documentation, 320–321, *see also*
     Technical documentation
capturing of embodied knowledge in, 2
creation of new forms of, 329
function of in hazardous environments,
     2, 324–328
incompleteness of, 199–203, 289
implications of gesture for, 282–283